I0885354

RESEARCH IN THE CREATIVE AND MEDIA ARTS

In *Research in the Creative and Media Arts*, Desmond Bell looks at contemporary art and design practice, arguing that research activity is now a vital part of the creative dynamic.

Today, creative arts and media students are expected to develop a range of research competencies and critical capacities in their creative project work. This book plots the basis for a research culture in the creative and media arts. It provides an illuminating genealogy of artistic research, revealing the intimate connections between art and science over the centuries and identifying some of the founding figures of practice-based artistic research. Bell explores the research that artists undertake through a number of case studies, talking to a range of contemporary artists and media makers about their work and the role research plays in this. He also traces the dialogues between art practice and a range of other humanity disciplines, such as history, anthropology and critical theory. His analysis reveals how contemporary art practice is now so locked into a set of interlocutions about process and purpose that it increasingly resembles a research practice in and of itself.

Research in the Creative and Media Arts is a comprehensive overview of the relationship between research and practice that is ideal for undergraduate and postgraduate students, as well as researchers in the fields of art and design, art history and visual culture.

Desmond Bell is a research fellow at the National College of Art and Design in Dublin, where he was previously Head of Research. He has held chairs in film, photography and media studies at a number of UK universities, combining his academic work with a career as an award-winning documentary film-maker.

RESEARCH IN THE CREATIVE AND MEDIA ARTS

Challenging Practice

Desmond Bell

Routledge
Taylor & Francis Group
LONDON AND NEW YORK

First published 2019
by Routledge
2 Park Square, Milton Park, Abingdon, Oxon OX14 4RN

and by Routledge
52 Vanderbilt Avenue, New York, NY 10017

Routledge is an imprint of the Taylor & Francis Group, an informa business

British Library Cataloguing-in-Publication Data
A catalogue record for this book is available from the British Library

Library of Congress Cataloging-in-Publication Data
Names: Bell, Desmond, author.
Title: Research in the creative and media arts : challenging practice / Desmond Bell.
Description: New York : Routledge, 2019. | Includes bibliographical references and index.
Identifiers: LCCN 2018055568 (print) | LCCN 2018056314 (ebook) | ISBN 9780429491948 (master) | ISBN 9780429959424 (adobe reader) | ISBN 9780429959400 (mobi pocket) | ISBN 9780429959417 (epub 3) | ISBN 9781138589056 (hardback : alk. paper) | ISBN 9781138589063 (pbk. : alk. paper)
Subjects: LCSH: Arts—Research. | Art and science.
Classification: LCC NX180.S3 (ebook) | LCC NX180.S3 B44 2019 (print) | DDC 701/.05—dc23
LC record available at https://lccn.loc.gov/2018055568

ISBN: 978-1-138-58905-6 (hbk)
ISBN: 978-1-138-58906-3 (pbk)
ISBN: 978-0-429-49194-8 (ebk)

Typeset in Bembo
by Apex CoVantage, LLC

CONTENTS

List of figures *ix*
Acknowledgements *xv*

Introduction 1

PART I
Critical parameters **7**

1 Creative arts and media research: key terms and perspectives 9
1.1 Introduction 9
1.2 Defining the key terms 9
1.3 The distinguishing characteristics of artistic research 15
1.4 The main types of artistic research 18
1.5 Modelling artistic research 29

2 The research that artists actually do 35
2.1 Introduction 35
2.2 Re-evaluating "background" research 36
2.3 Artist reflections and critical dialogues as research sources 42
2.4 The artist interview as a research tool 46
2.5 Case study: found footage film-making 49

3 A research turn in the visual arts? 58
3.1 Introduction 58
3.2 The dynamics of contemporary art practice 58
3.3 Conceptualism 59

3.4 Curational strategies 65
3.5 Relational aesthetics 73
3.6 Performativity 80
3.7 The practice turn in contemporary theory 84
3.8 Conclusion 88

4 The spectre of science 92
4.1 Introduction 92
4.2 The methodological drive in artistic research 92
4.3 Understanding the logic of scientific discovery 94
4.4 The specificity of artistic research 97
4.5 Against method (and for art) 101
4.6 Artists and science 106
4.7 Case studies of art–science collaboration 110
4.8 Conclusion 121

PART II
Antecedents and foundations **123**

5 The historical antecedents of artistic research: the Renaissance 125
5.1 Introduction 125
5.2 Renaissance vision 128
5.3 Leonardo, science and invention 137
5.4 Assessing the Renaissance contribution to the advancement of artistic research 142

6 A science of art?: theory and practice in the modernist avant-garde 145
6.1 Introduction 145
6.2 A science of art? 147
6.3 Founding fathers (and mothers) of twentieth-century artistic research 150
6.4 Theory and practice in the work of Wassily Kandinsky 162
6.5 Art as expression 168
6.6 Founding texts and axioms of modernist artistic research 173
6.7 Conclusion 178

7 Utopian visions and material realities: artistic research and the revolution 183

7.1 Introduction 183

7.2 Artistic research for a revolutionary moment 184

7.3 The auspices of non-objective art: from the analysis of style to language and system 192

7.4 Malevich as pedagogue: the UNOVIS experiment 200

7.5 The Constructivist challenge 206

7.6 From Suprematist painting to Constructivist imaging 214

7.7 The eclipse of the avant-garde 218

7.8 Conclusion 221

PART III
Interdisciplinary dialogues **225**

8 Art and the sciences of man: reframing history 227

8.1 Introduction 227

8.2 Method, craft and science in history 228

8.3 Material memory: exploring a family archive (case study) 231

8.4 The artwork as historiographical intervention: interpreting the Spanish Civil War (case study) 234

8.5 Imaging the past: The Enigma of Frank Ryan *(case study) 242*

8.6 Contemporary art and the engagement with history: Hito Steyerl's Der Bau *and* In Free Fall *246*

8.7 Conclusion 254

9 Artistic research and the human sciences: staging anthropology 257

9.1 Introduction 257

9.2 An ethnographic turn in the visual arts? 258

9.3 The tarnished legacy of anthropology 260

9.4 Lens-based ethnography: documenting Dublin's docklands (case study) 272

9.5 Susan Hiller: a coming to terms with anthropological method 277

9.6 Trinh T. Minh-ha: speaking nearby 283

9.7 The auto-ethnographic impulse in artistic research 290

10 Artistic research and critical theory 297
10.1 Introduction 297
10.2 Theory in the creative arts and media 298
10.3 Paradigms of the theory–practice relation 300
10.4 Exemplars of contemporary critical arts practice 307
10.5 Conclusion: artistic research and critical theory 323

Bibliography *328*
Index *340*

FIGURES

1.1	JCHP poster, *Critical Décor: What Works* Exhibition, Lanchester Galleries, Coventry, 2014	11
1.2	William Kentridge, *More Sweetly Play the Dance*, multiscreen installation, Luma Gallery, Arles 2016	14
1.3	*More Sweetly Play the Dance* close-ups	14
1.4	The creative arts PhD research route	21
1.5	Alison Pilkington, *Solo* (2013), oil on canvas 70 × 100 cm	22
1.6	Desmond Bell, *Cycle of Reflective Practice* (2018)	25
1.7	Desmond Bell, *Modelling Artistic Research* (2018)	30
1.8	Desmond Bell, *The Methodological Context of Practice Research in the Arts* (2018)	31
2.1	Stanley Kubrick, *Barry Lyndon* (1975), US/UK, 185 mins, colour, film still	39
2.2	Jean Rouch and Edgar Morin, *Chronicle of a Summer* (1960), film still	39
2.3	(left) Douglas Gordon and Philippe Parreno, *Zidane: A 21st Century Portrait*, National Gallery Scotland, 2008; (right) *Zidane* film still	41
2.4	William Kentridge, *6 Drawing Lessons* (2016), film still	44
2.5	Péter Forgács, *Wittgenstein Tractatus* (1992), film still	51
2.6	Bill Morrison, *The Film of Her* (1996), film still	52
2.7	Desmond Bell, *Hard Road to Klondike* (1999), film still	55
3.1	Joseph Kossuth, *One and Three Chairs* (1965)	61
3.2	Willie Doherty, *Ghost Story* (2007), film stills	64
3.3	Damian Ortega, *Altocumulus* (2016), photo by Ruth Clark	66
3.4	(above) William Kentridge *Tango for Page Turning*, installation view, Fruitmarket Gallery, Edinburgh 2016; (below) film stills (2013)	68
3.5	(left) *7 Reece Mews Francis Bacon Studio*, the Hugh Lane Gallery, Dublin, photo by Perry Ogden; (right) *Eduardo Paolozzi Studio*, Museum of Modern Art, Edinburgh	70

3.6	(left) Katerina Jebb, *Tablier*, from the Balthus Inventory; (right) *Cendrier*, from the Balthus Inventory (2017), Musée Rattéu, Arles 2016	72
3.7	(left) Rirkrit Tiravanija, *Bamboo Maze* (2017); (right) *Bamboo Maze* close-up	74
3.8	Exhibition view: Philippe Parreno, "Anywhere, Anywhere, Out of the World", 2011, Palais de Tokyo, Paris, 2 *TV Channel*, 2013, LED panel screen 768 × 576 cm – on screen: *The Writer*, 2007	77
3.9	JCHP Collective poster distributed as part of the *Critical Décor: What Works* Exhibition Lanchester Galleries, Coventry, 2014	78
3.10	(left) Jo Spence, *Libido Uprising* (1989), collaboration with Rosy Martin, archival pigment print, 105 × 70 cm; (right) *Photo Therapy: My Mother as a War Worker* (1986–1988), colour photograph mounted on board, 25.3 × 16.5 cm	81
3.11	Sophie Calle, *Venetian Suite* (Detail) (1980)	82
4.1	The hypothetical–deductive model	95
4.2	The deductive–nomological model	96
4.3	Damien Hirst, *Away from the Flock* (1997)	107
4.4	Robbie Collins, *64 Things to Worry About*, Science Gallery Dublin, 2017	109
4.5	Catherine Sarah Young, *An Olfactory Portrait of the Amazon Rainforest*, Science Gallery Dublin, 2017	111
4.6	Grace Weir, *Dark Room* (2015), (left) digital photograph, (right) calotype copy of digital print, Irish Museum of Modern Art	113
4.7	Grace Weir, *3 Different Blacks*, non-lightfast photographic ink on photo paper, 21 × 29.7 cm each, installation view and close-up, Irish Museum of Modern Art, 2015	116
4.8	Marc Didou, *Skull 1* (2007), black and white marble with anamorphic mirror (and close-up), Naughton Gallery, Belfast	119
4.9	Marc Didou, *Eco* (2007), statue in bronze located outside the library of Queens University Belfast	120
5.1	Students in Minnesota experiment with individual camera obscura	127
5.2	Flick book comprised of early cinema stills	128
5.3	(left) The Baptistry of San Giovani, Florence today, seen from the Duomo opposite; (right) Brunelleschi's demonstration of perspective (c.1420), line drawing	132
5.4	Diagram from Alberti's *Della Pictura* (1437)	134
5.5	Masaccio, *The Trinity with the Virgin, St John and Donors*, Santa Maria Novella, Florence (c.1425)	135
5.6	Leonardo da Vinci, *Superficial Anatomy of the Shoulder and Neck* (1452), Windsor Codex, Royal Collection	139
6.1	(left) Wassily Kandinsky, *Sketch for Achryka* (1901), oil and tempera on canvas board, 23.6 × 32.7 cm, Städtische Galerie, Lenbachhaus, Munich; (right) Wassily Kandinsky, *Old Town* (1902), oil and tempera on canvas board 23.6 × 32.7 cm, Georges Pompidou Centre, Paris	153

6.2 Wassily Kandinsky, *Riding Couple* (1906–1907), oil on canvas, 55 × 50.5 cm, Städtische Galerie, Lenbachhaus, Munich 154
6.3 Wassily Kandinsky, *Mountain* (1909), oil on canvas, 109 × 109 cm, Städtische Galerie, Lenbachhaus, Munich 154
6.4 (left) Wassily Kandinsky, *All Saints I* (1911), reverse glass painting, Städtische Galerie, Lenbachhaus, Munich; (right) Wassily Kandinsky, *All Saints I* (1911), oil on canvas, Städtische Galerie, Lenbachhaus, Munich 155
6.5 (left) Kasimir Malevich, *Birkenhain* (1905), oil on canvas, 34.5 × 49.5 cm, Albertine Museum, Vienna; (right) Kasimir Malevich, *Veil* (1908), gouache on cardboard, 61 × 26 cm, Tretyakov Gallery, Moscow 157
6.6 The Picasso Room in the Sergei Schukin Mansion, Moscow (1914) 158
6.7 (left) Kasimir Malevich, *Taking in the Rye* (1912), oil on canvas, 72 × 29 cm, Stedelijk Museum Amsterdam; (right) Kasimir Malevich, *Reaper on Red* (1911), oil on canvas, Nizhny Novgorod State Museum of Fine Arts, Nizhny Novgorod, Russia 159
6.8 Kasimir Malevich, *Painterly Realism of a Boy with a Knapsack: Colour Masses in the Fourth Dimension* (1915), oil on canvas, 71 × 44 cm, Museum of Modern Art, New York 160
6.9 (left) Kasimir Malevich, *Black Square* (1915), oil on linen, 79.5 × 79.5 cm, State Tretyakov Gallery, Moscow; (right) Last Futurist Exhibition 0.10, St Petersburg (1915), installation of the Black Square, St Petersburg, with Kasimir Malevich photographed seated beside fellow exhibiting artists Olga Rosanova and Kseniya Boguslavskava 161
6.10 Wassily Kandinsky, *Painting with White Border (Moscow)* (1913), oil on canvas, 140 × 200 cm, Solomon R Guggenheim Collection, NYC 165
6.11 (left) List of books published in the Bauhaus Books series from Malevich's *Non-Objective World*, Bibliotheque Nationale, Paris; (right) Wassily Kandinsky surrounded by students in the Bauhaus (1928), Bibliotheque Nationale, Paris 174
6.12 Photograph of the movement of the dancer Pelucca by Charlotte Rudoph (1926) and Kandinsky's abstracted drawing of this from *Point and Line to Plane*, Bibliotheque National, Paris 177
7.1 (left) Kasimir Malevich, pedagogical chart (1926/27); (right) Malevich demonstrating his research to his students in GINKhUK, St Petersburg (c.1926) 190
7.2 (left) Kasimir Malevich, design for UNOVIS publication (1922); (right) UNOVIS wall paintings, Vitebsk (1920 191
7.3 (left) Gustav Klutsis, design for loudspeaker system (1922), George Costakis Collection, Athens; (right) Vladimir Mayakosky, Civil War poster (1924) 192

7.4	(left) Kasimir Malevich, pedagogic chart indicating the evolution of modernist painting (1926); (right) Alfred Barr, front cover from his book, *Cubism and Abstract Art* (1936)	194
7.5	Kasimir Malevich, explanatory photomontage indicating the stylistic features of Cubism, Futurism and Suprematism (1925–1926), Museum of Modern Art, New York	195
7.6	(left) Front cover of Malevich's *Non-Objective World*, designed by Lazlo Moholy-Nagy (1927); (right) Students relax at the Bauhaus c.1931, Bauhaus Stiftung/Pompidou Centre, Paris	197
7.7	Malevich's painting class at UNOVIS (1920) (he can be seen by the blackboard)	202
7.8	(left) El Lizzitsky, *Proun 1A (The Bridge)* (1919), pencil and gouache, 15 × 19.5 cm, Berado Collection Museum, Lisbon; (right) El Lizzitsky, *New Man* (1923), from the series Figurines: "Victory of the Sun"	205
7.9	(left) Tatlin and students building a model of his *Monument to the Third International* (1920); (right) Tatlin's unfinished model in a street display (1920)	208
7.10	El Lizzitsky, *Lenin Tribune* (model for a publicity tower) (1920), State Tretyakov Gallery, Moscow	209
7.11	Aleksandr Rodchencko, *Spatial Construction*, with artist in background	210
7.12	Tatlin's flying machine on display at a glider show in Moscow (1933), State Tretyakov Gallery, Moscow, unknown photographer	211
7.13	Leonardo da Vinci, sketch for a flying machine (c.1500), Biblioteca Ambrosiana, Milan	211
7.14	(left) Stepanova at work in her studio (1923); (right) design workshop at Vkhuktemas (1925)	212
7.15	El Lizzitsky, *Footballer* (1921), illustration for Ehrenburg's "Six Tales with Easy Endings", photo collage, pencil, gouache, ink on cardboard, State Tretyakov Gallery, Moscow	216
7.16	Aleksandr Rodchenko, *Steps* (1930), gelatin-silver print, 12.4 × 12.7 cm	217
7.17	Lazlo Moholy-Nagy, *Sailors*, 1929, Metropolitan Museum, New York	218
7.18	Kasimir Malevich, *Arkhitekton* (1926–1928), State Russian Museum, St Petersburg	220
8.1	Margaret Fitzgibbon, *Dresser* (2011), appropriated 1950s bedroom dresser + original letters	233
8.2	Mathew Buckingham, *The Spirit and the Letter* (2007), film still	234
8.3	(left) Pauline Zenk, *Portrait of Marie* (2017), oil on canvas; (right) Pauline Zenk, *Maria coupe en deux* (2017), oil on canvas, 50 × 40 cm	235
8.4	(left) Pauline Zenk, *A Leap of Faith* (1917), aquarelle acrylic and crayon on paper, 29.7 × 21 cm; (right) *L'homme rouge* (1917), oil on canvas, 46 × 38 cm	237
8.5	(left) Trisha Ziff, *The Mexican Suitcase* (2015); (right) French Internment Camp, Le Barcarés, France, March 1939	239

8.6 Desmond Bell, *The Enigma of Frank Ryan* (2013), film still: Ryan meets Abwehr II Officer, Colonel Veseenmayer, in his Berlin office 245

8.7 (left) The Bridgehead building today in central Linz, photograph by Hito Steyerl; (right) The Wiener Graben Quarry Concentration Camp (c.1941), United States Holocaust Memorial Museum 247

8.8 Hito Steyerl, *In Free Fall*, 2010, video HDV, 32′, single channel sound, colour, film stills 253

9.1 Trinity College Dublin (TCD) researchers conduct cranial measurement and photographic documentation of an Aran islander c.1892, National Library of Ireland 261

9.2 Photographic portraits taken during the TCD survey: (top) full frontal and (bottom) side view of three Aran islanders 262

9.3 (left) Postcard of the Irish Village at the 1908 Franco-British Exhibition; (right) John Hinde postcard (1965–1970) 265

9.4 Sean Hillen, *Collecting Meteorites at Knowth* (1996) 266

9.5 John Kindness, *The Dealers* (1987), hand painted poster display, Dublin 267

9.6 Jean Rouch, *Petit à Petit* (1970), film stills: Damouré Zika conducts his own ethnographic research on the streets of Paris 268

9.7 Alberto Baraya, *Machu Pichu Expedition* (2013) 269

9.8 (left) Alberto Baraya, *Herbarium of Artificial Plants*, installation view, Berlin Bienalle, 2016; (right) Close-up of a "specimen" 270

9.9 Still from the film *The Forgotten Space* (2010), 110 mins, colour, by Allan Sekula and Noel Burch 273

9.10 (left) Moira Sweeney, *Rhythms of a Port* (2014), screen shot: "Amy cleaning the hold"; (right) installation view, 5 Digital HD screens in a dockland warehouse 275

9.11 Susan Hiller, *Fragments* (1978), installation, Hayward Gallery, London 281

9.12 (left) Still from the film *Naked Spaces* (1995), 135 mins, 16mm, colour, West Africa, by Trinh T. Minh-ha; (right) Still from the film *Reassemblage* (1982), 40 mins, 16mm, colour, Senegal, by Trinh T. Minh-ha 287

9.13 (left) Ruby Wallis, *My Brother Django* (2014), digital photograph; (right) *Dido* (2014), medium format photograph 292

10.1 Neil Beloufa, *Sustainable Development* (2017), installation view, Regional Museum of Contemporary Art, Serignan (MRAC) 308

10.2 Neil Beloufa, *World Domination*, 2012, shown at Artesmundi 7, Cardiff 2012 309

10.3 Neil Beloufa, *Ukrainian Monopoly* (2015), film still 311

10.4 (left) Lenin in Kremenchuk; (right) Lenin in Teplivka 313

10.5 (left) Lenin in Chernobyl; (right) Lenin in former Lenin Museum, Kiev 316

10.6 Mari Bastashevski, Erik Prince Frontier Service Group, view from the office in Hong Kong (2016), from *State Business* 318

10.7 Mari Bastashevski, SARACEN Uganda, new recruits waiting for transport to Mogadishu, Kampala, Uganda (2017), *State Business* 318
10.8 Mari Bastashevski, NATO Banquet, Pavlograd, Ukraine (2012), *State Business* 320
10.9 Mari Bastashevski, State Department conference room, Washington DC (2016); World Bank mailing room, Washington DC (2017), *State Business* 321

ACKNOWLEDGEMENTS

Many thanks to all the artists who responded to my queries and provided images of their work for the book. Hopefully I have represented their work in an accurate manner. The task of securing image permissions for such a broad range of work initially seemed a daunting task. However, the resulting dialogue, conducted in English, French, German and Russian with a range of artists, curators and gallerists from Moscow to Manhattan, Belfast to Mexico City (thanks to Edita Labanauskiene for translation help), confirmed one of the central tenets of the book, namely that not only do contemporary artists have an acute interest in research as a source of creative insight but also that through their work they make significant contributions to an emergent tradition of artistic research.

Thanks to Connor Clements for preparing the digital image files and to Natalie Foster and Jennifer Vennall at Routledge and Autumn Spalding at Apex CoVantage who went the extra mile in facilitating the sort of image production a book like this requires.

Marcus Banks in Oxford and Maruška Svašek in Belfast read the anthropology chapter with considerable insight, while Sam Cohn in Edinburgh commented incisively on the history chapter and the Renaissance one. Stephanie McBride and Luke Gibbon in Dublin provided invaluable comments on the completed manuscript. Thanks to all. Needless to say, none of the above bear any responsibility for the final text!

I wish to acknowledge the support of the National College of Art and Design in Dublin, which provided a period of research leave enabling me to devote my energies to this book. The Institute for the Advanced Study of the Humanities at the University of Edinburgh provided the hallowed space and intellectual fellowship that facilitated completion of the book. Louise Milne at Napier joined me in a series of late-night conversations which explored many of the key ideas that inform the text. Thanks are also due to all my doctoral students at NCAD with whom a range of methodological themes were initially explored, confirming another tenet of this book – that it is in the pedagogic encounter that the creative arts discover their capacity for theorizing from practice experience.

Lastly, thanks to my daughter Hannah B. She kept my feet firmly on the ground and any "pessimism of the intellect" firmly under control.

INTRODUCTION

What role does research play within the creative and media arts? How can it advance what we do as artists, critics and publics? This book seeks to plot the basis for a research culture for the creative and media arts, a composite field of practice within which lens and screen-based work has become central. It does so by seeking to elaborate the critical basis of contemporary *artistic research*. The purpose of the text is to raise problems, not to offer simple solutions. The arguments I make are largely unoriginal. They are drawn from existing philosophical reasoning and scientific case and from artistic example adapted to cast some light on the nature of creative arts research and celebrate art as a form of intelligence. Rather than offering a systematized handbook of methods, it seeks to provoke debate about the role of research within the creative arts. In any case, given the diverse nature of the disciplines within the creative arts area – encompassing art and design, film and photography, interactive media, music, drama and performance – achieving the former would be a tall order!

Research by artists? Surely artists make stuff, stage, exhibit and perform rather than research and theorize – activities that can be left to scientists and critics, respectively? Indeed, the traditional emphasis in our specialist art academies and conservatoires has been on "making work" rather than "talking up" process, method and rationale. Perhaps these justificatory activities are best left to critics. But, as we shall see, things are changing. Artists are talking back.

Actually, most of us have no difficulty in acknowledging that artists engage in serious research activity as part of the process of making their work. I survey a range of such work in Chapters 2 and 3. At various times in the past, whether in the Renaissance (examined in Chapter 5) or in the work of the radical avant-garde in revolutionary Russia (explored in Chapters 6 and 7), artists were also theorists and thought of their work in ways akin to science. The book offers a sort of genealogy of artistic research to which we might attend as we seek to fashion a contemporary paradigm of research practice for the creative arts. Indeed, the work of the modernist avant-garde in the first decades of the twentieth century increasingly looks like a foundational moment for artistic research. Perhaps we should celebrate it as such?

More recently, curators have responded to the public's interest in understanding the processes that artists employ in the making of their work. Increasingly, exhibitions are designed so that the research activities underpinning an artist's work can be shared with an audience. Curators are also interviewing artists about how their work is made and sharing this contextual information with a range of audiences. Such dialogue, I would suggest, is at the core of artistic research.

All well and good, but can artistic research ever be scientific in character? Should it have this aim? In Chapter 4, I look at the dominant model of science, which shapes discussion of research methodology across a range of disciplines, including the arts. I argue that if we have an open mind with regard to science and its methods, the creative arts researcher can be seen as operating within an expanded field of scientific investigation closely related to art practice. Some creative arts researchers have looked to the social sciences, particularly anthropology, to provide a model for artistic research. I explore this dynamic in Chapters 8 and 9, exploring firstly the work of artists who have engaged with history in their work and, subsequently, the relation between anthropology and the creative arts. In my final chapter, I explore the troubled relationship between critical theory and contemporary creative arts practice.

However, the key focus of this book is how practicing artists go about the business of research. Throughout the book, I draw on a series of interviews conducted with a range of contemporary artists. These focus on their creative process and research activities. Some writers have claimed that contemporary art practice is now so locked into a set of conversations about process and purpose that it increasingly resembles a research practice in and of itself. That is to say, artworks are increasingly produced within a broader framework of critical exchange and public engagement which has a strong investigative basis. Artists increasingly seek to integrate various research-led approaches into their work, often collapsing the barriers between research, project development, art process and exhibition practice and combining these within various forms of social engagement. That said, the focus for most artists when they articulate the research basis of their work tends to be on a description of process rather than an elaboration of method. Process is often project-specific, representing a signature approach by an artist. On the other hand, research methodology strives to achieve a degree of generic application. As we shall see, artistic research is predominantly concerned with questions of process, rather than issues of methodology. The logic of discovery is coeval with the work and the materials and media employed.

Rirkrit Tiravanija serves his audience a meal of curry in a New York gallery as an art inspired act of conviviality (p. 73). Tino Seghal, in his 2013 Turner Prize Exhibition piece *This Is Exchange*, directs the gallery assistants in Derry to quiz visitors about the market economy, rewarding some respondents with a one-pound coin. Neil Beloufa (p. 311) films a group of Kiev teenagers engaged in a perverse game of

Monopoly in which their troubled land is up for sale. The art object is now indissociably bound up with conceptual, communicational and action outcomes. Fine artists experiment with film-making; media makers grapple with conceptual questions. The "expanded practice" embraced by many artists is creating an opening for the forging of a research culture within the creative arts – perhaps for the first time.

In addition, the creative arts and media have demanded – and to some extent have been afforded – parity with other subject areas within the university system as areas of research. "We are also researchers", art educators assert. "Our art is an investigation like any other". The academy seems to have begun to accept the idea of artistic research. Moreover, the last twenty years have seen a massive growth in the number of courses in the creative arts and media in the UK and Ireland, across Europe and in the US. Much of this growth in creative arts provision has taken place outside of the art academy and specialist conservatory within the university sector. Most university courses in media, communication, cultural studies, drama, dance and performance, architecture and design incorporate a practice element.

They also foster research expectations. So, while advanced creative arts and media undergraduate students are expected to conduct a range of practice projects as forms of independent learning, such projects – whether driven by conceptualist concerns or following the codes of professional media production – are expected to demonstrate a research intent and underpinning. Research aids active learning. Postgraduate education in general is increasingly research-led in character, and postgraduate study in the creative and media arts is no exception to this. Students are expected to develop a range of research competencies and critical capacities in their creative project work and are assessed on this basis rather than solely on their creative outputs. Advanced creative arts and media students are expected to follow courses on research methods. Students are being enrolled in research degrees, pursuing PhDs by a practice route which combines the production of artworks with critical and reflective writing. Most creative arts academics now envisage art practice as research-led.

In the race to enhance an institution's research profile, the artworks produced by creative practice academics and research students – paintings, sculptures, films, photographic and computer imaging, performances – can be treated, for all intents and purposes, as equivalent to the print-based research publications of other disciplines. Considerable numbers of practicing artists and media professionals have been recruited into higher education to staff this expansion of creative arts and media teaching. Many seek to advance their careers through their research achievements. Given their professional backgrounds in art and media rather than in traditional humanities scholarship, practice-based forms of research seem like an attractive option for these new entrants. Such research facilitates a link to their teaching commitments and professional interests.

A new research-based, creative arts economy has emerged at a time when traditional forms of public funding for artists are in decline. New opportunities are there to be exploited. You can now undertake a funded PhD in creative arts subjects by practice, and large research grants are available to support various art-based investigations. With an expanded notion of academic publication that now incorporates exhibition and performance, the academic careers of artists can be advanced by research achievements. Hurrah!

However, in this rush to promote artistic research within the academy, certain confusions can arise as to how the creative arts might best be understood. For example, is research replacing creativity in a new administrative aesthetics which endangers artistic autonomy?

Another equally troubling question inevitably arises. The sciences have their research methodologies, which are often quantitative and replicable in form. These are capable of delivering law-like generalizations about the natural world. What is their equivalent in the creative arts and media, where the core activity is focused on the making of a singular artwork? Is artistic research primarily about a knowledge outcome or an aesthetic one? Surely it is about creative process rather than scientific method? And, if the creative arts have not quite resolved these matters, with no clear research tradition with definable methods emerging to date, where should we be looking for inspiration? To the natural sciences with their powerful models of verification and explanation, to the social sciences which seek to understand society and the social subject or to the humanities with their engagement with culture, history and meaning? Or, rather, can artistic researchers elaborate their own working methods? These may well be based on creative activity within the studio, field and gallery, and in artists' engagements with audiences and society, rather than adherence to a generic methodology.

As I shall argue, the creative arts and media, with their focus on "ways of making and doing", have their own distinctive modes of apprehending and understanding the world. A film-maker, photographer or performance artist sees and moves within the world in distinctive ways. The knowledge they acquire is rooted in their creative encounters. Artistic research needs to start from this specificity. Herbert Read, the empathetic historian of modernist art, talked about the need for a "philosophical approach that recognizes art as a means of conceiving the world visually". As he argued (1968:12),

> There are alternative methods of conceiving the world. We can measure the world and record our measurement in an agreed system of signs (numerals or letters); we can make statements about the world based on experiments. We can construct systems that explain the world imaginatively (myths). But art is not to be confused with any of these activities: it is an ever-living question, asked of the visible world by the visual sense, and the artist is simply the man who has the ability and the desire to transform his visual perception into a material form.

We can broaden the scope of Read's focus beyond the visual to include the sonic and performative aspects of art practice and to other ways of making and doing less concerned with the retinal. However, his general point about the distinctiveness of the arts as a way of knowing the world seems valid and is the premise from which this book proceeds. It is also, I would suggest, the basis on which a viable tradition of artistic research might be built. And, just as other disciplines celebrate their "founding fathers and mothers" and define themselves by reference to a history of key texts and projects, so the creative arts will have to attend to the provenance of its own distinctive research traditions (an idea explored in Chapters 5 to 7). At the same time, we need to acknowledge how the dialogue between the creative arts and a range of other disciplines and critical discourses has shaped artistic research practice (and this theme is pursued in Chapters 8 to 10).

PART I
Critical parameters

1

CREATIVE ARTS AND MEDIA RESEARCH

Key terms and perspectives

1.1 Introduction

We are, then, witnessing an increased role for research within the creative arts both within the academy and in the art world itself. The last ten years have in effect seen a reconceptualization of the creative practice project as a research-led exercise. Moreover, contemporary art practice and teaching are no longer tied to craft specifics and the mysteries of individual media. Conceptual activity and practices of research and development are now at the core of art and media education.

But how does artistic research fit into the broader field of academic research? What might be the basis for a systematization of its methods? For instance, can creative arts research methods be taught? Is it appropriate to try to systematize research methods in an area as singular as the creative arts? Does the emphasis within arts education on individual imagination and expressivity and on the creative development of the artist mean that artistic research is beyond the scientific pale?

1.2 Defining the key terms

Research is a portmanteau term with a number of meanings. For example, when we employ the internet and world wide web to conduct a series of informational searches for cheap air tickets, best web buys or possible partners, we now refer to this activity as "research". The dictionary definition of the term focuses more on the systematic character of research as a knowledge-gathering activity, characterizing it as "an endeavour to discover new or collate old facts etc. by the scientific study of a subject or by a course of critical investigation".[1]

This definition seems sufficiently wide to incorporate the creative arts. However, it is probably true to say that this and other standard definitions of research have at their core a privileged meaning, and this is related to notions of *scientific investigation*. Scientific research is held to follow certain methodological procedures involving observation, objective data gathering and quantification. It provides generic explanations rooted in substantiated bodies of theory. These *empirical* methods underwrite the validity of science's knowledge claims (the logic of scientific discovery is examined in Chapter 4). It is argued that there is a unity of scientific method. So, ideally, all research activity should adhere to this model, including artistic research.

In addition, for research to be judged worthwhile in character, it should have a measure of *originality* and contribute something new and innovative to the field of investigation. It should come up with fresh empirical or theoretical findings which advance our knowledge in a field or apply existing knowledge to generate new practical outcomes of value, such as the technological sciences and design research do. Artistic research actually scores pretty well on the criteria of originality. Though, as we might expect, it is the originality of the artwork produced, whether painting, performance, film, composition or design, that is highly valued in arts research rather than any general conclusion or knowledge claim.

Of course, when we invoke the criteria of systematicity, artistic research seems to immediately run into problems. Research in the arts tends to follow an open discovery path (though often in the area of media arts, this is guided by professionally established codes of production). Its focus is primarily on the making of a unique art object or media artefact and only secondarily on reflecting upon this process with a view to arriving at any set of abstract generalizations.

The exception is perhaps advanced art education. Here, as we shall see, students undertaking creative project work are encouraged to treat the studio as a laboratory of sorts. And, as we shall see, historically it has been the pedagogic challenges associated with modernist art education that have provided the basis for an emergent research tradition within the visual arts. Thus, while there has been much talk within art discourse about the experimental nature of contemporary art, and a corresponding fondness for conceiving of the studio as a laboratory and the art process as an investigative activity, this use of scientific terminology within the creative sphere is largely metaphorical. The mode of understanding in the arts tends to focus on the specifics and materiality of the art process and on the production of an artefact, proposition or performance, rather than on the sort of law-like generalizations to which the natural sciences and some of the social sciences aspire. Yet to count as research within the academy, artistic research has to be seen as conforming – to some extent – to the established protocols of science.

This privileging of objectivist scientific method in the conduct of research and conceptualization of knowledge is often referred to as *positivism*. This powerful, indeed, "imperial" theoretical position (examined in depth in Chapter 4) plays a dominant role within the broad research community. With its roots in philosophy's early twentieth-century accommodation to the rise of the natural sciences, positivism seems to want to legislate methodologically for other domains of enquiry, such as the arts and social sciences. Often it appears to have the institutional power to do so, despite the fact that a field like the creative arts does not seem at all to conform to the rationalist and nomological (law-like) character of the natural sciences.

Moreover, this totalizing position is often what gets us into trouble when we seek to model artistic research. "Stupid as a painter" goes the age-old denigration of artistic intelligence. Certainly, in the past the work of visual artists was often unfavourably compared with the forms of rationality deployed by scientists and writers. But artists, I will argue, have a special way of seeing and understanding the world through their practice. Artistic research seeks to recognize and formalize the work of the "thinking eye" and the artist's sensory engagement with the world. To quote the motif displayed on a poster prepared by the artwork collective Jeffrey Charles Henry Peacock for their 2014 Lanchester Gallery Projects (Figure 1.1) "our practice testifies that things sensed cannot be immediately understood by us and that only things which are understood can be still more profoundly sensed."

By *artistic research*, I mean the research that artists do as artists, rather than the research that is done on them and their work by critical scholars. I prefer this term to the clumsier couplets *practice-based* arts research or *practice-led* research or, indeed, to the term *practice as research*. The notion of artistic research seems to

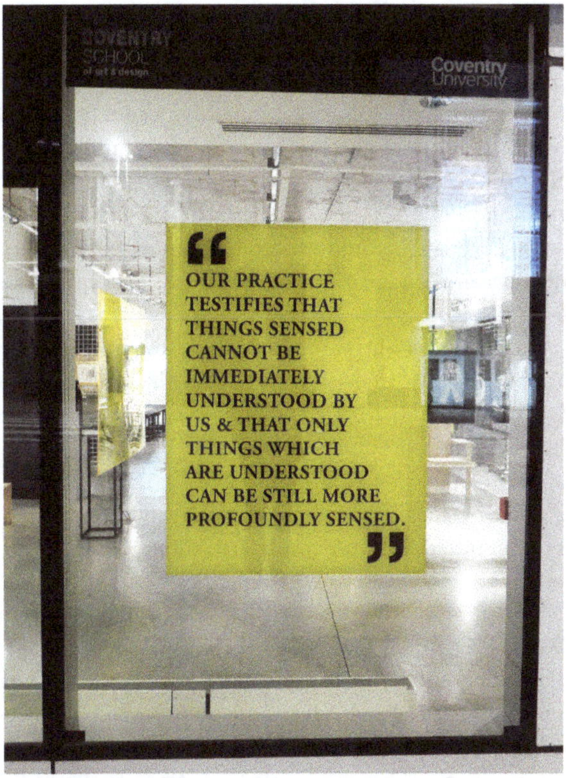

FIGURE 1.1 JCHP poster, *Critical Décor: What Works* Exhibition, Lanchester Galleries, Coventry, 2014

Source: courtesy of the JHCP Collective

acknowledge art activity as a fusion of creative and critical elements. It rejects any attempt to drive an analytical or institutional wedge between the acts of making and of *theorizing*. In the first instance, by theorizing I mean the process of providing a critical context for one's creative work so that both its creative intent and communicational achievement can be understood and evaluated by others (in part by placing it alongside comparable work in the field and in part by setting it into conversation with various critical discourses and interlocutors and giving the work a social and historical context). This text champions the idea that practice can provide both an empirical and critical basis for theorizing arts and culture.

In part, the notion of "theory" employed in creative arts research and pedagogy refers to the importation of general sociological concepts, models of mass communication process and forms of cultural analysis into the fields of media, communication and cultural studies. But it is also used in a more specific manner within media arts and in studio-based art education to refer to the analysis of cultural representation and of semiotic and ideological processes *and* to privilege this field of investigation over other approaches to researching the creative arts. In this latter case, the term *critical theory* is often invoked to qualify the notion of theory employed. And, in this guise, the notion draws upon a varied body of sophisticated critical and philosophical theory, much of which is derived from French poststructuralist and deconstructionist sources (Roland Barthes, Jacques Lyotard, Julia Kristeva, Michel Foucault, Jacques Derrida, Gilles Deleuze, Helene Cixous – to list but a few of the luminaries whose work is often drawn upon).

However, the notion of *critical* invoked in artistic research is a complex one with at least three, often overlapping, meanings.[2] Firstly, the term alludes to the classic Enlightenment concern with the rational foundations of scientific and other forms of knowledge. In the case of the creative arts, the focus is often on the basis of aesthetic judgements. Philosopher Immanuel Kant sought to identify the conceptual foundations of Newtonian physics and expose the limitations of metaphysical thinking in his *Critique of Pure Reason* (1781/2007). Kant was also concerned with the rational basis of moral reasoning and of aesthetic judgement.

The second use refers to the radical critique of political economy and capitalist society advanced by Karl Marx. This approach envisages a form of social and political theory committed not only to understanding the structure and dynamics of capitalist development but also to providing the scientific basis for transforming society. A later generation of social theorists, associated with the Institute for Social Research in Weimar Germany, sought to extend Marx's analysis to the broader study of social domination. They began to use the term *critical theory* to discriminate their radical approach to social and cultural theory from positivist accounts of social science method.

Lastly, the term critical has been used to refer to traditions of exegetical scholarship and cultural evaluation, that is to say, to practices of *criticism* or *critique.*[3] With its origins in biblical interpretation, from the eighteenth century the term spread to the assessment of cultural works – books, plays, performances, exhibitions – by specialist intellectuals operating within the sphere of public culture. The critic sought to understand and evaluate the merits of a text or form of discourse or artwork, performance or proposition in a reading shared with an interested audience to whom they offered a considered judgement. As Raymond Williams points out (1976:75), such judgements became associated with an ability to "discriminate" in matters cultural. Thus, the term became inextricably connected with questions of taste and distinction:

> **criticism** in its specialized sense developed towards TASTE (q.v.), *cultivation*, and later CULTURE (q.v.) and *discrimination* (itself a split word, with this positive sense for good or informed judgment, but also a strong negative sense of unreasonable exclusion or unfair treatment of some outside group – cf. RACIAL).

Creative art and media students are pretty familiar this third use of "criticism". After all, since the end of the nineteenth century, the "crit" has been institutionalized within the art school as a key pedagogic tool central to studio-based learning. The crit, or critical review, is a one-on-one tutorial meeting between student and tutor – or a group tutorial encounter between tutors and a group of students. Work is presented to a tutor for comment and in some cases to industry practitioners (common in architecture, design, film and photographic education). The aim is to strengthen the development of the student's work and their insight into art and design processes more generally.[4] The crit is designed to provide feedback and evaluation of a student's creative work and to more generally foster a spirit of self and peer criticism on the part of students. Students are expected to communicate the creative intent of their work and identify its artistic and intellectual auspices to their tutor, who, in turn, seeks to objectively comment on work presented.

I use the term *creative art*[5] to refer to the broad range of *studio-based* arts disciplines concerned with "making and doing" – the visual arts and design; the lens and screen-based arts forms, such as film and photography; performance, literary and sonic art activities, such as drama, dance, music making and creative writing; and all the hybrid forms that mix and merge these practices. The couplet *creative arts and media* acknowledges a common field of professional and educational practices concerned with invention and creation. Indeed, that commonality at least in part rests on the widespread use of a range of lens and screen-based media across a range of arts disciplines. It also rests on the *conceptualist* character of much contemporary art. Ideas, rather than the use of specific media or production techniques, now drive most contemporary art projects. The notion of *studio* I use here is a broad-based one incorporating the traditional artist and designer's atelier, but also the photography and film studio, the

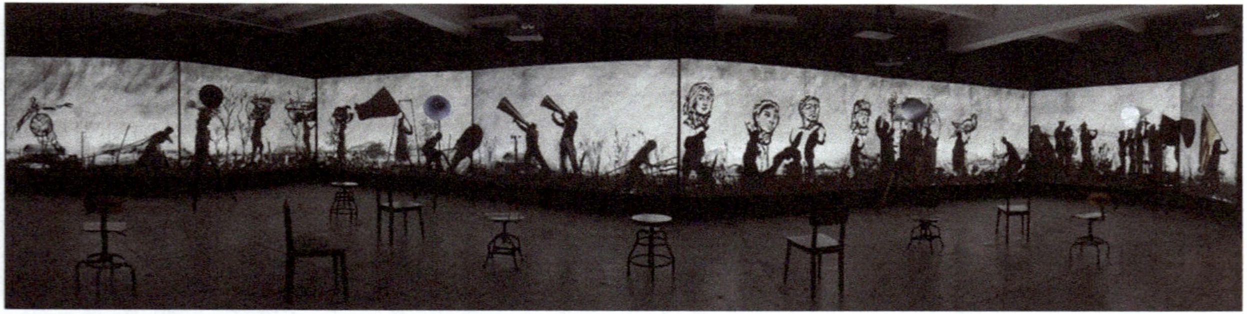

FIGURE 1.2 William Kentridge, *More Sweetly Play the Dance*, multiscreen installation, Luma Gallery, Arles 2016

Source: courtesy of the artist

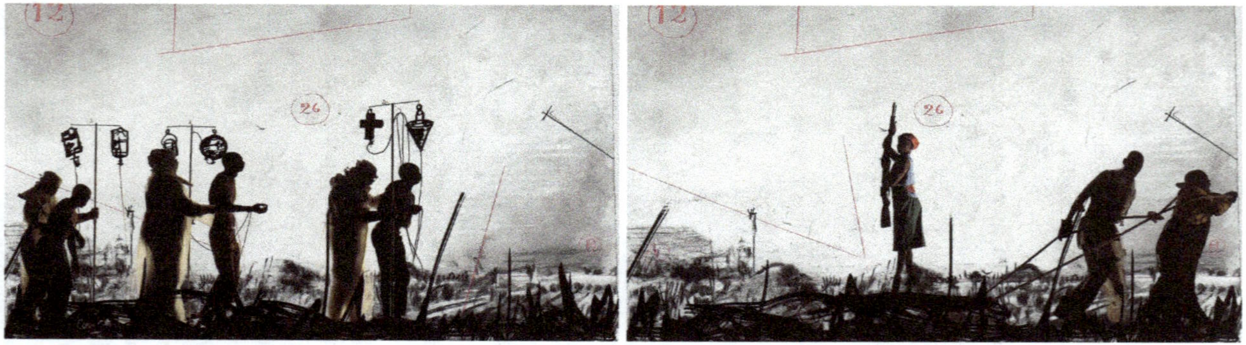

FIGURE 1.3 *More Sweetly Play the Dance* close-ups

Source: courtesy of the artist and the Marion Goodman Gallery

performance workshop and adapted gallery space, the writer's study and the multimedia workspace – in short, anywhere creative projects and practices are originated, developed and delivered and ideas and creative strategies are explored and implemented to produce creative work.

Again, I do not wish to draw any hard and fast lines between fine and applied art, or between the visual arts and the media and performing arts. Today the creative arts are increasingly interdisciplinary in character with elements of performance, installation, lens-based imaging, blending with the traditional concerns of the painter, sculptor, musician, film-maker and photographer.

For example, in the summer of 2016 the South African artist–film-maker William Kentridge exhibited a multiscreen installation *More Sweetly Play the Dance* at the Luma Foundation gallery in Arles in the south of France (Figures 1.2/1.3), a piece previously shown at the EYE Film Museum in Amsterdam and reported on in a book with the same title (Kentridge 2015).

This tableau vivant portrays a caravan of dancers, marching bands and animated figures that winds its way past the viewer across eight assembled cinema size screens over 40 metres in length in a riveting *Danse Macabre*. Kentridge employs his trademark animated charcoal drawings overlaid with a range of video elements, including live action performance of dance and dramatic gestures (a clip from a video of the exhibition can be seen at: https://youtu.be/pA7uob5PIPw).

Kentridge's ceremonial parade/funeral is part sorrowful, part joyous and can be read as a moving catalogue of the winners and losers thrown up in the course of history.[6] It alludes specifically to the victims and malefactors in South Africa's recent political struggles. Categorizing this brilliant and evocative piece of contemporary art in medium-specific or disciplinary terms is all but impossible. Its research content is ever present but hard to methodologically divine. And, that's where we are at.

1.3 The distinguishing characteristics of artistic research

What the different arts disciplines concerns have in common is their concern with *invention*. This is a notion which, as we shall see, had particular purchase during the Renaissance (where it was closely related to artisan-based, design and scientific activity and thinking). Much later, the notion becomes central to the artistic culture of the modernist avant-garde. Making art involves a projection of a creative imagination realized and given public form through a medium. As art theorist Paul Carter (2010:15) argues,

> in the process of invention, the heterogeneous interests of the poet, choreographer, the hip-hop deejay, the AutoCad designer, and the landscape architecture display their common interests. The condition of invention – the state of being that allows a state of becoming to

> emerge – is a perception, or recognition, of the ambiguity of appearances. Invention begins when what signifies exceeds its signification – when what means one thing, or conventionally functions in one role, discloses other possibilities.

What, then, distinguishes artistic research from other fields of enquiry? We can, I think, identify two core characteristics common to artistic research which differentiate its methods from the research and scholarship conducted by disciplines like art history, visual cultural studies and film and media theory (critical approaches that also have artworks and cultural matters as their objects of analysis). Firstly, artistic research employs a research process executed through the making of artwork – paintings, sculptures, films, art propositions, events, photographs, sonic and performance activities and recorded outputs, etc. It is, in other words, practice-based.

Secondly, such research is "discovery-led" rather than hypothesis-driven like much research in the dominant scientific tradition. French philosophers Gilles Deleuze and Felix Guattari (2013:421) identify a "nomadic" conception of science that can be distinguished from the forms of institutionalized or "royal science" closely associated with state and corporatist interests. Nomadic science is not so much concerned with reproducing existing knowledge through formalized procedures concerned with confirming theories but rather with exploring various concrete lines of enquiry. These emerge hesitatingly out of engagements with materials, professional practices and real-world problems. In this investigative approach, historically firmly rooted in artisanal experience, the researcher subordinates "all their operations to the sensible conditions of intuition and construction" (ibid.:435). As we shall see, intuition and construction are core to artistic research.

One of the best descriptions of the process of artistic discovery within the studio has been provided by William Kentridge. In the course of an illuminating public lecture on the "natural history of the studio" given in Tel Aviv in 2012, he traces the discovery path he usually follows as an artist (www.youtu.be/Vm1juXIaCsg). Kentridge sees the studio – in his case the place where his drawing and film animation activities are conducted and reflected upon – as a "protected space" for the construction of artistic meaning. This he sees as closely related to the physical and material processes involved in producing his work – the charcoal drawing, photographing and animation activities that make up his film practice. He talks about the artist's discovery path involving a "giving over to the medium". His particular interest as an artist is in the relationship between these physical processes of making and the forms of thinking they can release.

As he notes, the initial impulse in the art process may be an irrational one and subject to the ad hoc constraints intrinsic to the medium and materials employed. In his terms, it takes the form of a "provocation". However, what follows as the artist responds to this spur is a rigorous and reflective process through which they learn to explore and master "what the provocation suggests". As he concludes,

> The idea, the first provocation, is very quick. What take the days, the hours, is trying to train yourself to understand what has been shown to you . . . what a random or ad hoc chance or whim unleashes. So, this is the work of the studio the constructing, filming, performing, rewinding, drawing redrawing . . . re-looking.
>
> The idea is never enough . . . it's always the embodiment of the idea, how it finally is turned into a piece of film, a piece of paper . . . that is the question.

The studio (like the scientist laboratory) provides a vital protective space for experiment, uncertainty and, indeed, failure. Kentridge wants to valorize this exploratory aspect of the studio over "a studio of good ideas, of things worked out in advance then executed", that is to say, a studio of hypotheses, scripts, storyboards and received ideas.[7] For it is "from the making and from the very imperfections of technique that parts of the world and parts of us are revealed that we neither expected nor knew before". Studio activity both mirrors and shapes the creative process that goes on in our heads. In the studio, this process gets externalized and "demonstrated".

> The finished drawing or the finished film is a demonstration of the process, of the completion of the process. The physical activity – the months, the weeks, the hours – of doing the drawing are a slow-motion version of how it is we appropriate and understand the world.

However, for Kentridge the studio, while providing a protected space for experimentation, also resonates with the outer world. It can be seen as "a kind of membrane between the outside world (. . .) and the activity of making the image, of making the artwork inside the studio". For him, "the studio becomes an emblematic demonstration of what it is when we go through the world". What, he asks, is it that penetrates into the studio?

> It's obviously history, it's politics, it's the newspaper you've read in the morning, it's images coming in from television, it's a memory of a conversation from a week before. . . . The raw material which enters the studio is transformed and gets disintegrated into fragments and then reconstituted as a drawing, as a film, as a text, as a lecture.

So, artistic research is primarily pursued via practice. Its outcomes are predominantly concerned with specific artworks, propositions, performances and engagements, rather than with an abstract knowledge object or a generalizable set of theoretical arguments or understandings expressed in formal academic or scientific language. Artistic research may be more concerned with facilitating imaginative and critical thinking and practice than in the generation of systematic knowledge or the search for an elusive universal methodology. The artwork is a complex of meanings, some intentionally striven for by the artists, some imputed by their audience. Artistic research both assists in the development and production of the artwork and critically addresses the processes of making meaning through art activity. As such, its "results" may be based as much on intuition, professional *nous*, openness to a myriad of intellectual and cultural influences and a capacity to respond intelligently to contingency as on the application of any rigorous methodology. That said, as arts academics we anxiously ask – are the array of ad hoc professional research practices employed within the creative arts and concerned primarily with making work comparable in any way to those found in the more established areas of the natural and social sciences, or indeed to those found in humanities scholarship? And if not, does this matter?

1.4 The main types of artistic research

Arts academic Chris Frayling (1994) has attempted to delineate the major types of artistic research. His schema provides a good overview of the activities which cluster under the general title of arts research and which are present in any art college or university department of creative arts where critical scholars work alongside practicing artists.[8] He does not, however, identify any specific set of methods to guide artistic research.

Frayling plots three modes of artistic research:

- Research into Art and Design
- Research through Art and Design
- Research for Art and Design

1.4.1 Research into the creative arts

This focuses on the critical interpretation of artworks and remains the predominant tradition of visual arts research in higher education. Indeed, it is the one most widely accepted within the academy. It involves scholarly investigation into art, design and media works, into creative processes, institutions and cultures, whether employing historical, critical or empirical methods. This academic approach draws predominantly upon the broad hermeneutical and critical traditions of the humanities and has a prominent place within teaching and research in the visual arts sector through the teaching of art history and visual culture.

For want of a better term, I will label the methodology of the research conducted in these fields as *theoretico-critical*, signalling that the key methods employed are exegetical, conceptual and evaluative. Sometimes within the creative arts, this sort of research is simply called "theory", reflecting the impact of complex bodies of poststructuralist critical theorizing on art and cultural discourse. In the field of media arts, scholars working in a tradition of film and media studies have fused humanities and social science concerns to produce a body of interpretative research on film and media works. They also analyse media processes and communicational flows, examining the ownership and control of media institutions and how these structures facilitate certain political interests. Some call this *media theory*, and we now have a range of textbooks with deal with research methods within media studies (Priest 2009; Anderson 2012) and, indeed, more recently, with particular reference to social media (Sloan and Quaane-Hase 2017).

My interest in this form of research is predominantly how it might inform creative practice. Art, design and media students of course write academic essays and dissertations related to the broad field of art history and visual cultural study. That aspect of their programmes is often described as "critical", "contextual" or "complimentary" studies. However, they are also expected in their studio work to be able to draw upon perspectives and concepts from these discourses to provide a critical and coherent methodological and evaluative dimension to their studio work.

Take, for example, an advanced student who is interested in innovation within the documentary film form around the notion of "performativity". They plan an experimental, authored piece of an autobiographical character dealing, for instance, with the topic of parent–child relationships in dysfunctional families. They have access to and plan to use a range of family home movie material and recorded communications with their parents as key visual and sonic sources. Their tutors will expect them to review a body of relevant film work in order to situate their own studio and field plans – work that addresses this specific topic or which perhaps experiments with autobiographic voice and other performative elements in the narrational process. They may be directed to a film such as Jonathan Caouette's 2003 creative documentary *Tarnation*, which drew from hundreds of hours of old Super 8 family footage, VHS videotape, photographs and answering machine messages, collected over twenty years, to tell the story of his life and his relationship with his mentally ill mother Renee (for the trailer of this movie, see: www.youtube.com/watch?v=mLDQL23nutw).

They will also, however, be expected to be aware of a range of critical issues which may have bearing on their work, for example, the problematization of the boundary between the factual and fictive within contemporary documentary film, as explored by authors such as Bill Nicholls (1992) and Stella Bruzzi (2000). Engaging with such work, it is hoped, will facilitate a critical contextualization of their practice. Their finished submission might typically involve a ten-minute film together with documentation of the film's development and production and an analytical summary and self-critical evaluation of the project. All of these elements are seen as constituting the research – although it is somewhat telling that the theorizing is often thought to reside in the written part of the student's submission.

In the past, the discipline of art history provided many of the analytical tools for visual arts students to "theorize" their practice. Similarly, film studies and critical theories of photography have facilitated the contextualization of imaging practices. Again, advanced students are expected to situate their imaging work within a critical context constituted in part by theoretically inflected readings of their work and in part by a consideration of exemplars of leading practice in the field. I explore the relationship between theory and practice in more depth in the next chapter, as well as in my final one, where I identify some of the limitations of current formulations of theory within the creative arts.

1.4.2 Research through creative arts practice

Frayling's second category (often referred to as "practice-led" or "practice-based" research) identifies art process as the key method employed in an arts investigation. Accordingly, the focus is on the making of the creative work *as a vehicle for an enquiry*. The research outcome usually involves the production of an artefact and the critical appraisal of this in a body of writing. Such research is usually sited in a university and tends to be shaped by the broader research norms of the academy. In this model, the artist–researcher is encouraged to elaborate a research design within which various research questions will be identified and investigated and subsequently to systematically reflect on their creative process and product, arriving at some set of conclusions with regard to the research questions initially posed and refined throughout the life of the project. This approach perhaps finds its most coherent expression in the practice-based PhD, and it may be useful to map the research terrain (Figure 1.4) that a doctoral student is expected to navigate.

As can be seen from this chart, the demands made of the doctoral student are considerable. Not only must they originate, develop, deliver and exhibit a body of creative work of an advanced and original character, but they must also provide a written exegesis detailing and defending the creative approach taken. Their written dissertation must explicate a range of existing sources on their chosen topic via a literature review, also identifying and evaluating cognate practice in the field in order to clarify the artistic auspices of their work. Candidates are expected to achieve levels of scholarship and methodological rigour in their writing. These should be equivalent to those achieved by other doctoral students in terms of the coherence of the arguments they advance and the evidence provided to support their thesis, and should conform to an academic style of citation and layout.

To give an example, painter Alison Pilkington embarked upon a PhD programme in Fine Art by practice at the National College of Art and Design in Dublin. She had completed her undergraduate study in Fine Art in the college and had subsequently gained an MA in Film and Television Studies at one of Dublin's

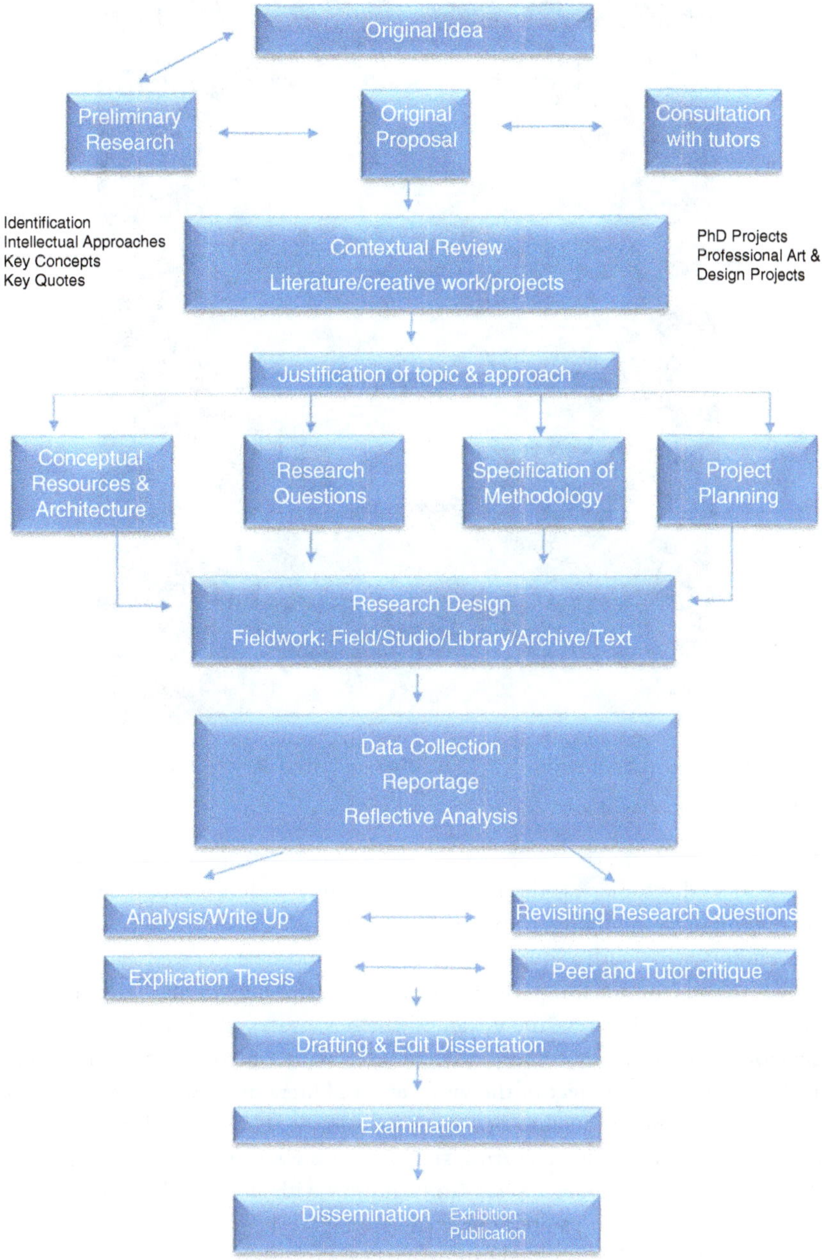

FIGURE 1.4 The creative arts PhD research route

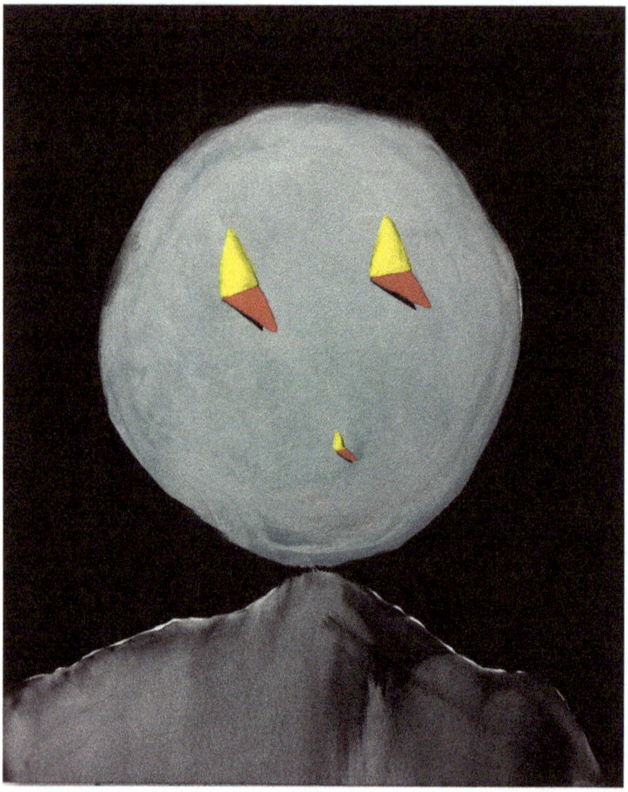

FIGURE 1.5 Alison Pilkington, *Solo* (2013), oil on canvas 70 × 100 cm

Source: courtesy of the artist

universities. Her topic was the role of the uncanny within painting, in particular, in contemporary figurative painting. As she tells us in in her dissertation (2015), "my enquiry starts with a simple premise, that it is possible to investigate the highly subjective and emotional experience known as the uncanny through the practice of painting". As she observes, "the uncanny, which is associated with a feeling of disorientation, mild panic or confusion when faced with something strangely familiar, has been a frequent subject of the visual arts and literature". She chose to explore this theme via the production of a body of painting. Her work, delicately poised as it is between figurative picturing and abstraction and with its appropriation of childish symbols and sources, seems to furnish us with an example of the uncanny at work in and through painterly practice (see Figure 1.5).

In turn, she sought to provide an interpretative context for the work produced. This involved excavating the notion of the uncanny – as present in the work of Freud – and tracing how a number of post-war artists, such as Francis Bacon and

Philip Guston, and more contemporary ones, like Glenn Brown, Angela de la Cruz and Andrew Glassie, have grappled with the psychic and creative dynamic around the uncanny in their work.

Her preliminary research centred, then, both on matters of conceptual clarification and art historical review and on the documentation of her own studio process. In the latter task, she experimented with journal writing, photographic documentation, auto-ethnographic reportage and, indeed, creative writing strategies. Her methodology, in other words, involved the making of a body of work and documentation of this, exploring a range of memories and emotional sensibilities through the materials and practices of painting. As a reviewer of the exhibited work commented,

> the paintings and collages are the fabric woven around an occurrence lodged in the midst of other more banal psychic workings. Referencing now the backdrop of a mundane day that proved worthy of repression, they pinpoint a confused moment extracted from the onslaught of visual stimuli that is childhood.
>
> *(Rainsfort 2014)*

Pilkington reports that the original idea for the project came from her own unnerving experience of a morning viewing of Giovanni Bellini's *The Frari Triptych* (1488) in Venice, the work backlit by the enormous nave windows of the church of Santa Maria Gloriosa dei Frari with her viewing of the images inflected by this explosion of light. Given the highly personal nature of this quest and its expressive dimensions, she experienced some difficulty in formalizing the research questions which her tutors thought should guide her investigation. Indeed, like many artists, she was resistant to the idea that her research journey could be pre-mapped in objectivist language of "research design" rather than discovered in and through painterly processes guided as much by instinctive concerns and provocations than by ratiocination. As she argued (op. cit.),

> my research demonstrates how art practice as a mode of research, demands a "*sui generis*" approach. *Sui generis*, meaning that it engages processes and procedures that produce outcomes and knowledge that are particular to art and embedded in the unique knowledge that art can produce.

That said, she had to navigate a number of methodological culs de sac before arriving at an appropriate academic organization of her work. Early aspirations to invoke a communicational model for painting which might sketch the encoding and decoding of cultural meaning by artist and spectator had to be abandoned, as did plans to include an element of reception study to gauge how spectators might view and respond to her paintings. The research at times threatened to turn into bad social science – though, like any artist, Pilkington retained a keen interest in how members of the public reacted to the exhibited work. Similarly, early attempts to locate her reflective activity within a phenomenological philosophical framework gave way to a more art-grounded approach to understanding the act of reflection within creative practice. Art is not philosophy, though it can be informed by philosophical thinking. Slowly and surely, the discovery path began to centre on the exploration of the inherent qualities of painting as a medium and how its materials and processes provided the basis for a "thinking through" of painting's mode of display and how specifically this might support certain representations of the uncanny. This "data" was in effect generated by close attention to her studio process and through a set of analytical reflections supported by her extensive documentation. And, throughout her investigation, she continued exhibit work, including a number of her doctoral pieces in the Royal Academy 2012 Summer Show, the 2013 RHA Annual Exhibition in Dublin and the Marmite Painting Prize touring exhibition (2012–2013). Her PhD exhibition at the Pallas Gallery in Dublin in February 2014 presented a completed body of work to her examiners and later to the public. This, together with a 55,000-word dissertation, constituted her successful doctoral submission.

1.4.2.1 Reflective practice

In an attempt to elaborate a methodology for artistic research through practice that remains faithful to the demands of art process yet is capable of some measure of formal modelling, a number of authors have offered a methodological prospectus based on the notion of *reflective practice*. Reflective practice has been understood as the ability of a professional or team of professionals to reflect on an action as an individual or group, so as to engage in a process of continuous learning, "leading to developmental insights and the refinement of that practice".

This model has its origins in practice-based professional settings such as architecture and design (the initial formulations were by Donald Schön (1930–1997), an architect, educationalist and pragmatist philosopher). Schön (1983) defines reflective practice as the research process by which professionals become aware of their implicit knowledge base and learn from their experience, constantly testing and refining this with each new project.

In the field of design research, such reflective practice is often combined with a variety of other observational and data-gathering methodologies, such as the use of ethnography, focus groups and sample surveys. Designers develop and test products and services through observing people and asking questions about user need. Their approach is inclusive: they search for information about materials, users and functionality; they test product ideas and refine prototypes through various iterations and analytical reflections.[9]

Indeed, many design academics now see the research process as being at the core of intelligent design activity. As Alex Milton and Paul Rodgers argue in their textbook *Research Methods for Product Design* (2013:18),

> typically, the creation of a new product begins with an idea and ends with the production of a physical artefact. Design research methods support the design and development of new products by helping provide invaluable data, expertise and knowledge through observing, recording and analyzing how consumers interact with the designed world.

The focus of the design researcher is not, then, on arriving at an abstract knowledge outcome, nor indeed a purely technical one, but on generating an informed design solution to a problem that has been presented to a design team and which has been rethought and refigured. Researchers in a design enquiry learn from their own professional experiences, through the processes of goal setting, reflection, reportage and design iteration that take place in a project-based, problem-solving, professional environment. Design process is in effect a cycle of knowledge gathering, visualizing, building and testing, or, rather, it is a series of such iterative cycles through which a product or design solution emerges. Reflective practice is identified as the methodological basis of this research dynamic.

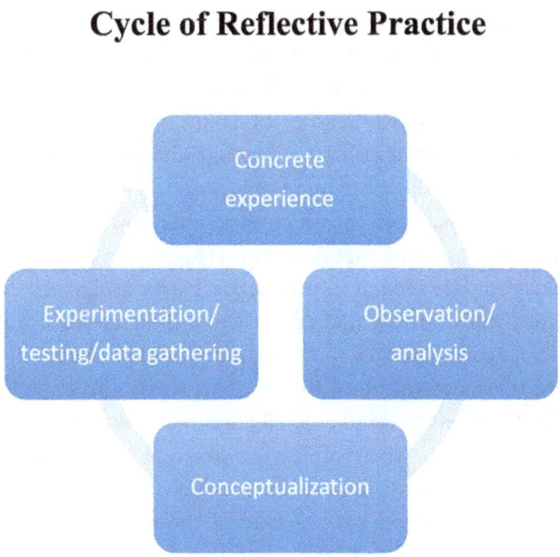

FIGURE 1.6 Desmond Bell, *Cycle of Reflective Practice* (2018)

This reflective approach has now found its way into the modelling of artistic research process well beyond its traditional application in architecture and design studies (Figure 1.6). The adoption of this approach within art and design research rests largely on the popularity within art education of a *constructivist* model of learning. As Grey and Malins note (2004:2),

> Constructivism is based on three key principles; the first being that learning is constructed as a response to each individual's experiences and prior knowledge; the second is that learning occurs through active exploration; and the third principle is that learning occurs within a social context – interaction between learners.

The artistic research process can be represented as a cycle of experiential learning forged through the concrete tasks of artistic production. When accompanied by processes of reportage and reflection, this facilitates an analysis of art and design process and its reconceptualization within a knowledge dynamic. This enables the artist–researcher to develop further processes of active experimentation, providing the basis for a further refinement of insights into art practice. So, a methodology of artistic research is erected on the site of a pedagogic theory, which in turn rests on a particular *epistemology* or theory of knowledge.

The US philosopher John Dewey (1859–1952) was one of the main intellectual influences on the development of Donald Schön's ideas. Dewey advances a pragmatist theory of knowledge closely related, in turn, to an account of active learning (1998). Dewey's epistemology asserts that rather than knowledge being a repository of fully formed facts and theories based on objective scientific activity, learners (and researchers) in fact construct knowledge for themselves through experience, experiment and reflection. There is no knowledge independent of the meaning attributed to experience by the learner or community of learners. Each learner individually, or in a group, constructs meaning and insight as he or she learns. Learning is understood as the refinement of practice. Accordingly, experience, practice, experiment, learning, research and constructed knowledge form a continuum. Dewey's epistemology is *both* a theory of knowledge and one of learning. In it, art, science and education are seen as in productive harmony. Dewey conceives of the forms of invention and learning that take place in the creative arts as akin to those found in science, as each is rooted in the primacy of experience and practice.

Indeed, in historical terms, the auspices of art theory lie in pedagogic engagement. Very often the need to think radically about art education and to encourage a new generation of artists to rethink their practice has encouraged a range of ambitious theoretical work which aspires to scientific status. This, as we shall see, was particularly the case during the period of the radical avant-garde in Russia and Germany. Between 1910 and 1930, figures as diverse as painters Wassily Kandinsky (1866–1944), Kasimir Malevich (1879–1935) and Paul Klee (1879–1940); photographers and designers Aleksandr Rodchenko (1891–1956) and Lazlo

Moholy-Nagy (1895–1946); and film-maker Sergei Eisenstein (1898–1948) all made substantial theoretical contributions to their disciplines, motivated, in part, by their pedagogic commitments (we explore this important legacy in Chapters 6 and 7). New art institutions sprang up, like the VKhUTEMAS in Russia (Higher Art and Technical Studios) and the Bauhaus in Germany, offering courses in the new art and codifying the language of this new visual sensibility via both curriculum provision and in a series of published texts. Transformed art practice, reformed pedagogy and theoretical insight chimed in unison in a period of social and political transformation.

However, the model of the reflective practitioner employed by today's art educationalists has its limits. It differs in significant ways from the pedagogical and theoretical concerns of the radical avant-garde. Today, a generation of creative arts students are being encouraged by tutors committed to constructivist learning approaches to experiment with reflective journals, mind maps, task-oriented iterative models and self and peer evaluations. A capacity on the part of an artist to reflect on, and critique, their artistic practices and the auspices of these is, of course, to be highly valued. And, of course, this is one of the pedagogic outcomes we strive for in arts education through encouraging robust practices of project documentation and self and peer evaluation. But *reflection* is not *reflexivity*. The latter term refers to the process by which a researcher seeks to both acknowledge their subject position and the partiality of their knowledge stance, whether set by the research apparatus employed or by the social relations governing artistic production and distribution (the art system) or by the ideological stance of the researcher. Nor can practice be reflected upon in a critical vacuum. In the art college, criticism is generally deployed retrospectively after work is completed and presented. As some art activists have noted (Jeffrey Charles Henry Peacock 2017),

> this encourages students towards a situation of trying to justify their artwork from the position of being its responsible authority. In reality, the work may have come about through a feeling of obligation or confusion. This kind of introduction to so-called criticism is likely to negate any interest in being puzzled for the sake of it and only to reify the notion of the artist as a self-affirming centre of truth.

Criticism, including self-criticism, has public–critical requirements. We engage with our experience – which is always a socially mediated one structured by various relations of production – via various theoretical problematics and public discourses. These structures colour how we make sense of that experience as artists. In a creative project, practice needs to be open to various critical interlocutions.

Research practice has a broader base of critical awareness than experiential learning itself can provide. It is sharpened by theoretical engagement, critical interlocution and the deployment of various proven methodologies. For reflexivity to be achieved within a research practice, not only has the partiality of the artist as subject to be acknowledged, but a critical context has also to be brought to bear upon a body of work in addition to any reflective gesture.

Moreover, the artist may or may not be able to provide a full critical interrogation of their work.[10] Many are not disposed to do so. Indeed, some have made refusal to provide a reflective account of their work, or to evaluate its meaning, as either a badge of professional pride or a grand modernist gesture.[11] But as we shall see, this is changing, and contemporary art practice is much more likely to regard the art process as interwoven with a research trajectory and to facilitate critical discussion with a public (even if the artist does not always want to join the dialogue they have initiated).

I would suggest that there are also philosophical limitations to the application of this approach more generally within the creative arts in respect to the degree of criticality that can be achieved within it.[12] Simply facilitating an iterative loop within studio and field practice whereby the artist–researcher/artist–practitioner is encouraged to develop a dynamic of experiment, reportage and reflective analysis captures only aspects of the artistic research process. The "reflective practitioner" approach needs to be supplemented by forms of critical dialogue with other interlocutors outside of the reflective dynamic. Moreover, a triangulation of the experience of the artist–researcher with a range of theoretical sources and forms of evidence which lie in broader cultural understandings of the art process and its social context is necessary to complement the reflective dynamic and anchor it in a viable research tradition. It is these forms of critical dialogue and social and political engagement which characterize the work of the radical avant-garde artists whose work and pedagogic experiments John Dewey had a chance to observe at close quarters during his trip to Russia in 1927.

1.4.3 Research for art and design

Frayling's third category takes us back to the idea of artistic research as a grounded activity primarily harnessed to the needs of creative production – an approach often dismissively called "background research". As Frayling describes it, this approach involves research conducted by an artist to directly assist them in the making of an artwork – whether a film, performance, sculpture, design product etc. The artefact, proposition or performance produced, rather than a body of reflective or critical writing about the work, is the research output. This third mode, as Frayling admits, is the approach closest to what artists actually do when they engage in research. Moreover, this sort of research is conducted by professional artists and media makers and not just by university-based academics.

Frayling hints at the academy's reluctance to recognize "applied" artistic research, suggesting that this may be due to the fact that it is less inclined to acknowledge the knowledge status of such an approach, "where the thinking, is so as to speak, embodied in the artefact, where the goal is not primarily communicable knowledge in the sense of verbal communication" (147). From the academy's point of view, such practice-embedded investigation may simply be the "wrong sort of research".

Arts academics have come under pressure to identify the "research value added" of their projects. What, it is asked, has to accompany the art object to make the art process a research one? Does it need, for instance, a written exegesis to explain and evaluate the work?

Indeed, the research body responsible for public funding of art and humanities research in the UK, the Arts and Humanities Research Council, has made it clear that an artistic research project, if it is to be judged eligible for public funding, must have a clear "research focus" and not just be about "making work".[13] For them, this means the investigative concerns of a project should be expressed in a series of formalized research questions informed by existing research in the field and capable of empirical corroboration. The research pursued should employ a coherent and, ideally, replicable methodology (where in principle anyone else could repeat the research and arrive at the same results).

I have to say, that this seems a patently absurd requirement for a field of research concerned with creative originality. Yet conformity to this investigative logic is held to distinguish practice-based arts research from simply art practice. The unfortunate consequence of this ruling has been to encourage some creative arts academics and graduate students to shroud their studio projects in theoreticist verbiage, or to employ the instrumental language of project management with its quasi-positivist logic of enquiry. Researchers find themselves adopting what has been dubbed an "administrative aesthetics"[14] based on project management tools, which may well damage the creative dynamic at the core of practice-based arts research. They are required by funders to identify "milestones" and "deliverables" within their projects. This institutionally dominant notion of research may be expressed in the language of scientific enquiry; however, it now functions primarily as a tool of bureaucratic accountability.

1.5 Modelling artistic research

It might make sense to resist the temptation to divide the field of creative arts research into three distinct categories, as Frayling does. Maybe we need to explore the dynamic that operates between these approaches. In reality, most artistic research involves elements of each approach and profits from the interactions between critical, creative and professional considerations. Indeed, it is precisely these synergies that interest me in this book.

We can, I think, identify a dynamic relation between three elements of art and media practice which, when brought into productive alignment, might sustain a research culture for the creative arts: firstly, autonomous creative arts and media activity conducted in the studio and field, responsive to its own inner voice and studio discipline, and fearlessly committed to innovation in form and social purpose; secondly, a body of critical and contextual theory, elaborated in and around the core tasks of art practice, but responsive to a range of interdisciplinary encounters and research methodologies; and lastly, an arts pedagogy acknowledging the central role of research, professional considerations and forms of social engagement in the conduct of the creative arts. We can represent this model schematically (Figure 1.7).

So, this book seeks to plot the basis for a grounded research culture for the creative arts – one that is attentive to the demands of creativity but informed by critical considerations. One that is true to its own lights, but open to a range of cross-disciplinary affiliations, inspirations and methodologies as well as to pedagogical and professional concerns. The relation between these critical and methodological considerations and the dynamics of creative practice can be represented diagrammatically (Figure 1.8). Of course, the specific critical sources and methodological approaches employed by arts researchers tend to vary considerably across the various creative arts and media disciplines, though a degree of commonality seems to exist across the field.

For some creative arts researchers, the theory and methods of art history and the study of visual culture (via film, photographic and imaging studies) remain the critical axis shaping practice-based visual arts research. For others, as we shall see, anthropological and socially engaged concerns have replaced this focus with

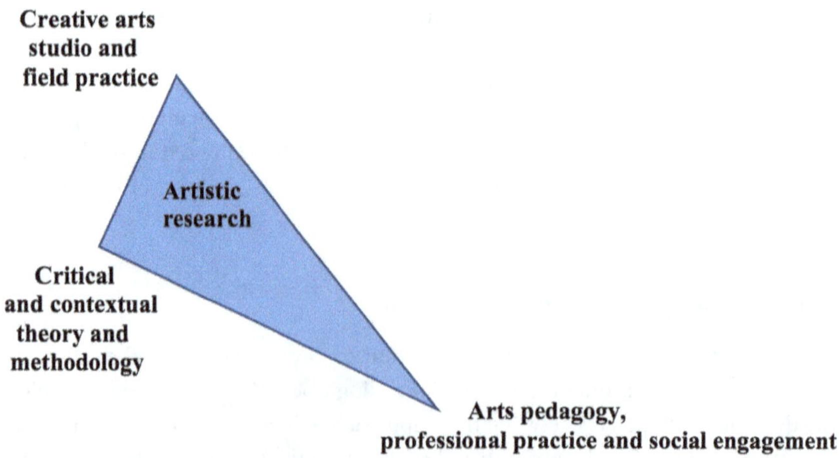

FIGURE 1.7 Desmond Bell, *Modelling Artistic Research* (2018)

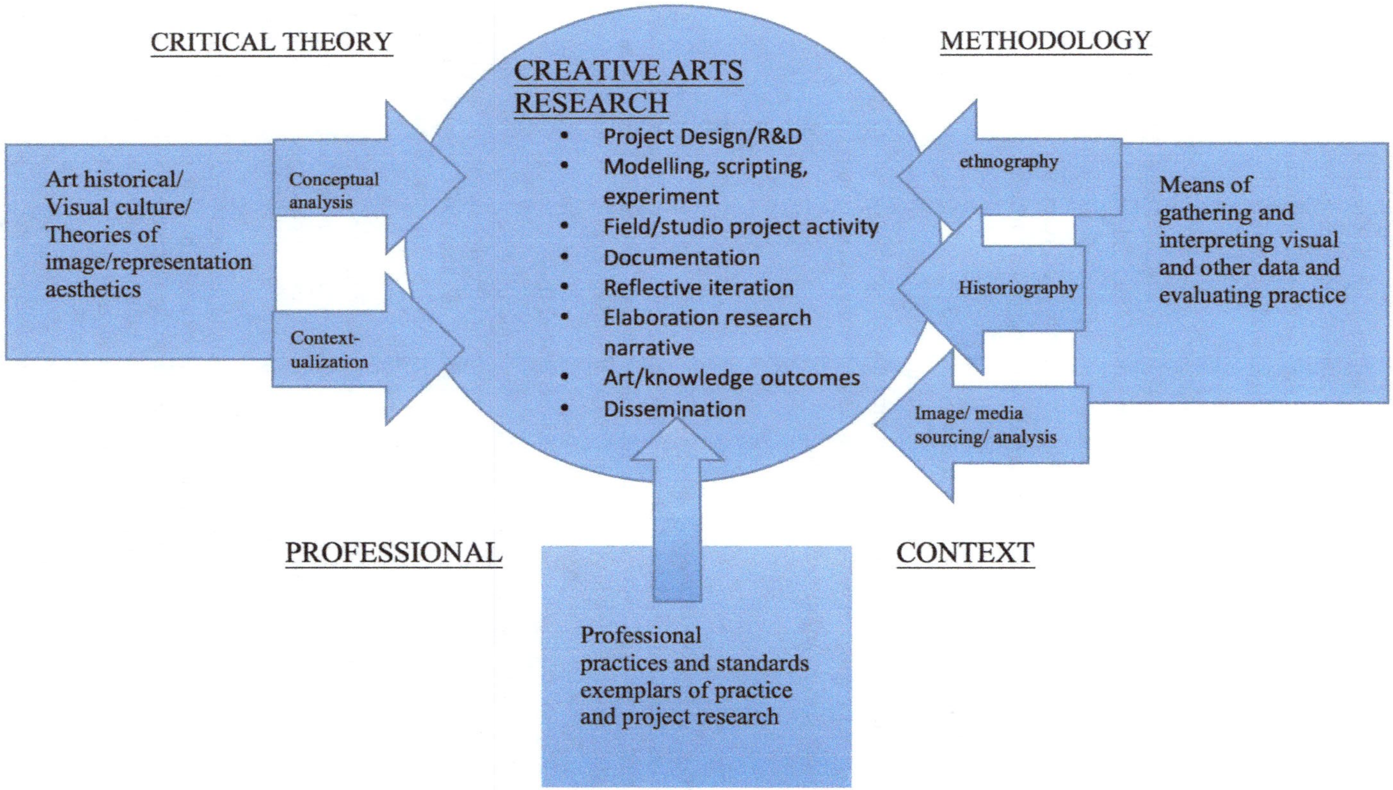

FIGURE 1.8 Desmond Bell, *The Methodological Context of Practice Research in the Arts* (2018)

new field methodologies, such as ethnographic research (explored in Chapter 9). Others seek to revisit the archive and to re-engage with the past, including the recent past (see Chapter 8). Still others, as we have seen, privilege the forms of grounded research and learning that take place within professional practice, which they believe can, in part, be formalized and transmitted in a set of reflective methods.

As we shall see in our next chapter, some seek to identify a "research turn" in contemporary art practice itself as the basis for a methodological prospectus for the creative arts that is true to its object of study. Others, like myself, search in the history of the avant-garde for an elusive science of art with its own analytical language and methodological concerns which might found a research tradition for the creative arts that goes beyond the current semiotic-based forms of enquiry (this is detailed in Chapters 6 and 7).

Yet, whatever the extrinsic methodological and critical considerations which shape artistic research, the creative arts research project has its own core methods and dynamic. These focus on the development of the work as a set of creative possibilities to be tested in various studio, field and professional settings and in relation to various critical arguments and publics.

Notes

1 *The Concise Oxford Dictionary*, 1990 edition.

2 Kant's later enquiries focused on the basis of moral and aesthetic judgements, or what he terms "practical reasoning" and "subjective universal" judgements, respectively. The former is addressed in his 1788 book *Critique of Practical Reason* (2015) and the latter in his 1790 *Critique of Judgement* (2009).

3 As Raymond Williams tells us in his *Keywords* (1976:74),

> The word came into English in eC17, from critic and critical, mC16, fw criticus, L, kritikos, Gk, rw krites, Gk – a judge. Its predominant early sense was of fault-finding: 'stand at the marke of criticisme . . . to bee shot at' (Dekker 1607). It was also used for commentary on literature and especially from lC17 for a sense of the act of judging literature and the writing which embodied this.

4 The tutor's evaluation of the work presented may or may not be part of the formal assessment of the student's creative work for examination purposes. Such criticism differs from literary criticism, for example, in that it is usually conducted verbally in an immediate face-to-face manner and is primarily concerned with pedagogic outcomes. For discussion of this pedagogic tool, see Healy (2016).

5 Raymond Williams reminds us that the association of the word "creative" with the arts did not happen until the end of the eighteenth century. Before that, the word was predominantly used in a theological manner to refer to the act of creation. The new association between the arts and creativity seems to reflect the increased status of the visual and performing arts (and the artist) in a secularizing bourgeois society and the corresponding academization and intellectualization of artistic practice.

6 As Raymond Bellour argues (quoted in Peucker 2007:66), the use of tableau in film "makes cinema lean in the direction of photography, towards its power to inscribe death".

7 As Kentridge freely admits, in his film practice he works largely alone in his studio rather than with technicians. This gives him considerable creative freedom in comparison with a director working within commercial film production with its elaborate scripting and production planning strictures.

8 Frayling's classification is largely applicable to the visual arts and was elaborated from his perspective as rector of the Royal College of Art and Design in London. His thinking on practice-based research was developed in his contributions to the Working Group convened by the UK Council for Graduate Education on practice-based doctorates in the creative and performing arts, a forum which reported in 1997.

9 Milton and Rodgers provide us with an example. In 1990, the kitchen utensil manufacturer OXO set itself the task of designing a range of easily gripped utensils for use by those with restricted manual capabilities impaired, in particular, by arthritis. The starting point was not a functional agenda concerned with developing an assistive design technical specification. Instead, the research of the design team focused on firstly exploring actual users' needs and existing experiences in the kitchen, employing a number of observational, ethnographic and empathy tools with actual arthritis sufferers. The resulting ergonomically fashioned kitchen tools facilitated the needs of manually restricted users. However, the stylish final product achieved a successful cross-over to a mainstream consumer market. Many designers attempt to address a range of social and ecological issues in their work, invoking concepts like "inclusive design", "user design", "ethical design" and "sustainable design" to signal the concerns of the design community with a broad range of user, societal and environment issues.

10 Poet Rainer Maria Rilke, at the time serving as the secretary to sculptor August Rodin, in 1905 reports Cezanne's acute discomfort when asked to give an account of his work in a personal correspondence with a critic:

> Cezanne was provoked (. . .) to make some pronouncements about painting; but when you see the few letters the old man wrote: how awkward this effort at self-explication remains, and how extremely repugnant it was to him. He was almost incapable of saying anything. The sentences in which he attempted it became long and convoluted, they balk and bristle, get knotted up, and finally he drops them, beside himself with rage.
>
> (2002:76)

11 Abstract Expressionism is often identified as the quintessential art movement which, conceiving as it does of "painting [as] a struggle between the artist and the canvas", preferred existential encounter and performative flourish to any mission to explain.

12 Moreover, given the reservations within contemporary philosophy about the limitations of the reflective self as the basis of knowledge, we have to be circumspect about the claims of reflective practice. For example, we cannot really be critiquing the structural role of the Cartesian ego in the visual domain (Crary 1990; Jay 1999) or deconstructing the idea of the author as the sovereign creative subject, only to reintroduce the figure of self-hood unproblematically into our core methodological premises.

13 Moreover, the funding guidelines for practice-based research drawn up by a number of research funding bodies exclude support for "work which is purely a development of an individual's professional practice (and which might be more appropriate for support from an arts funding body purely concerned with the creative output)" (AHRC 2004). As the AHRC intones, "work that results purely from the creative or professional development of an artist, however distinguished, is unlikely to fulfil the requirements of research".

14 The term "aesthetic of administration" was used by art historian Benjamin Buchloh (1990) in his critique of conceptual art. It acknowledges critical theorist Theodore Adorno's notion that 1960s capitalism was evolving towards a "totally administered world".

Buchloh argues that far from Conceptualism challenging this real-world development, it becomes part of it. As the features of artistic practice which underwrote its claims to autonomy – skill, respect for the image, memory, tradition and cultural vision – are stripped away, art is opened up to the vicissitudes of the marketplace, to the games of the PR industry and to limitless bureaucratic control.

2

THE RESEARCH THAT ARTISTS ACTUALLY DO

2.1 Introduction

So, what sort of research do practicing artists actually do? From the outset, let us admit – and we seem to be confirming the worst fears of the academy – that the majority of research conducted by artists is not systematic in form. Such investigative activity is predominantly concerned with professional considerations rather than the disinterested pursuit of original knowledge. Indeed, much of such exploratory research is not of a primary nature, nor original in itself. The originality, as I have argued, lies in the creative work it facilitates. The artist collates and makes use of existing sources of knowledge to advance a creative project. The new knowledge quotient lies in the making of work and the forms of understanding which arise in this process. The research undertaken is harnessed to the demands of what we could call "informed creative production". It focuses on the production, exhibition and to a lesser extent, reception and interpretation of an original work. This sort of "background research" is often marginalized in discussions of artistic research, but perhaps we need to re-evaluate such research. For, despite the recent attempts to formalize a practice-based research paradigm, this form of research activity continues to be the primary one engaged in by artists.

Thus, while the academy may get itself tied up in knots about the methodological auspices of university-based arts research (Bell 2006), no one actually doubts that artists engage in substantial research activity in producing their work. Film treatments test and develop creative ideas. In turn, scripts and other bodies of creative writing rest on detailed research activity into the cultural and historical location of events; into the language, habit, psychological motives of characters; and into the visual elements that make up, for instance, a film's "look". Models, maquettes and prototypes are used to explore the design of work in the art studio and to facilitate planning for its execution. Sketches and photographs prepare the way for paintings and sculpture. Ethnographic encounters and social enquiries facilitate documentary film and photography work and a range of other art projects. Choreographic and performance ideas and approaches are workshopped and tested in practice. Improvisation facilitates experiment and innovation in music and drama. And so on.

In this chapter, I examine some examples of such artistic research where the import of the work seems to extend well beyond the immediate production concerns of the artists who undertook it. I also explore some of the ways that artists have articulated an analytical understanding of their work, whether via bodies of reflective writing; in interviews, lectures and teaching engagements; or, indeed, through intellectual and creative encounters with other artists and writers.

2.2 Re-evaluating "background" research

Some academics are reluctant to ascribe research status to these exploratory activities. They argue that such research can be hard to evaluate *per se*, as its outcome is the artwork produced rather than any supplementary written account that usually figures in peer-reviewed academic research. Of course, we have long-established practices and protocols for evaluating artworks – whether in the judgements of the critic, the peer group, the public or the marketplace. However, the assessment of creative work as an academic research contribution remains problematic. What are we assessing in addition to the quality of the artwork produced? Can a body of accompanying writing and documentation about the making of the work tease out the research significance of a creative project? Can such ruminations have a research status separate from the work itself? And who exactly constitutes the tribunal of peers assessing such research – one's fellow artists, art critics and the general public, or one's academic colleagues and a broader community of scholars? Maybe it is because of such complexities that background research is often dismissed by academics tasked with assessing research performance. Many of these proto-managers work with a positivist modelling of knowledge and rather crude performative criteria driven by bureaucratic imperatives. Sadly, many such reviewers have little direct experience of the creative arts.

Background research is, then, part and parcel of the art process. It is not overly concerned with the generation of propositional knowledge within an academic discipline. This is not to say that knowledge cannot be abstracted and elaborated via acts of reflection and critical dialogue which may accompany the act of making. Surely facilitating this dialogue is one of the key tasks of practice-based artistic research. However, it is not the primary concern of the artist. As I have said, it seems strange that this set of professionally grounded background research activities receives so little attention from academics, who have for their part expended so much effort trying to delineate the character of practice-based arts research across a range of texts.[1] There are very few books dealing with the conduct of background research in the creative arts. What texts there are tend to be in fields like film and photography or in journalism and creative writing and usually deal with the topic within a more general discussion of project development within the medium concerned. The approach adopted is often technical and instructional in tone.

For example, in a broad range of media production textbooks (many published by Focal Press, the leading publisher in the field), we find detailed discussion of how to conduct research to advance documentary film projects (Rabinger 2014) and to edit them in an effective and ethical manner (Vaughan 1999); how to develop fiction film ideas through project research and the scripting process (Dancyger 2006; Rabinger 2006); and how to edit story film in a creative manner (Dancyger 2010; Crittenden 2010). Within the specialist area of ethnographic film, we do have discussions of the contribution of fieldwork research to factual film project development (Barbash and Taylor 1997). Similarly, we now have texts which examine the use of photography as a research method and resource within the social sciences (Banks 2001; Holm 2014). However, these latter publications are predominantly concerned with social science considerations rather than creative ones.

Some of this lack of academic attention to background research activity in the creative arts maybe due to the fact that the process of creative development within the arts is generally poorly understood. Indeed, it is often all but obscured by surviving romantic notions of the artist as an individual genius whose work springs from deep within their soul! However, today most artists see such primary investigative activity as a vital part of making original work and improving the acuity and impact of this. Any account of artistic research will have to give a central place to this basic reality.

An expansion of our notion of artistic research, if it is to include background research such as takes place in film and media production, will also have to acknowledge the status of the professionally grounded practices of project research within the wider family of arts research. Many scholars of the media have justifiable reservations about the lack of criticality which can on occasion characterize such research, focused as it is on pragmatic concerns with the development and delivery of a specific project. Of course, we have the profession of the media researcher. This is someone attached to a television or web production who provides information, contacts, archival sources, location support and general production development assistance for the director and the production team. The methods they employ tend to be *ad hoc* and governed by professional considerations. There are a small number of texts which seek to provide guidance for aspiring professionals in this field. For example, Kathy Chater's overview of professional media research *for media production* (1998) is a slim volume which, while full of advice and concrete tips for future media professionals, is short on any specific account of research methodology and makes little attempt to relate the tasks of professional media research to the extensive critical literature on mass media and popular culture.

Professional practices of research can also be identified in design and architecture. Milton and Rodgers (2013) and Vaughan (2017) have sketched these for design. In the field of architecture, RIBA (2017) published the text *How Architects Use Research*. Via a series of interviews and case studies involving leading architects, this explores the role of research within architectural practice – although, as the editors acknowledge, the professional organization of architecture sets limits on the development of research capacity within individual architectural practices (p. 7).

> The project focused nature of architectural practice tends to mean that research is mainly undertaken in relation to individual building projects and is largely funded through marginal elements of project fees.

This constraint is to be found across a range of other creative arts disciplines, such as theatre, dance, musical composition and performance. Often the background research that informs the creative and conceptual development of a creative project fuses with project management considerations and professional practices conducted under commercial conditions. These are research considerations which are specific to the creative disciplines and indeed to the institutional and professional context within which a project is developed (a television, film or theatre commission, a film or television studio, an architect's office or design practice).

For example, television and film producers working in the factual field hire specialist picture researchers to locate relevant archival images or to identify and investigate the background and circumstances of potential interviewees. Fiction film producers call on the services of specialist academics such as historians to develop scripts which accurately deal with the past in terms of factual accuracy and realistic depiction. Directors read avidly and widely around the topics they propose to address. Cinematographers pour over visual sources and experiment with lenses, filters, stock and lighting to develop a film's "look". Indeed, every film student is aware of the apocryphal story of director Stanley Kubrick (originally a photo-journalist) scouring the world looking for ultra-fast lenses capable of handling extremely low levels of light available in the candlelit scenes he envisaged for his 1975 film *Barry Lyndon*. Kubrick located a Zeiss still camera lens with a focal length of 50mm and a maximum aperture of f/O.7. for use with an 'antique' Mitchell BNC camera he had acquired from a US studio store. The lens had been developed by Zeiss for NASA for use in satellite photography. This set-up proved capable, with some technical adaption, of achieving the desired aesthetic outcome – the painterly images we see on screen (Figure 2.1), which strive to invoke the salon life of the eighteenth-century landed society. We all understand this as an artistic research process. The results were apparent to all. Kubrick let others provide the narrative.[2]

A less well-known historical example of such production-oriented technical and aesthetic research, which has had significant import for how we think about practice-based arts research, is the research undertaken by the film-maker and anthropologist Jean Rouch into how best to adapt the 16mm film camera and lightweight sound recorder for use in documentary and ethnographic film work. The immediate context was Rouch's production planning for his film of life in Paris in the summer of 1960, *Chronique d'un été*. The film, produced together with sociologist Edgar Morin (who immediately recognized the research potential of using documentary film within social science), was shot by Jean Luc Godard's favoured cinematographer,

Raoul Coutard.[3] Like Rouch, Coutard was keen on experimenting with the new lightweight camera equipment, with a view to expanding the visual vocabulary of cinema (Figure 2.2).

We can list their technical innovations: the use of a lightweight, mobile, easily reloadable and relatively silent 16mm éclair camera used in synchrony with an independently operating portable Nagra sound recorder; the use of tiny Lavoiser

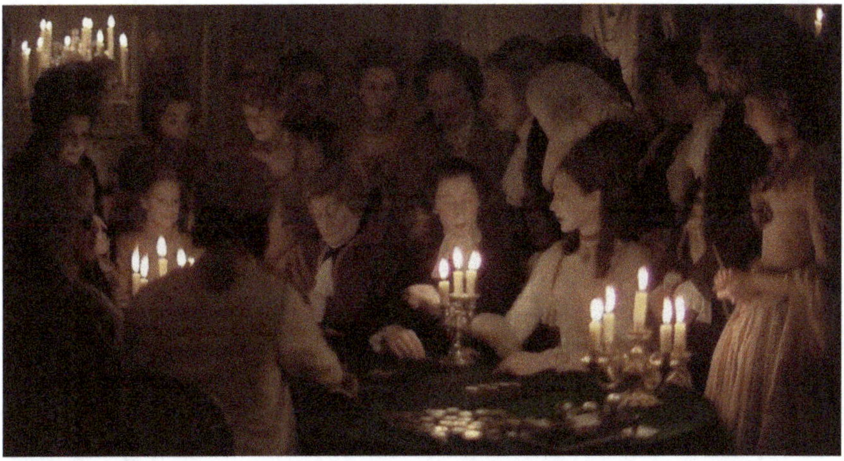

FIGURE 2.1 Stanley Kubrick, *Barry Lyndon* (1975), US/UK, 185 mins, colour, film still

Source: courtesy of the Criterion Collection

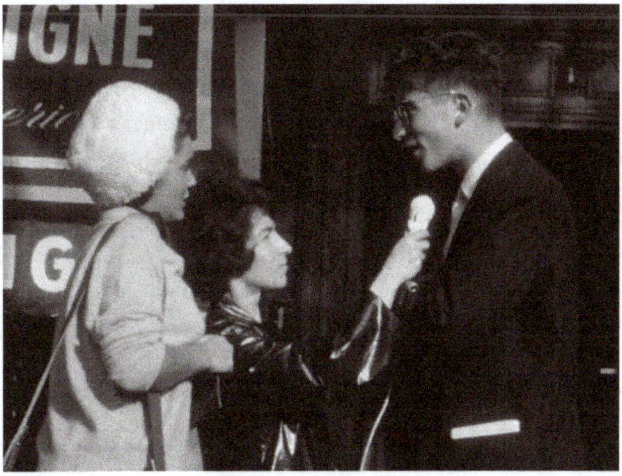

FIGURE 2.2 Jean Rouch and Edgar Morin, *Chronicle of a Summer* (1960), film still

Source: courtesy of Criterion Collection

microphones attached to interviewees; the employment of a wide lens that facilitated hand-held shooting. With this new toolkit, the documentary film-maker seemed for the first time to have the technical means to penetrate deeper into everyday life. A number of these innovations had already been experimented with by the French-Canadian film-maker Michel Brault, who Rouch invited to Paris to demonstrate his use of the new technology in 1960. André Coutant, the éclair company designer, provided the camera for the production and supported the unit's experiments. In Rouch's "cine-provocations" on the streets of Paris and in his pattern of collaboration with his non-actor participants, a new realist template for a socially engaged film practice was emerging. This was immediately heralded by *Nouvelle Vague* directors like Jean-Luc Godard and Jacques Rivette and seen as the rebirth of French cinema. Technical research, project development, political engagement and aesthetic innovation went hand in hand, and the results soon had an impact across European cinema.

More recently, in 2006, Scottish artist Douglas Gordon, in collaboration with the French artist Philippe Parreno, made an experimental documentary feature, *Zidane: A 21st Century Portrait.* The film received a cinema release as well as figuring in various multiscreen gallery installations (Figure 2.3). It follows the performance of the Algerian-French mid-fielder Zinédine Zidane in a single game of football he played for Real Madrid in 2005. Gordon and Parreno scoured the world for camera lenses capable of providing the magnifying power needed to achieve the extreme close-ups of Zidane on the field of play that they would employ in the film. Eventually they secured the loan from British military sources of an experimental zoom lens not yet in commercial production. They experimented with complex strategies for positioning and using the seventeen synchronized 35mm and high definition video cameras used to capture the action. They are reputed to have also marched their Director of Photography, Darius Khondji, and his camera crew through the Prado Gallery in Madrid so they could explore a range of forms of painted portraiture as 'prep' for the shoot, hoping that the camera operators might find inspiration among the Goyas and Velasquezes for their photographic framing and portrayal of Zidane on the field of play (Kubrick did something similar during the preproduction for *Barry Lyndon*, seeking to familiarize his design team with the visual sensibility of English eighteenth-century portraiture and landscape).

Such research is of course unoriginal in an art historical sense, and its outcome is inextricably bound up with the aesthetic objectives of a professionally bounded creative project. But, like scientific research, it has a cumulative dimension in so far as novel creative developments are often built on a refinement of earlier achievements and understandings.[4]

And indeed, as we shall see, this sort of research involving experiments with lens use and optics; the manufacturing of maquettes, models and prototypes to test design options; and scripting and scenario planning to manage productions can look rather like the sort of research that 'hard' scientists and, in particular, technologists do in their labs and field testing. It certainly resembles certain sorts of

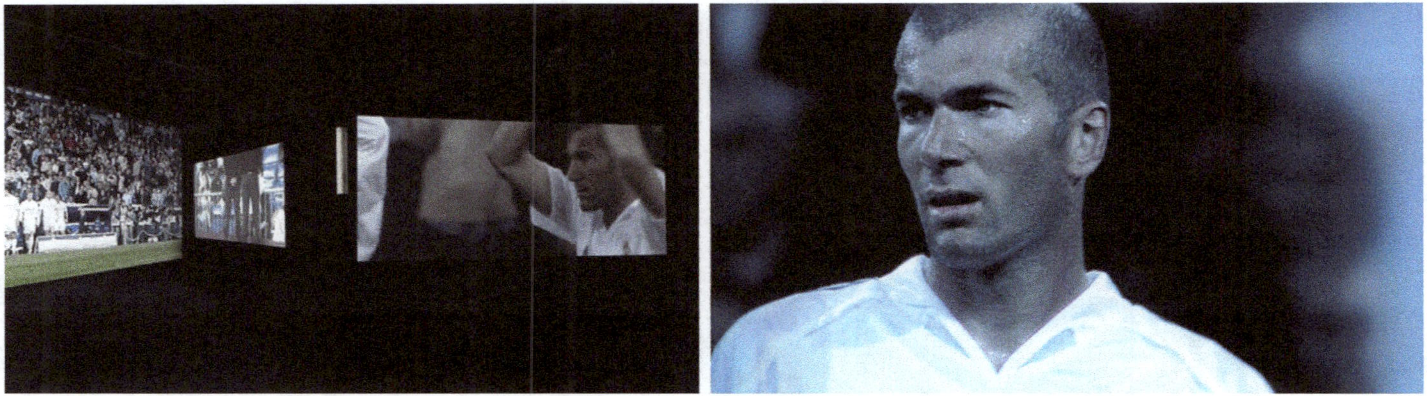

FIGURE 2.3 (left) Douglas Gordon and Philippe Parreno, *Zidane: A 21st Century Portrait*, National Gallery Scotland, 2008; (right) *Zidane* film still

Source: courtesy of the artist

technological investigation in that it is focused on project-based problem-solving approaches that are reliant on professional skill and experience, including input from technical experts. In addition, it also often employs lateral thinking, inspired guesswork and persistent curiosity rather than adherence to methodological exactitude. As we shall see when we look in Chapter 4 at the investigative logic of natural science, the processes of art executed in the studio and the science done in labs where researchers also often employ 'fuzzy logics' and creative leaps of the imagination don't look a million miles apart. Such background research, I would argue, demonstrates the application of an acute research intelligence in its handling of sources and testing of strategies of creative innovation and, in turn, adds to the stock of knowledge and critical insight in our arts disciplines.

2.3 Artist reflections and critical dialogues as research sources

Creative practice can also occasion critical insight on the part of the artist and their *interlocuters* (collaborators, critics and publics). Under certain circumstances, this insight can be abstracted and generalized to produce a valuable knowledge outcome. To develop a research culture for the creative arts, we need to be able to recognize those circumstances and to encourage the forms of critical dialogue and reflective practice which bring to the fore the research and ideational components of creative art production. Today this is one of the key tasks of art educators. This intellectual journey is perhaps best pursued not via the narrow pathways of methodological refinement but rather by the consideration of a range of case studies, particularly of contemporary arts and media practice, that can raise core methodological, critical and ethical issues.

What do creative artists come to know as a result of making their work? Students of film, drama and music confronted with the abstractions of cultural theory, and unsure how such theorizations of art and media activity might relate to their own creative concerns and professional formation, often turn to accounts of creative practice provided by esteemed practitioners who have set down their reflections on their work in print or expressed these in interviews. What can they learn from such sources?

There is a rich tradition of artists – painters, film-makers, writers, composers or architects – critically commenting on their own work in a series of measured reflections.

Famously, film editor Walter Murch, in dialogue with a range of film directors and theorists of sound, like Micheal Chion, has written illuminatingly about the processes of film and sound editing (Chion 1994). Similarly, the documentarist Frederick Wiseman has been able to draw upon symbolic interactionist sociological perspectives to illuminate his work as a chronicler of American social institutions (Luccia 1994). More recently, film theorist Bill Nichols (Nichols and Renov 2011) conducted a series of email interviews with Hungarian found footage film-maker Péter Forgács co-editing a book with Micheal Renov which explores both the creative and critical dimensions of Forgács's work as chronicler of the fate of central European Jewry and to which a number of other film and cultural theorists make contributions, identifying Forgács's filmic approach as a form of historical research.

William Kentridge has, as we have seen, cogently discussed his working methods and the forms of reflection and creative thinking that emerge in the studio. In his public lectures and interviews and books (2014), he provides us with an analysis that, as he puts it in a 2013 lecture,

> starts from work in the studio and then goes through a kind of "reverse engineering" to try to understand a possible interpretation of work that has been done and to try to find the thinking inherent in what the hand and the eye is doing in the studio.

He talks engagingly about the bifurcation of self – playfully portrayed in one of his video pieces (Figure 2.4) – that occurs as the artist moves from the active subject at work drawing and photographing, to the observing self that stands back from their work to assess its development and emerging meaning (2012).

> Each time you step back from the drawing you see it in a different way and give instruction to yourself, a set of mental notes which you reapply when you return to the drawing. This split between oneself the viewer, and the maker, becomes essential in the gathering and construction of sense in the studio. . . . This split of the self in the studio becomes emblematic of the split one does in the work itself.

So, the playful trope of creative self and interfering, interrogating other gives way to the idea of the canvas or sheet of drawing paper or filmic image as a membrane resonating to the beat of the world and acting as a condenser that registers the impact of external forces. This "data" is then creatively re-configured within the protected space of the studio. So, the conversation is not just within an artistic self – between a maker and observer – conducted through an act of ekstasis, but also a dialogue with that external world and the impulses and discourses that flow into the studio from it. In his lectures and interviews, Kentridge takes another pace backwards, establishing a further remove that reframes the artistic process within an ongoing critical dialogue. And while this reframing can be thought of in terms of the model of reflective practice sketched in the previous chapter, it moves beyond this. Kentridge embraces a dialogical notion of the studio and of the artistic process that opens a pathway to a grounded artistic research practice.

Let me explore another example of such artistic reflection and research impetus in a non-visual field. Contemporary South African-born composer Kevin Volans, a student of the German avant-garde composer Karl Heinz Stockhausen, has developed an interest in the encounter of modernist music with a range of native African musical cultures. I have a particular interest in Volans's work, as I have used elements of his music in my own films. Over a thirty-year period, Volans has developed a

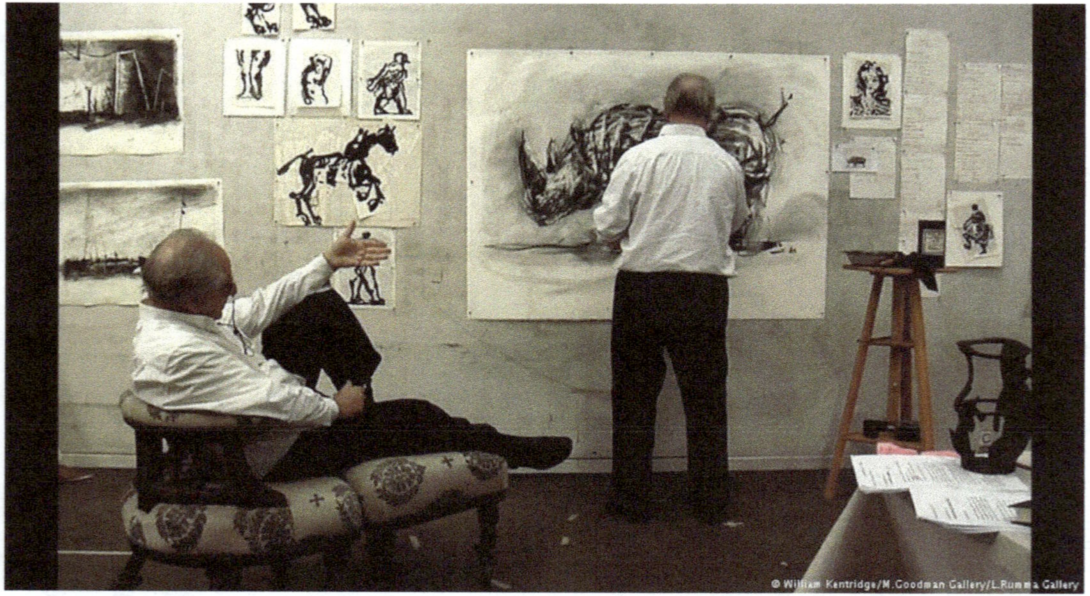

FIGURE 2.4 William Kentridge, *6 Drawing Lessons* (2016), film still

Source: courtesy of the artist and the Marion Goodman Gallery

body of work significantly influenced by traditional African musical forms, instrumentation, pitch and tuning, timbre and colour. He has also written illuminatingly about this work, addressing how he as a composer working within the western classical tradition has navigated these ethnic influences. His auto-critical framework combines ethno-musicological considerations with cross-cultural concerns. It seeks to make sense of the hybrid character of his own music, recognizing in this a strategy that uses "cross-fertilising techniques of different musical traditions to arrive at a new musical perception" (Volans:1986). This approach challenges the euro-centricity of the classical tradition but refuses to retreat into "hands off ethnomusicology".

For example, his now classic piece *White Man Sleeps* (1982/1986) was initially scored for two harpsichords, viola da gamba and percussion. His composition drew upon a range of ethno-musicological research conducted by South African academics on the music of native South African peoples, and it incorporates Venda dance elements together with San hunting pieces, each reworked for western classical instruments. The keyboards were specially re-tuned to acknowledge his debt to original African musical sources. As Volans acknowledges in his 1986 artist's statement about the piece in *White Man Sleeps*, Venda, San, Basotho, Nyungwe, Baroque and contemporary Western music are all represented in roughly equal measure, all played in a tuning system derived from Shona Mbira music (and modified to suit the harpsichords). *White Man Sleeps* was subsequently rewritten as a string quartet (for the Kronos Quartet in 1986: www.youtube.com/watch?v=uENZhNnP2Mk). This involved a further translation of the music. As he argues (ibid.),

> it was obvious to me, that by re-casting the source material onto new instruments, it stripped it of its original significance, and gave it a new one. (However accurately transcribed, a folk tune played by a symphony orchestra is no longer a folk tune.) This does not imply, however, that one is free to do anything. What would be the point of working with original source material, if one did not wish to **translate** some of the values implicit in it? By working with the right instruments, in the right register, at the right dynamic, one can infer a world foreign, or rather new, to the traditional world of the instruments themselves.

His written reflections chronicle a complex musicological journey and, in turn, open up a critical line of enquiry that in some ways parallels the film practice and theorizations of Jean Rouch.

> By introducing some strictly non-Western aspects of African music into the European concert repertoire I hoped to gently set up an African colonisation of Western music and instruments and thus preserve some unique qualities, albeit in a new form. It was a bit like introducing an African computer virus into the heart of Western contemporary music.
>
> *(ibid.)*

I used some of Kevin Volans's music in my found footage film *Hard Road to Klondike* (1999), including his suite *White Man Sleep*; *Mbira*, his Shona- inspired harpsicord piece; and material from his string quartet *Hunter Gatherer*. The film (discussed in more depth in the next section of this chapter) narrates the journey of Irish migrant labourer Micí Mac Gabhann through 1890s America in search of work. Volans's music is used heavily in the new world section of the film, together with a range of contemporary Irish sean-nós singing present throughout the film. A passage of driving music from the second movement of *White Man Sleeps* (played by the Kronos Quartet in the version we use) accompanies the sequence representing Mac Gabhann's rail journey from the steel mills of Pennsylvania, through the Rocky Mountains to Butte, Montana, where he becomes a hard rock miner in what was then the western frontier. The sequence is visually comprised of a montage of early silent cinema material. Volans's music is interwoven with various sound effects (train whistles and locomotive sounds) and with the voice of the narrator, Irish actor Stephen Rea.

Volans's composition, with its transcription of African music for western instruments, seems to resonate with our transfiguring of Mac Gabhann's folksy migrant's tale into a contemporary film work sensitive to the postcolonial character of his story (Gibbons 2005). This element is revealed in the hard rock miners' confrontation with the native Americans whose land they are encroaching upon. Mac Gabhann, while in part identifying with the Indians due to his own experience as an Irish peasant on the receiving end of colonial expropriation, finds himself involved in the oppression of the Blackfoot people (this section of the film can be viewed at: http://youtu.be/9nIunjiI6mA). The sonic fabric, with Volans's music central to this tension, is as a result able to contribute to "the film's over-all plastic conception", and to its critical sensibility. Moreover, in my own subsequent attempts to discuss the use of music and sound in my films (Bell 2015), I have found Volans's writing about his cross-cultural practice invaluable. This is so particularly with regard to the value he places on both the hybridity of contemporary cultural work and the need to respect the sources one draws from when fashioning one's own work, a respect demonstrated not in a fetishizing of tradition but rather in the creative transfiguration of such sources. So, we have a case here of a body of critical reflection by an artist facilitating a further set of artistic reflections on work conducted in a different medium, forming, in effect, a chain of practice-based research events.

2.4 The artist interview as a research tool

Self-explication, once frowned upon within some strands of modernism (significantly, not, as we shall see, in the art activity and pedagogic proposals of the early twentieth-century avant-garde), has now become *de rigueur* for many artists. Sometimes these reflections are auto-critical in character. Often, they are elicited in critical dialogue with scholars and critics.

A classical case is David Sylvester's series of interviews conducted with the Irish-born painter Francis Bacon (Sylvester 1987). These in-depth interviews, conducted between 1962 and 1978, were originally filmed by the BBC for a series of television arts programmes on the painter's work. They were subsequently transcribed and published by Sylvester (elements of the interview can be seen at: https://youtu.be/xoFMH_D6xLk). In time, this material was to provide the basis for French philosopher Gilles Deleuze's "dialogue" with Bacon (2003), setting off another chain of critical events.

Deleuze's work consists of seven tentatively linked essays of philosophically inflected criticism. Each of these provides a different formal analysis of an aspect of Bacon's practice as an artist (the actual meeting between the philosopher and the painter was by all accounts somewhat less successful). Deleuze seeks to identify a common "logic" determining Bacon's choices in relation to his use of colour and figuration and with regard the mobilization of expressive force within his work. For him, Bacon's capacity to move between figuration and abstraction in his portrayal of the pulverized human body seems to echo his own interests as a philosopher in the fluidity of corporal form and its irreducibility to any fixed identity. The Sylvester material also informed the evolution of John Maybury's script for his 1998 film *Love Is the Devil: Study for a Portrait of Francis Bacon.*

Indeed, Sylvester's work as an interviewer (the methodology of which is discussed in Finch 2017) seems to have provided a model for the more recent research practice of Hans Obrist, the energetic director of the Serpentine Gallery in London. Obrist has embarked on an ambitious project over a number of years to interview a range of contemporary visual artists. In so doing, he has demonstrated the utility of the in-depth interview as an artistic research tool. He identifies (Obrist 2008:72) the influence of Sylvester's work on his own research practice as well as that of Pierre Cabanne, who in 1968[5] conducted a series of interviews with the artist Marcel Duchamp shortly before his death. As Obrist acknowledges,

> these books somehow brought me to art. They were like oxygen and were the first time that the idea of an interview with an artist as a medium became of interest to me. They also sparked my interest in the idea of sustained conversations – of interviews recorded over a period of time, perhaps over the course of many years; the Bacon/Sylvester interviews took place over three long sessions, for example.

Sometimes an artist's reflections can be occasioned by their engagement with other bodies of creative work, as in Rilke's encounter with Cezanne's paintings in Paris in 1907. Rilke reports (2002:54) the "conflagration of clarity" he experienced in front of Cezanne's paintings. He acknowledges how the artist's exact use of colour and his engagement with the quotidian (a "silent telling of an objective strength") and with the problem of abstraction all were to impact on his own approach to poetry, language and image.

We also know of Soviet film-maker Sergei Eisenstein's engagement with the work of James Joyce, reported in his 1934 lecture on the Irish writer delivered at the Russian State Institute for Cinematography (GIK). Eisenstein had read *Ulysses* in English with some difficulty in 1926. In his notes on a planned film to be based on Marx's *Capital*, he talks about finding a useful "methodology" in Joyce's work for his own thinking about film form. Joyce's work seems to provide for him an important exemplar of how to handle the relationship between the concrete and the abstract within cinema. Eisenstein observes (1976:7) that in Joyce's work, "the maximum abstractness of an expanding idea appears particularly bold when presented as an offshoot from extreme concreteness – the banality of life". Eisenstein not only acknowledges Joyce's celebration of contemporary urban life as a literary theme but focuses on the modes of exposition the writer employs to chart the odyssey of his character Leopold Bloom through Dublin. He finds in *Ulysses* a literary approach very much in tune with cinematic sensibilities. The final chapters of the book – "Ithaca" and "Penelope" – seemed to particularly inspire Eisenstein. The former is assembled in question and response form like a religious catechism to provide a documentary narrative of the perambulations and deliberations of Bloom. The latter consists of a long soliloquy, or internal monologue, delivered by his wife Molly. In Eisenstein's well-thumbed copy of *Ulysses*, Molly Bloom's soliloquy had been marked up to indicate material for use for a possible shooting script. Eisenstein valorizes this type of voice with its capacity to plot a concrete relation between literary language and consciousness.

The director had hoped to film *Ulysses* and indeed met Joyce in Paris in 1929 to discuss this possibility.[6] The planned film project never materialized. There was, however, a meeting of minds and fascination with each other's creative methods. As Eisenstein recalled (Werner 1990:495) "he has just asked me – entreated me – to show him my films; he seems interested in my experiments with film language that I am carrying out (just as I am interested in his literary experiments of a similar nature)". Eisenstein continued in his teaching and research into cinema to identify the importance of Joyce's literary tropes for the filmic form. In his directing classes, students were invited to "translate" elements of *Ulysses* into "filmic language". Joyce, for his part – ever the cinephiliac – acknowledged the role of cinematic practices like montage in facilitating a new way of handling space and time within the novel and of engineering dramatic collisions within material and story lines. Here we seem to have an "elective affinity" between two artists that occasions an illuminating critical dialogue and, in turn, facilitates autocritique.

As I have said, such auto-criticism is often most effectively elicited in concert with other interlocuters (including one's students) and often within a framework of public culture. Through dialogue, creative work is deconstructed, its auspices examined and its claims evaluated. And this is the stuff of artistic research. When an artist strives to impart a generalized understanding of their work, and that of others, to their students, their discourse often takes a more systematic form. And as we shall see with regard to the modernist avant-garde, it was often in the context of their educational activities, both teaching and curriculum development, that artists moved to theorize their thinking and more generally to develop an analytical take on their art. So, French film historian Jean Luc Comolli convincingly argues that for Sergei Eisenstein, the teaching encounter was "the true site of theoretical construction" (quoted in Aumont 1987:32). The same seems true with regard to Wassily Kandinsky and Kasimir Malevich.

2.5 Case study: found footage film-making

To illuminate how "background" artistic research, primarily concerned in the first instance with creative project development, might in time lead to more general trajectories of research and theorization, I now examine the area of found footage film-making and plot my own experience of research in this area. This is a film practice where artists draw upon a range of archival and found images to fashion new film works, often reframing, slowing down, speeding up, freezing or colourizing the footage used and employing music and sound effects to achieve expressive effects. There is to date very little research on found footage film-making within academic film studies. Certainly, studies of early Russian film (Taylor and Christie 1988) have identified Esfir Shub as the pioneer of the compilation film and drawn our attention to Eisenstein's pedagogic experiments, which required his students to recut existing footage to generate new narratives and filmic meanings (a strategy which has always seemed more than a pragmatic response to the unavailability of film stock at the time in Russia). Indeed, coming out of this historical interest, Jay Leyda published his 1964 book *Films Beget Films: A Study of the Compilation Film*. However, this remains one of the few texts directly on the topic.

The research has largely been left to documentary and experimental film-makers themselves. Their focus, of course, has largely been on the onerous task of locating, cataloguing, copying and creatively employing a vast range of found images within their own films, most of which seek to engage with history in one form or another. Many found footage film-makers effectively combine an archival research practice with a creative one. Like much background research, this sort of archival research process can at first glance look like a form of "mindless empiricism" involving the endless search for and collection of filmic fragments – many of them "orphans" without directorial attribution. Then there is the perpetual play with image, sound and text as new creative strategies combine with original material. But a certain type of film intelligence is at work here, and a distinctive process of appropriation and refashioning is being employed which seems to have resonance right across the creative arts and beyond to the fields of historiography and indeed philosophy. Let me elaborate based on my own experience.

For a number of years in the 1990s, I laboured in the vineyard of television documentary production directing a number of films for Channel 4 in the UK. These sought to address the complexities of political identity in trouble-scarred Northern Ireland. In 1993, during a trip to Budapest, I was exposed to the work of the avant-garde film-maker Péter Forgács. In the offices of Magyar Television, I was shown his 1992 serialized found footage film *Wittgenstein Tractatus: 7 Video Paragraph* (https://vimeo.com/1607404). I was amazed. His free association of image, sound and narration, in what might be best called "memory work", threw my rather prosaic understanding of factual film form into disarray. Here was a film-maker who was doing something really interesting with moving image archive, sound and narrative voice. Forgács's films have been predominantly, but not exclusively, concerned with the fate of central European Jewry after the ascent of Nazism. He employs archive footage not simply as historical evidence to support a documentary assertion but also to achieve expressive and critical outcomes concerned with questions of memory and loss. He encourages us as an audience to look closely at the images he has assembled for our eyes . . . and then to look again, for the world they portray has disappeared, enveloped by historical disaster.

For example, his Wittgenstein film is not "about" the philosopher Wittgenstein, or his work. It is neither a bio-pic nor an arts documentary. Rather, it is a visual essay on film's capacity to make us see and think and to reflect on the fragmentary nature of memory. It is in a sense directly "doing" philosophy, though that is not its primary concern. Forgács's piece involves a narration largely drawn from Wittgenstein's thesis, the *Tractatus Logico-Philosophicus*. This work was completed in 1918 and published in 1921 in German and a year later in English. It is a text every bit as difficult as Marx's *Capital* to "film". As a former philosophy student, I was acquainted with this classic work. But here was a creative artist translating this most abstract and austere of texts into the visual and sensuous language of film. Wittgenstein's epigrams are voiced by Forgács himself (and reproduced in text on the screen). They are cut against a compilation of archival images portraying Hungarian rural and urban life in the interwar period in footage shot by amateur film-makers. Forgács had unearthed a collection of home movies shot by a Budapest bourgeois family from the 1930s to the 1950s. Across a range of projects, he returns to this treasure trove of images. The power of the film seems to reside in the capacity of the film-maker to set into motion a series of resonances between, on the one hand, the philosopher's nostrums on the nature of language, the world and how we picture it, and, on the other, the filmic assembly of image, sound and text. The associations established between word and image are often simply uncanny and give the work a brooding philosophic presence. As Whitney Davis argues (in Nichols and Renov 2011:210):

> The film tries to show (though it would be difficult or impossible to speak or discursively describe) the contingency of the depicted world or to show the world as a picture, specifically a film-like picture in which its unpredictable reversibility can be shown.

In its capacity to "show" rather than "tell", the film-maker demonstrates his capacity to address both philosophical and historiographical questions. Forgács's *Wittgenstein Tractatus* explores the philosopher's ideas about the relation between how we picture and speak about the world through a filmic medium. Indeed, his approach adds something distinctive to Wittgenstein's original philosophic project. As Davis argues (ibid.:209), "uncertainty, ambiguity, and indeterminacy underlie much of

Wittgenstein's understanding of language; Forgács uses found footage to a make a parallel argument". As Wittgenstein himself had argued in the *Tractatus* (2.224), "it is impossible to tell from the picture alone whether it is true or false" (1961). This of course is particularly the case as regards the filmic image, and Forgács's images have a dream-like quality further enhanced by his manipulation of them in the digital editing suite. The images seem to slip their immediate referent (the recreational activities and lived world of a middle-class Jewish family) and begin to figure a broader historical narrative. The film, and all of Forgács's film work, is not just "doing philosophy", as Davis claims, but providing an inspired telling of history.

For example, towards the start of the film, we see a striking shot of a pig writhing about in mud with a jackbooted foot prodding it (Figure 2.5). Over this image, Forgács voices a quotation from Wittgenstein. The lines this time are from the book *Cultural and Value*, a collection of the philosopher's writing (1998) on ethics and personal life gathered together and published after his death. This is a text which can only but be read as framed by the Holocaust. Forgács voices the philosopher's anguish in his own rendering,

> No cry of torment can be greater than the cry of one man
> Or again, no torment can be greater than what a single human being may suffer
> Man is capable of infinite torment therefore
> And so, he can stand in need of infinite help

As viewers, we associate the scene and these spoken lines with the fascist brutality that is soon to overwhelm the Jewish family whose family movies we are

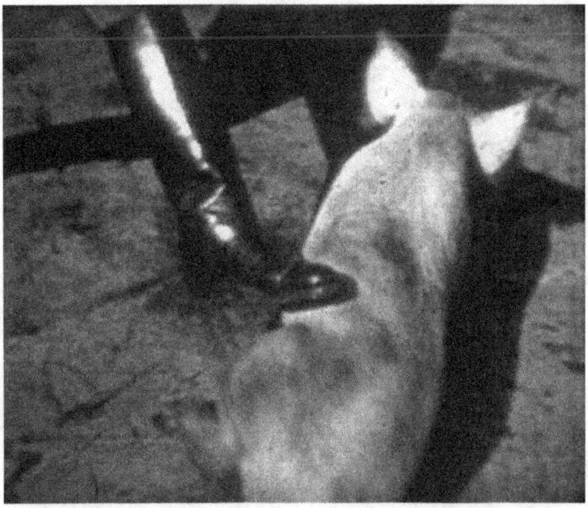

FIGURE 2.5 Péter Forgács, *Wittgenstein Tractatus* (1992), film still

Source: courtesy of the artist

sharing. In fact, as Forgács readily admits, the specific footage of the suffering pig is from a veterinary science film and his use of the image to hint at authoritarian brutality is largely figurative. As in a number of Forgács's films, as viewers we feel a certain apprehension when confronted with a range of images that, while having no immediate historical referent (the footage is largely of the everyday life and recreational activities of families), seem to allude to historical dangers lying ahead for those captured in the footage. We know what will befall them, as they almost certainly did not. They are captured in carefree pose. As he later intones over shots of Jewish people waiting at a rail line, "what is thinkable is also possible" (*Tractatus* 3.02), an interpretation of Wittgenstein's line which stiches the philosopher's abstract considerations on thought and action into the fabric of history.

I returned to Ireland determined to rethink my own documentary practice.

I then came across the 1996 short film *The Film of Her* (Figure 2.6) by Bill Morrison and realized that a number of film-makers, Morrison[7] in the US, Forgács[8] in Hungary, Peter Delpeut[9] in the Netherlands and Edgardo Cozarinsky,[10] the Paris-based Argentinian writer and film-maker, were simultaneously engaged in making experimental found-footage works, appropriating and creatively reworking early silent cinema or news reel footage. These are artists who employ historic film archives in a creative way to produce experimental works musing on history, the fragmentary nature of memory and the materiality of film process. All of them work closely with composers and regard the musical and sonic track fashioned for

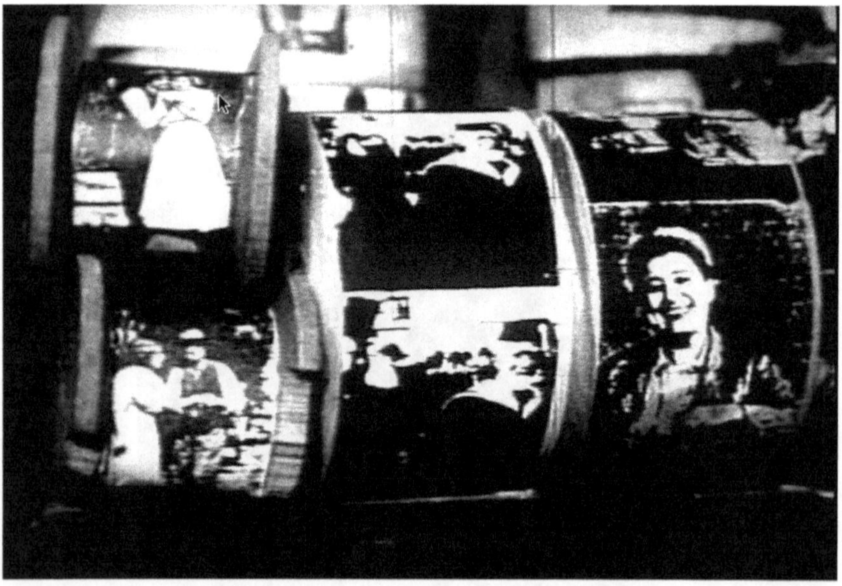

FIGURE 2.6 Bill Morrison, *The Film of Her* (1996), film still

Source: courtesy of the artist

their films as central to the power of the work. They use archive in different ways, but each transforms this material in a plastic manner to refigure the archival image for new filmic purposes. Each transforms the nature of documentary narration, whether by poetical inflection, fictive surmise or by problematizing the truth status of that narration, in part through acknowledging the figure of the "unreliable narrator".

Bill Morrison's *The Film of Her* concerns the rescue of cans of early cinema paper prints[11] held by the Library of Congress Motion Picture Department. These had been scheduled to be dumped but were saved from destruction by a diligent clerk in the 1930s, and this tale provides the scenario for Morrison's found-footage film. Morrison employs a first-person voice-over loosely based on an interview with Howard Walls, the long-forgotten curator who rescued the paper prints in 1939. Among the footage retrieved were images of an unknown erotic dancer appearing in what seems to have been an early pornographic film (Figure 2.6). From these fragments, Morrison fashions a story centred on an imagined Library of Congress clerk/curator toiling in the film copyright section who is entranced by a fleeting image of a young woman that flashes across the screen of his flatbed editing machine. We follow Morrison's search through the retrieved footage to locate this image. This quest allows the film-maker to share a range of early film material with his audience and explore the archival process in an affecting way. Morrison offers the viewer a montage of scratched and degraded archival images which dance before our ideas. The assembly is testament to not only the electrifying character of the early cinematic image but also the fragility of historical memory.

Morrison, in a recent interview conducted by Christian Lund for the Louisiana Museum of Modern Art in Denmark (2016), describes his practice:

> I would define myself as a film-maker . . . as someone who has chiefly used archive footage – and archive footage in various states of degradation – to make new films and usually new films with original footage.
>
> I came out of a painting and animation background. So, the idea that films had already gone through this plastic medium and were being re-mediated frame by frame meant that there were several different layers of reality happening. There was the reality from the turn of the twentieth century and then everything that happened when it (the film) was committed to paper. Then everything that happened since then. So, there are these archeological layers of time and that was fascinating to me.

I then discovered that this filmic approach had come to the attention of certain film theorists wrestling with the problem of indexicality[12] in relation to the filmic image. For example, Catherine Russell (1999:240) observes that found footage film-making approaches the archive in a distinctive manner: "in the process of being appropriated, the original image gives over its meaning to the new text and is manipulated by the new film maker on the level of the signifier". In so doing, the film-makers offer their audience the indexical richness and link to the past that the filmic image as photographic trace provides. At the same time, they undermine that testimony by both revealing the fragmentary nature of the photographic record and setting free these images as signs, which now take their place within new narratives and are thereby transformed in their meaning and affect.[13]

In 1999, I set about making an archive-based film of my own. This drew on the book *Hard Road to Klondike*, a classic story of late nineteenth-century Irish emigration to America based on the life of Donegal migrant worker Micí Mac Gabhann (MacGowan 1962/2014). The resulting film employed a range of silent cinema archive, both factual and fictive, and a narration by Oscar-nominated actor Stephen Rea. My script drew from the original drafts of the book, re-telling Mac Gabhann's story for a different era and seeking to critically re-examine the story of Irish migration to America.

The production of this film involved extensive "background research": moving and still image archival searches in the Library of Congress and US National Archives and in a number of other film and photographic collections in Ireland, Scotland and the US; script research involving re-translation from the original Irish sources to English with a view to retaining the poetical character of the former; historical background research on patterns of Irish migration to Scotland and the US, some of it library based, plus interviews with leading historians (clips from which were eventually employed in the completed film); location scouting and shooting in Ireland, Scotland, Pennsylvania, Montana and Alaska; postproduction experiments with the use of contemporary dissonant music (Kevin Volans) used against archival image sequences and with contemporary Irish music that has revisited traditions of traditional sean-nós song in the context of interactions with other varieties of "world music"; experiments with the multi-layering of video images to produce collage effects (Figure 2.7) – including slowing down and freezing moving images to lift archive out of its usual illustrative and evidential role within documentary film and to reposition it in with a more experimental and expressive framework. All of this, needless to say, was hard work and not overly concerned with any programme of academic research within the field of film studies or indeed history or Irish studies.[14]

Indeed, it was only several years later, with the film completed, broadcast and exhibited at a range of international film festivals and subsequently reviewed and written about by others, that I had the time and inclination to draw some critical lessons from this experience. A number of colleagues in Ireland and in the US were using the film in their undergraduate film studies teaching (or specialist Irish studies seminars) and I found myself being regularly asked to screen the film for groups of students and to explain how it was made. I published several articles (Bell 2004, 2012a and b) locating the film within a tradition of found footage film-making and in relation to critical debates about new documentary practices.

My reflections were certainly sharpened by these pedagogic encounters and were now framed within the discourses of film studies and critical theory (rather than being the bar room or festival anecdotes of the film-maker). They became concerned with the theorization of contemporary documentary film practice and

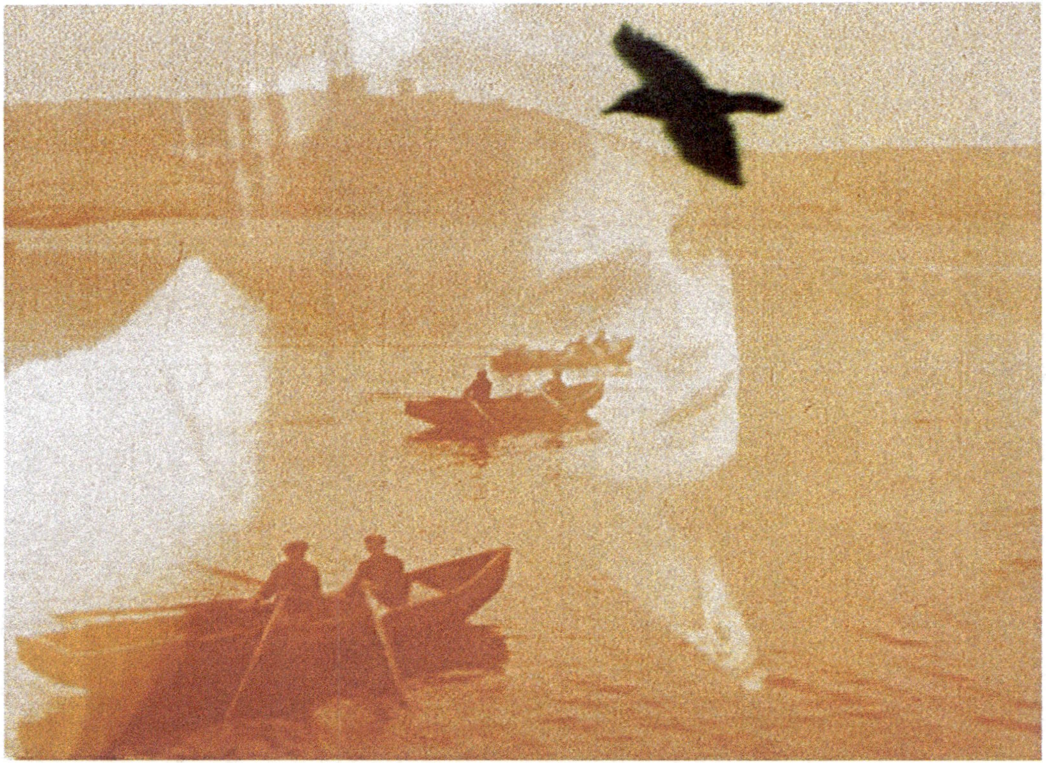

FIGURE 2.7 Desmond Bell, *Hard Road to Klondike* (1999), film still

Source: courtesy of glass machine productions

its engagement with history, rather than purely with the pragmatics of production. Understanding this sort of trajectory – project research > scripting and creative experiment > artisanal production > editorial process > exhibition and review > reflective analysis and critical contextualization – is, I would argue, key to developing models of practice research in the creative arts. This is just one case where the conduct of artistic research is mapped onto existing cycles of creative production and where the project, whether film, photographic venture, musical composition, dramatic work or performance, provides the "well-founded lab" within which research into arts practice can successfully take place. And yet, this template of practice-based research remains largely unrecognized, marginalized by more formalized but less grounded methodologies of research and criticism.

Notes

1 See for example: Barret and Bolt (2007), Macleod and Holdridge (2007) Smith and Deane (2009), Wilson and van Schelte (2013).
2 Interestingly enough, Stanley Kubrick was able to break free from the shackles of the Hollywood studio system and fashion his own artist's studio at his premises at Childwickbury Manor in Hertfordshire, England, complete with extensive picture and sound editing facilities. This gave him the sort of the sort of autonomy and creative freedom more typical of an artist's atelier.
3 Coutard moved effortlessly from fiction work to documentary projects, and similarities can be seen in his treatment of the streets and crowds of Paris in Godard's *À Bout de Souffle* (shot 1959) and in *Chronique d'une été*, shot a year later, or indeed in the film he later shot in Ireland for director Peter Lennon, *Rocky Road to Dublin* (1968). In this, the closing shot of a posse of kids from a Dublin housing estate chasing the lorry from which Coutard is shooting is a reprise of the tracking technique used in *Chronique*, where Marceline walks in a large warehouse towards Coutard's camera, also placed in the back of a truck.
4 As film critics never fail to remind Gordon and Perrano, their filmic strategy for *Zidane* had already been pioneered by German film-maker Hellmuth Costard in his 1978 documentary portrait of George Best, *Football as Never Before*. Costard used eight 16mm film cameras to follow Best in real time for the course of an entire game playing for Manchester United against Coventry City.
5 As Motherwell (in Cabanne 1987:7) argues,

> these conversations are more than mere interviews. They are Marcel Duchamp's "summing up," and constitute as vivid a self-portrait as we possess of a major twentieth-century artist, thanks to Duchamp's intelligence, scrupulousness, and disdain for the petty.

Certainly, a methodology of sympathetic elicitation is in play here in tandem with an ethics of documentary exactitude in this talk of scrupulousness, modesty and objectivity. One can see why this exemplary dialogue with the iconic "father" of conceptualism would have made Obrist sit up and take notice.
6 We have a description (Werner 1990) of the film-maker being ushered into a darkened room in Joyce's Parisian apartment to meet the esteemed writer, whose eyesight by this time was almost gone. The two sat opposite each other, in the gloom. It was, Eisenstein recalled, "a ghost experience" (1983:214).
7 Bill Morrison trained as a painter and animator. He has had a prolific output as a film-maker, producing an extensive series of short films, interspersed with a number of features. The most important of these is undoubtedly his 2002 work, *Decasia*, produced

together with composer Micheal Gordon (the trailer hosted on YouTube gives some idea of his visual approach: www.youtube.com/watch?v=hDa-mmSldDg).

8 Péter Forgács trained as a fine artist, working mainly in video, and subsequently practiced as a film-maker. He is perhaps best known for his "Private Hungary" series of found footage films based on a collection of Hungarian home movies from the 1930s and 1960s. He has drawn upon this material to pursue what has been called a "postmodern re-orchestration of history". In 2005, he also produced *El Perro Negro – Stories from the Spanish Civil War*, a film that engages with the Spanish Civil War through the newsreel footage of the period. Forgács also has an installation practice. He represented Hungary at the 2009 Venice Biennale, producing an ambitious installation piece, *Col Tempo*.

9 Peter Delpeut, an arts educator and film-maker, is best known for his films *Lyrical Nitrate* (1990) and, in particular, *Diva Dolorosa* (1999), which was screened alongside my own *Hard Road to Klondike* at the 56th Venice Film Festival in 1999. Like the other found footage film-makers discussed here, he works closely with a range of film archives and archivists. He is perhaps more faithful to his sources than Morrison, Forgács and myself, but again his use of specially composed music and cutting of his archive footage into montages which have their own narrative and formal construction significantly transfigure the material compiled.

10 Edgardo Cozarinsky studied literature in Argentina and aspired to a career as a writer before becoming involved in film-making in Paris. He is perhaps the most literary-minded of the film-makers discussed here. His most significant film, *One Man's War* (1981), has a distinctive modernist literary feel. Cozarinsky employs a narration based on the wartime diaries of maverick fascist writer Ernst Jünger, who had entered Paris in June 1940 as an officer with the occupying German army. This is dubbed over French and German newsreel footage from this period. This counterpoint produces a tension between the discourse of its unreliable narrator and the matter-of-factness of the documentary film record of Paris and the Nazi occupation.

11 These paper prints were used as a medium to preserve early film works so they could be deposited in the Library of Congress for copyright purposes. The Library has more recently used these paper prints as the source for making new celluloid prints of this early silent cinema material.

12 Within linguistics, indexicality is the notion of a sign pointing to (or indexing) some object in the context in which it occurs. The photographic image is often seen as an objective documentation, an index or copy, of what appeared before the camera. The concept became a significant one within film theory largely due to the work of French critic and theorist Andre Bazin (1967/2005), who sought to assert the realist character of cinema.

13 For a fuller account of found footage film-making and its use of film archive (including a discussion of my own films), see my 2004 essay, "Shooting the Past? Found Footage Film Making and Popular Memory".

14 It is probably worth noting that *Hard Road to Klondike* was made as a television commission involving all the public service broadcasters in Ireland (BBC:NI, RTÉ and TG4). From the outset, it set itself the task of appealing to a popular broadcast audience rather than the specialist audience for avant-garde film work.

3

A RESEARCH TURN IN THE VISUAL ARTS?

3.1 Introduction

In this chapter, I suggest it is possible to identify a "research turn" within contemporary visual arts practice. Indeed, understanding this development might be a good place to begin in elaborating a research tradition for the creative arts and media today. My focus in this chapter is largely on what is often referred to as fine art practice (as opposed to applied arts, design or media activity). But, as we shall see, the interdisciplinarity at the core of such practice and the extensive use by visual artists of a range of media, including photography, film and video, mean that the old divisions between fine and applied art have all but broken down. And while film and media students who approach practice from a media industry and popular cultural perspective may be less aware of the visual art debates I consider here, the issues raised seem to have import for how we as practitioners think about research across the entire field of art, media and cultural production. So, bear with me!

3.2 The dynamics of contemporary art practice

Art practice today is less concerned with the individual art object and more focused on art process and purpose. Today, art is increasingly seen as a series of events "where things can happen", rather than a thing in the world. This shift means that the role of research activity within visual arts practice has been enhanced. Under the sway of Conceptualism, creative practice involves a sharpening of focus on what might be called the "ontology" of the artwork – what art is or can be in this era. The advent of mass communication played its part in dismantling a cultural hierarchy which valorized high culture over popular forms of cultural expression, but Conceptualist argument did the critical spadework in levelling the ground.

But if in the new critical dispensation anything can be considered an artwork, how can we make valid judgements regarding the status and value of a piece of art? As art historian Benjamin Buchloh (1990:118) has accurately observed, conceptual art effaced "all residues of representation and style, of individuality and skill". This created a situation where the attribution of aesthetic value to the artwork became difficult. Who really knew how to make sense and evaluate the new work?[1]

One of the ways we cope with this dilemma is by focusing on art practice as a creative research process. This shift from object-hood, aura and authorship to an artwork that is dematerialized and *relational* creates a new engagement between art practice and research. The artwork is often now conceived of as an investigation with its own experiments and methodologies which can be shared with an audience. Thus, the "expanded practice" embraced by many artists primarily is exploratory in character.

We can then identify a number of major developments within contemporary visual arts practice which are encouraging an enhanced role for research across the creative arts:

1. The dominance of Conceptualism and the dematerialization of the artwork
2. The enhanced role of curation in reshaping the form of the artwork
3. The adoption of a relational aesthetic by many artists and curators
4. The increasing role of the performative in art practice
5. New interactions between critical theory and creative practice, conditioned by what has been called a "practice turn" within contemporary theory

Can we get to grips with these developments and tease out their significance for understanding the character of artistic research? An "immanent critique"[2] of these developments can, I think, contribute to the elaboration of a model of critically informed research for the arts.

3.3 Conceptualism

Art historians[3] usually trace the origins of Conceptualism to Marcel Duchamp's staging of the "readymade". The eponymous R. Mutt (the signature that adorns Duchamp's 1917 *Fountain*)[4] invites us to consider a mass produced, ceramic urinal as an artwork. This anti-aesthetic gesture indicated that nothing was to be definitively excluded from the domain of art and initiated a displacement of painting from its core status within traditional art practice. In turn, it paved the way for a subsequent critique of the "retinal". The emphasis now fell on the conceptual dimension of the art process rather than on the visualization process which preoccupied painting.

Pierre Cabanne, in his classic interview with Duchamp (1987), probes the emergence of a conceptualist sensibility in the artist's work (43):

CABANNE: Where does your anti-retinal attitude come from?

DUCHAMP: From too great an importance given to the retinal. Since Courbet, it's been believed that painting is addressed to the retina. That was everyone's error.

> The retinal shudder! Before, painting had other functions: it could be religious, philosophical, moral.

Duchamp presented the readymade as a conceptually framed artistic statement.[5] As he was to conclude with regard to the modernist movement (39), "everything was becoming conceptual, that is depended on things other than the retina". His anti-art stance (56) was "principally a matter of questioning the artist's behaviour, as people envisaged it. The absurdity of technique, of traditional things".

Art practice became focused on the ideational content of the artwork and no longer primarily concerned with exploring the logic of a medium or the visual elements of a work. This encourages the idea of art practice as a critical intervention within aesthetic discourse that is driven by conceptual considerations and rests on a core investigation.

As Sol LeWitt, one of the first self-proclaimed conceptual artists, was to argue in 1965,

> in conceptual art, the idea or concept is the most important aspect of the work. When an artist uses a conceptual form of art, it means that all of the planning and decisions are made beforehand and the execution is a perfunctory affair. The idea becomes a machine that makes the art.
>
> *(p. 1)*

His contemporary, Joseph Kossuth,[6] identified the rupture in art practice wrought by Conceptualism (1993:56).

> Art before the modern period is as much art as Neanderthal man is man. It is for this reason that around the same time that I replaced the term "work" for *art proposition*. Because a conceptual *work* of art in the traditional sense is a contradiction in terms.

The term proposition used here has philosophical allusions. In the late 1960s, American artists were becoming aware of structuralist approaches to understanding language. Roland Barthes had published his ground-breaking *Mythologies* in 1957. In this text, he suggests that the visual could be treated as a system of signs rather than as a realm of psychological experience, launching a critical strategy that set into motion the cultural studies approach to understanding art and popular culture.

Kossuth pioneered the use of words within the artwork in place of visual imagery. In his 1965 piece *One and Three Chairs*, he presents us with a found object, a wooden chair (Figure 3.1). This is flanked on one side by a black and white full-size photograph of the same chair and on the other side by a dictionary definition of the word "chair". The photographic representation, linguistic term and "real" object are transformed in their significance by their inclusion within an art context. All three are symbolic codes that the artist plays with, encouraging us to think

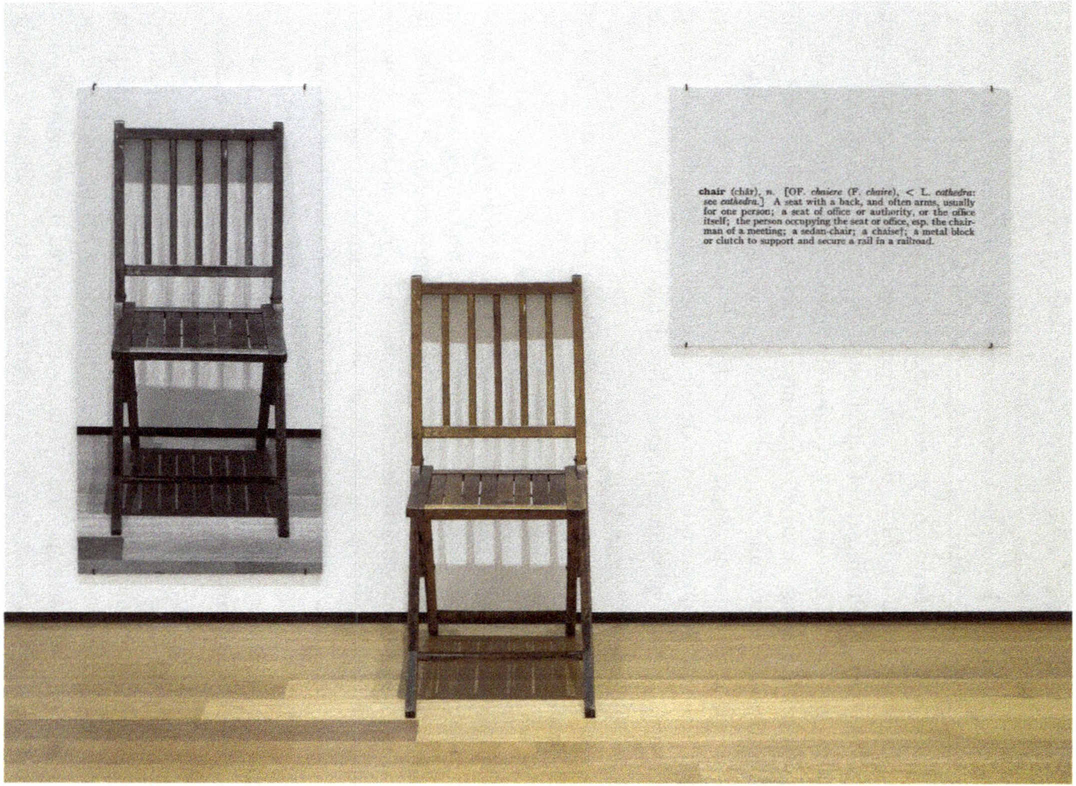

FIGURE 3.1 Joseph Kossuth, *One and Three Chairs* (1965)

Source: courtesy of Museum of Modern Art, New York

about the nature of the conceptualist artwork. Much of Kossuth's work at this time explores the relationship between ideas and the written word, which he treats as an image rather than an abstract signifier.[7]

Non-objective art was an art of concepts, propositions and problematic linguistic signs. Conceptualism quickly became aligned with a structuralist analysis of language and cultural form and in many ways anticipated the subsequent assertions of the "death of the author" by theorists such as Barthes (1972) and Michel Foucault (1977). With the assertion of the conceptual character of all artistic endeavour, the cognitive assumed a larger role in contemporary art – as does the research and development aspect of art practice. If the idea is the "machine that makes art", then the research and conceptualization process is the fuel that drives this mechanism.[8]

Conceptualism certainly led to the *dematerialization* of the art object. Of late, the focus has been on the impact of digitalization on the practices of artists. As a result of the ephemeral, networked and infinitely reproducible character of digital art, the artwork is radically dematerialized and art practice resembles a mass media process. As the art object can increasingly take the form of a digitalized image file, such work has neither uniqueness, spatial location nor medium specificity. It exists in a precarious realm dependent on a technological apparatus for the quality of its relayed image. Its exchange value is threatened with the eclipse of the singular and auratic quality of the artwork, and its use value remains unclear.

However, as Lucy Lippard has argued (1997:42), this dematerialization of art long pre-dates the digital revolution.

> During the 1960's, the anti-intellectual, emotional/intuitive processes of art-making characteristic of the last two decades have begun to give way to an ultra-conceptual art that emphasizes the thinking process almost exclusively. As more and more work is designed in the studio but executed elsewhere by professional craftsmen, as the object becomes merely the end product, a number of artists are losing interest in the physical evolution of the work of art. The studio is again becoming a study. Such a trend appears to be provoking a profound dematerialization of art, especially of art as object, and if it continues to prevail, it may result in the object becoming wholly obsolete.

Somewhat ironically, an art movement which sought to free art from commoditization and maximize its autonomy may have enhanced the role of the marketplace as the ultimate arbiter of value. Paradoxically, the academy has also assumed a role in the validation of the artwork. Some critics have suggested that many artists now seek validation for their work initially within the knowledge economy (certainly a tempting strategy for the creative arts doctoral student). Academic valorization provides an alternative to the harsh judgements of the market (at least in the first instance). As Simon Sheikh notes (2009:5),

> often, but not always, in such dematerialized, post-conceptual and, perhaps more accurately termed, re-contextualized art practices, there is of course a

> notion of research invoked. Research has even, to some extent, superseded studio practice. Artists are increasingly researching projects, not only to make site-specific works, but also time and content specific works.

Conceptualism has also transformed the nature of art education. Art academic Steven Henry Madoff (2009:ix) observes that,

> from the 1980s on, the influence of conceptualism has affected art schools all over the world. Many schools have erased the boundaries between disciplines, as the expression of a concept in this post-Duchampian epoch rides across all material means – photography, video, painting, drawing, sculpture, or any of them and more joined in an installation. Somewhere between philosophy, research, manual training, technological training, and marketing, an evolved profile of contemporary artistic practice has pressed the art school as a pedagogical concept itself to address what an artist is now and what the critical criteria and physical requirements are for educating one.

The creative arts field today is characterized by a high level of interdisciplinarity, which has overturned traditional hierarchies within art education. Today, within both the academy and the art world, painters and sculptors – or dancers and architects – are just as likely to be employing photography, video and digital processes in their work as film-makers or media students.

To give some examples: Irish artist Willie Doherty, a sculptor by training, employs a lens-based practice to explore the fallibility of memory and the troubling residues of past violence in Northern Ireland and elsewhere. His short film *Ghost Story* (2007) utilizes traditional filmic means[9] to explore the forms of trauma and amnesia associated with unresolved political conflict. Now permanently displayed in the Ulster Museum in Belfast (a clip can be seen at: www.youtu.be/ee_7ZS1vhna), his film employs a range of cinematographically composed tracking shots provided by Oscar-nominated Director of Photography Seamus McGarvey. Over these images (Figure 3.2), an ethereal voice-over by Stephen Rea is dubbed, narrating a ghost story that hints at real-world political terror and unresolved trauma, situations to which Doherty is sensitive given his upbringing in a troubled Northern Ireland. Doherty invokes the figure of the "daylight wraith" or revenant who, burdened with a traumatic past, continues to haunt the present.

As he told me in an interview conducted at the Venice Biennale in 2007,

> I have a love–hate relationship with the idea of story-telling. And, I guess I have a very unorthodox idea of process. The work is never script driven. I usually start with a location, which is how *Ghost Story* developed. I had an idea of the sort of places I wanted to shoot this piece in, and a number of characters, but beyond that I didn't really know what it was about. After I shot the footage I didn't look at it for several months . . . hated it . . . didn't know what it was about.

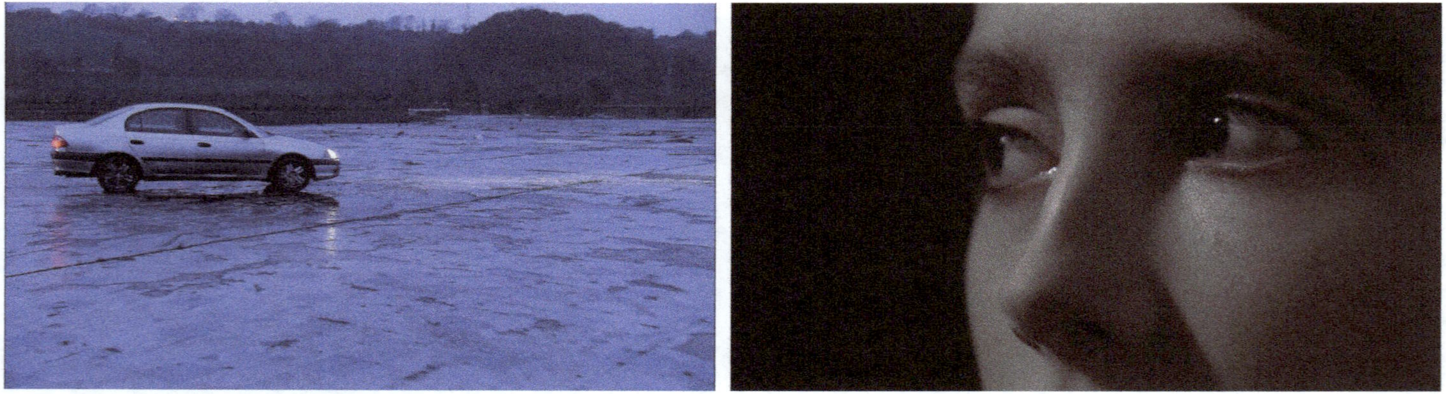

FIGURE 3.2 Willie Doherty, *Ghost Story* (2007), film stills

Source: courtesy of the artist and the Kerlin Gallery, Dublin

> The idea that the piece was a sort of ghost story emerged out of the way the camera was moving through space in a kind of floating, slightly disembodied feeling (in a long tracking shot along a deserted country road in the evening gloom). The form developed out of the process. On the other hand, I must have known at some level what this piece was about, but I wasn't able to articulate it. I always wanted to create a piece that was about loss and our inability to deal with memory.

For artists who employ lens-based media, the emphasis is likely to be on process rather than script or abstract concept. Form emerges from creative process and experiment, as do articulation and research insight.

3.4 Curational strategies

The impact of Conceptualism has also been found in the worlds of curation, exhibition and gallery practice. As the criteria for comprehending and judging artworks have become less clear, curators have stepped in to supplement the meaning of the artwork. In part, they are responding to the public's interest in understanding the processes that artists employ in making their work. Indeed, some have identified a more general "curatorial turn" in contemporary art. As Paul O'Neill and Mick Wilson argue (2015:12), "the 'curatorial' and certain understandings of 'research' have become aligned with each other, perhaps as a means of moving beyond an understanding of exhibitions as the main outcome of curating as production".[10] In the last few years, curators seeking new ways to present the work of key visual artists have on occasion gone to considerable lengths to provide glimpses of the artistic processes and research activities employed by these artists.

For example, the Fruitmarket Gallery in Edinburgh mounted an exhibition of the work of a leading Mexican artist, Damian Ortega, in October 2016, titled "States of Time" (one of the pieces, *Altocumulus*, is seen in Figure 3.3). This was described as "site specific" work. And, indeed, some of the work exhibited had been produced in the gallery during a residency. As Damian expressed it in an email to me,

> the gallery was very generous in letting me experiment with this project without asking for a particular result. The whole team in Edinburgh was very process-focused and didn't freak out when we decided to make a few additional artworks for the show at the very last minute during the installation and after some interaction with the gallery visitors.

The audience were invited to observe this process close up, and as Damian reports,

> I think I had a very good communication with the Fruitmarket Gallery from the preproduction stage onwards and they were incredibly receptive and engaged both with my ideas and with how their audience might interact

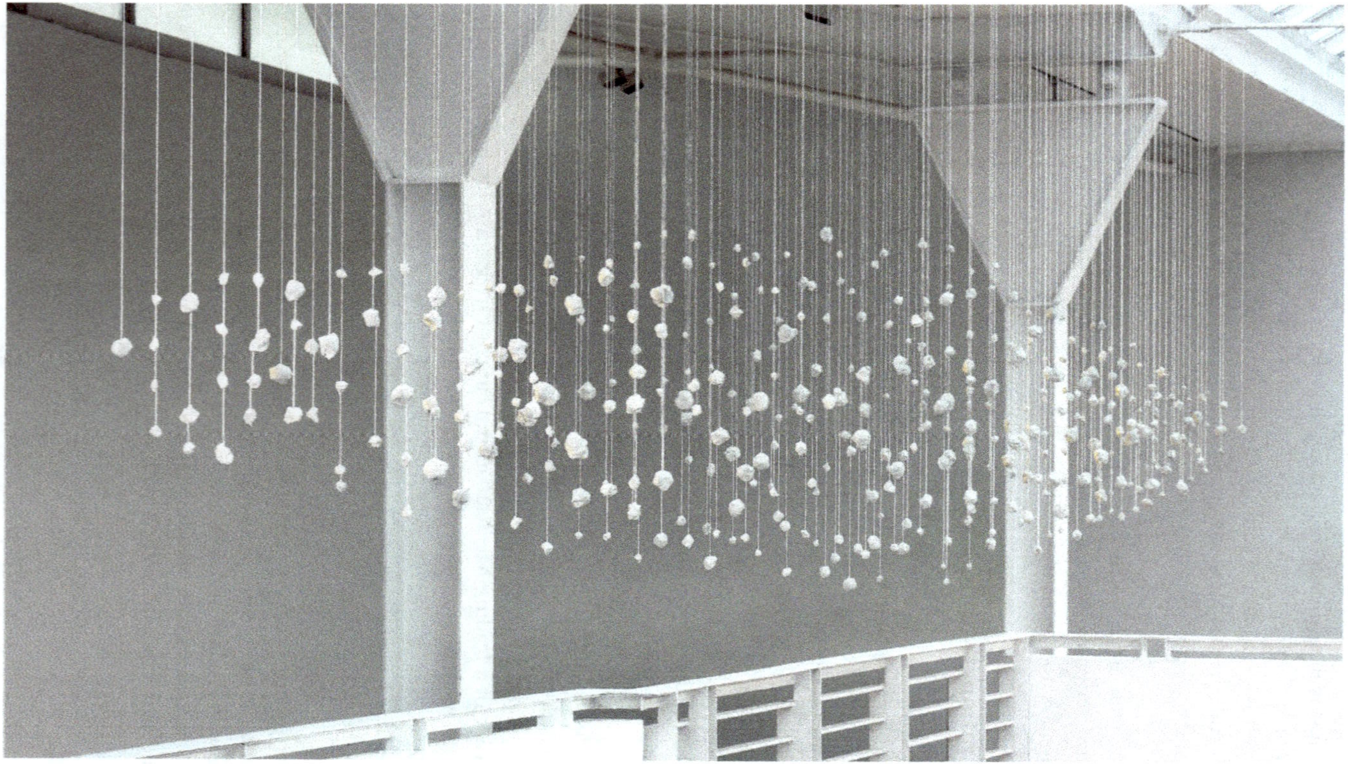

FIGURE 3.3 Damian Ortega, *Altocumulus* (2016), photo by Ruth Clark

Source: courtesy of the artist and the Fruitmarket Gallery, Edinburgh

> with them. As it turned out, the Edinburgh public had a very good reaction to the show and with the production of the exhibition book and videos, and the holding of a series of public workshops, I think we fostered a closer relationship for the audience with these works.

The Fruitmarket describes itself as "a kind of laboratory – a space in which artists are encouraged to experiment, to try things out, to reach out to audiences" (2016). This is a self-description to which many other galleries of contemporary art would subscribe.

As part of this mission to explain and engage, it has become a standard practice in many contemporary art galleries to commission short, interview-based documentary films on the artists whose work they are exhibiting.[11] For example, William Kentridge, the South African visual artist whose project *More Sweetly Play the Dance* we discussed in Chapter 1, provides an illuminating videotaped interview (made in tandem with his co-exhibitor Vivienne Koorland) to accompany their 2017 joint exhibition in the Fruitmarket. This exhibition involved a number of Kentridge's animated films employing his charcoal drawings, with their characteristic marks of erasure and reworking, together with live action performance. In *Tango for Page Turning* (Figure 3.4; https://vimeo.com/257386772), the artist dances with a female figure against the background of an open book. Each figure draws closer to the other as the pages turn, until they disappear into the fold of the book. These films engage with aspects of recent South African history, particularly during the Apartheid era, but seem to have a broader existential resonance. Kentridge's films complement Koorland's paintings and tapestry works, which explore South African history and landscape, both real and imagined.

The Fruitmarket interpretative film takes the form of a video conversation between the two artists and curator and academic Tamar Garb about the work in the show and how this evolved (an element from the interpretative film can be seen at: www.fruitmarket.co.uk/archive/kentridge-koorland/). The piece explores "the formal and thematic links between the work of Kentridge and Koorland, mapping their artistic friendship through shared artistic strategies and expression of a common sense of the urgency and agency of art". Garb sets out the critical context for the discussion.

Tamar Garb: Both of these artists understand history through the representations and material practices of the past. So rather than trying to picture history as an event one tries to imagine afresh, they use the residues of history, the material traces of history – whether these are books, or maps or scripts, or drawings . . . it's not about a recording of history, it's about a personal relationship to history and about situating of the self in relation to those historical events.

Kentridge responds in a more personalized and concrete manner.

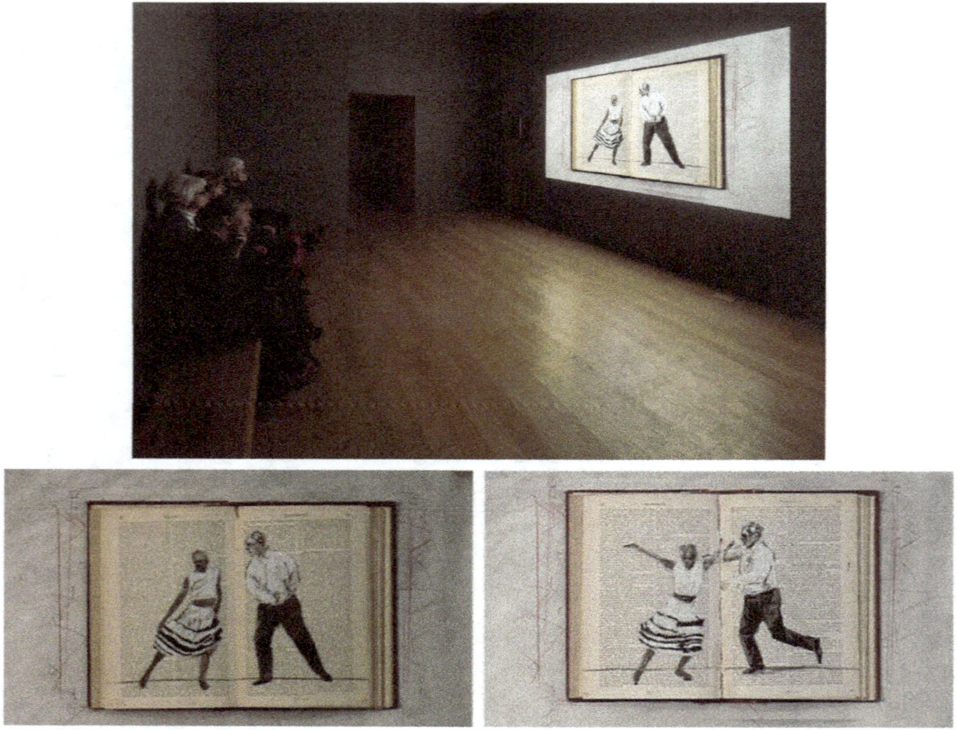

FIGURE 3.4 (above) William Kentridge *Tango for Page Turning*, installation view, Fruitmarket Gallery, Edinburgh 2016; (below) film stills (2013)

Source: courtesy of the artist and Marion Goodman Gallery

William Kentridge: The need to make art is different from the need to make art about history. The need to make art is primary and that gets fed by so many things around you, history, memory, your own thoughts and desires are there. The world comes into the studio in all its different forms, whether in newspaper articles, things you've read, things you are enraged about, things you're fearful of . . . and they get cut up and rearranged in the studio as a painting, as a film, as a projection and then sent out. And, that's not the same for me as saying this is a historical responsibility, it's a much less coherent need but it feeds on everything around it.

The two artists subsequently published a book (2017) which extends the critical conversation between themselves and Garb to include other contributors, such as the art historians Griselda Pollock and Briony Fer. Such books and films aim to assist audiences in their engagement with work, providing a context for the work exhibited and insights into the working methods and creative aspirations of the artists.

On occasion, this interpretative process has involved reconstructing artist's studios – for example that of Francis Bacon in the Hugh Lane Gallery in Dublin (Figure 3.5 left), a display which now has assumed a permanent and indeed monumental presence in that gallery. In Edinburgh, the reconstructed studio of sculptor Eduardo Paolozzi (Figure 3.5 right) has been afforded a similar status in the city's Museum of Modern Art. It is not clear what insight, if any, either of these reconstructed studios provide into the work of Bacon and Paolozzi. Do these installations purport to be artworks or merely artefacts assembled in a simulacrum to document the artists' work and lives? Both seek to portray a working environment of sublime disorder for each artist in an attempt to authenticate these installations as "sites of genius" and claim them for the national culture (the original studios of these artists were in London, not in Dublin and Edinburgh).

On other occasions, this process of excavating an artist's work has involved assembling a range of records, including photographic documentation, personal communications and notebooks, with a view to revealing an artist's experiments and the thinking and sources that may have shaped their work.

Indeed, over the last few years artists themselves have developed an interest in curatorial and archival activities, exploring the relationship within photography between its capacity for documentation and the artist's concern with expression. English artist Katerina Jebb has a practice that combines costume photography and portraiture, but her abiding interest has perhaps been in a series of ambitious and innovative art projects that explore the creative potential of high-resolution scanning technologies. While describing herself as a "scannographer" rather than photographer, her work clearly lies in a long tradition of research within photography which has explored the relationship between photographic practice and a range of other imaging technologies – such as X-ray imaging, use of infra-red techniques and MRI scanning.

In 2011, she was commissioned by the Balthus Foundation to document the contents of the artist's studio in the village of La Rossinière high in the Swiss Alps,

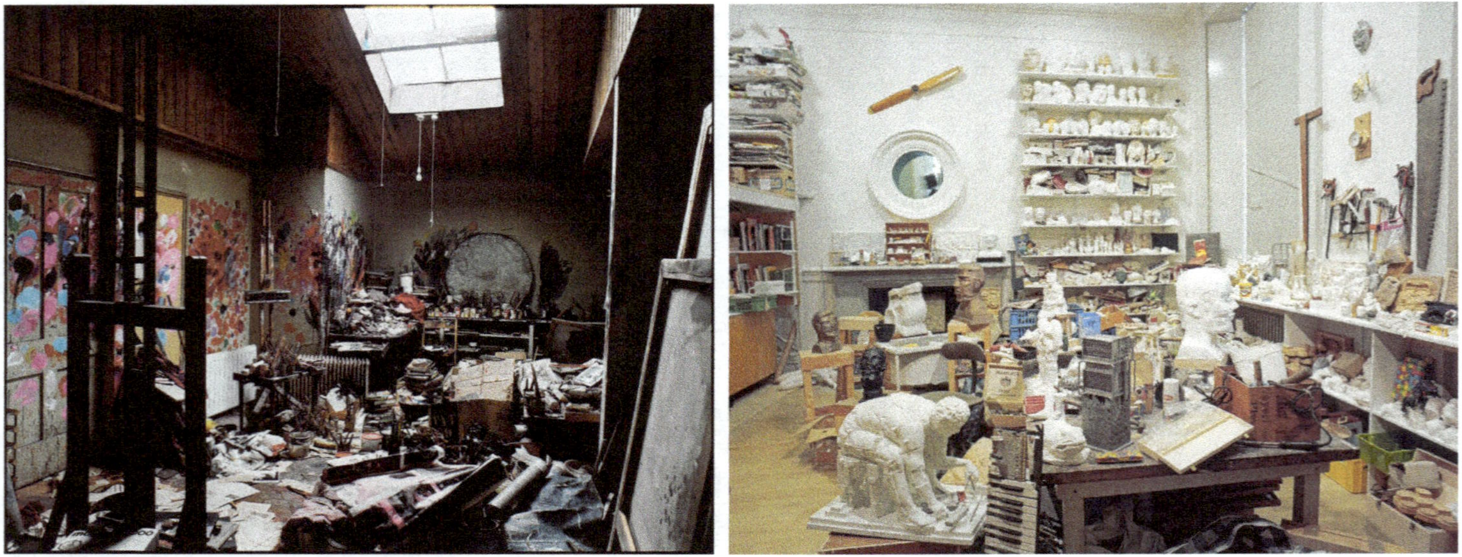

FIGURE 3.5 (left) 7 *Reece Mews Francis Bacon Studio*, the Hugh Lane Gallery, Dublin, photo by Perry Ogden; (right) *Eduardo Paolozzi Studio*, Museum of Modern Art, Edinburgh

Source: (left) © The Estate of Francis Bacon. All rights reserved, DACS; (right) courtesy of the National Galleries of Scotland

in which the reclusive Balthus had painted for over twenty years. In a 2018 email interview with me, Jebb traced the origins of her Balthus project:

> Balthus died in 2001 and nothing in the studio has been moved since, so in essence it's like a large still life painting. Most of my work is made with a high-resolution scanner and so it followed that I would document the subject using this medium. Balthus wore the same smock to paint in for over 60 years and it was the central piece of his working life. He was obsessed by it, and it seemed to hold more for him than just an ordinary object of clothing. I started by scanning the smock, which is known as "The Tablier." As the size of the scanner is A3 the work is made in sections that are rendered to form a life-size composition and then digitally retouched to eliminate the traces. The viewer could think the result is one image but in fact it can be as many as 50. What evolved from this beginning was that I began to make an inventory of his entire studio which has continued for 7 years.

Her image of the *Tablier* can be seen in Figure 3.6 left. Through the high-resolution scanning process she employs, the Balthus smock is separated from the jumble and detritus of his mothballed studio and afforded an elevated documentary status. Indeed, the garment seems to take on the form of a funerary shroud, reminding us of that intimate link between photography and death which so fascinated Roland Barthes. Similarly, Jebb scanned, then reconstructed, a cigarette butt-filled ashtray found in the studio (Figure 3.6 right). Separated from its quotidian surroundings and "embalmed" via the scanning process, the object also has an uncanny feel, serving as it does as a *memento mori*.

She exhibited the *Balthus Inventory* first at the Tokyo Metropolitan Art Museum in April 2014 and subsequently at Musée Rattéu, in Arles, in 2016. In this latter exhibition, "inventories" on Balthus and on the futurist artist Francis Picabia are shown alongside more traditional photographic portraiture and films documenting her interest in costume and identity. In my email interview with her, I asked her about this "expanded" aspect of her work.

DB: Do you see yourself performing an archival and curatorial role in this inventory work?

KJ: I try to examine and represent the object with truthfulness and precision. There is a forensic leaning to my practice, indeed I consider my work to primarily be about the close study of objects. The fact that the object and the glass plate of the scanner physically make contact is vital in exposing the material basis of the subject matter.

I think that it is essentially this contact which distinguishes the medium of "scannography" from traditional "light borne" photography. The exactitude of the scanning process enables me to make mechanical compositions whereby eliminating the need to paint or draw the subject.

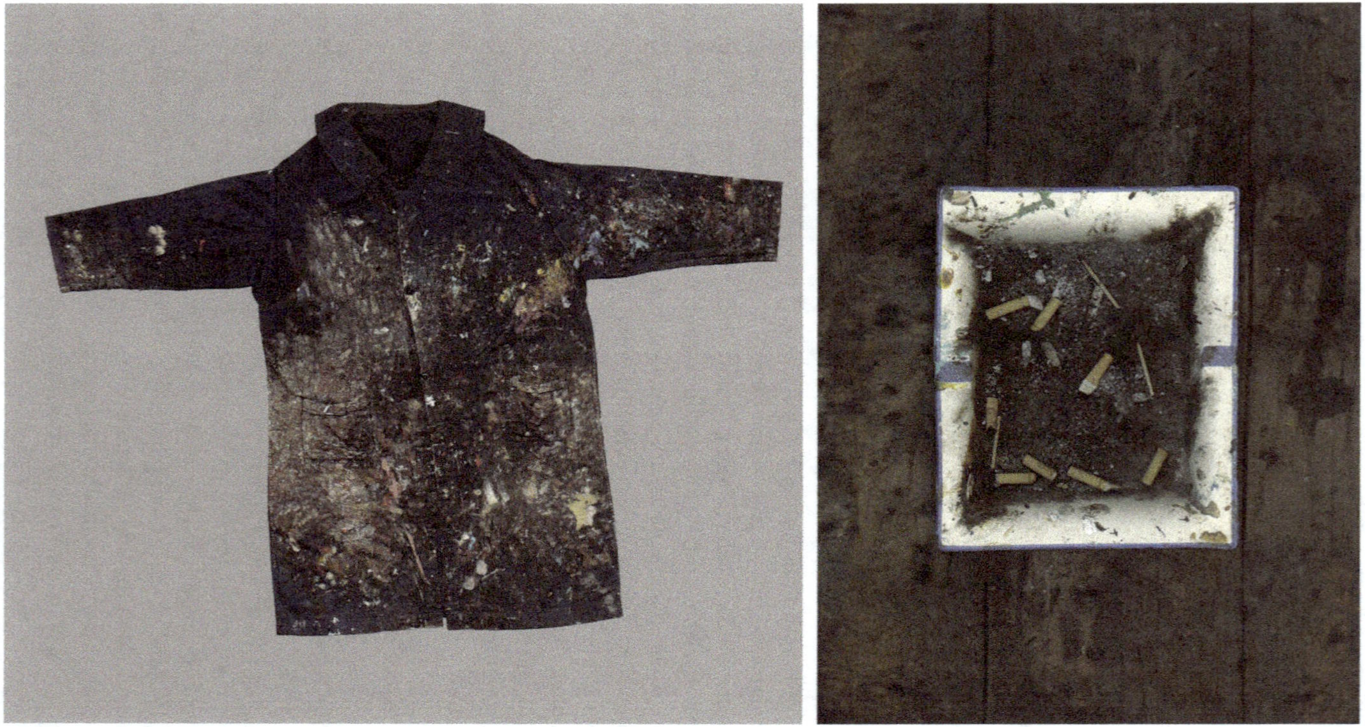

FIGURE 3.6 (left) Katerina Jebb, *Tablier*, from the Balthus Inventory; (right) *Cendrier*, from the Balthus Inventory (2017), Musée Rattéu, Arles 2016

Source: courtesy of the artist

The objectivity belongs to the machine and I assume the role of the programmer and so the result is a collaborative effort between a machine, myself and an object.

DB: What does the artist add to this process of documentation and preservation?

KB: I curate and produce works based on my personal obsessions and aesthetic choices. In the case of the Balthus studio I could sense that the space was still charged with his presence. Time seems to be suspended and it is a contemplative place where one asks questions and makes a personal inventory of the mind. It's very cold inside but there is an overwhelming feeling of peace and harmony. Moreover the space holds meaning which is a difficult feeling to try to describe.

Jebb's Balthus work captures, I think, elements of this embodied meaning much more successfully than the faux artist studios reconstructed in galleries that we have discussed.

Objects are lifted out of their original material context and deconstructed in the scanning process that she employs, then re-assembled in a composite image. A distinctive and insightful artwork emerges. For her, as for many contemporary artists, research, documentation, re-appropriation and re-presentation are now seen as central to the art process.

3.5 Relational aesthetics

In 1992, artist Rirkrit Tiravanija mounted an exhibition entitled *Untitled (Free)* at the 303 Gallery in New York. He converted the gallery space into a makeshift kitchen. There he served rice and Thai curry to the visitors without charge, later recreating the event on the second floor of MoMA with help from the museum's kitchen staff. The scenario can be glimpsed in an image of the staging of the original exhibition (https://arteatart.wordpress.com/2010/03/10/rirkrit-tiravanija-untitled-free-19922007/).

As the publicity for the later presentation of the work at MoMA show puts it,

> in this deceptively simple conceptual piece, the artist invites the visitor to interact with contemporary art in a more sociable way and blurs the distance between artist and viewer. You aren't looking at the art, but are part of it – and are, in fact, making the art as you eat curry and talk with friends or new acquaintances.

Here we seem to have art as a form of what is sometimes called "action research". The art facilitates a social engagement with an audience (in Tiravanija's case, an invitation to eat together) and documents an interactive process between artist and their public.

In a more recent work, *Bamboo Maze* (2017), Tiravanija builds an elaborate wooden maze on a museum rooftop in downtown Singapore and invites visitors to explore the maze structure, which leads them to a tea house located in its centre where refreshments are available and lectures and events held (Figure 3.7; a video "taster" for the project can be seen at https://youtu.be/EzQESWM2Imo).

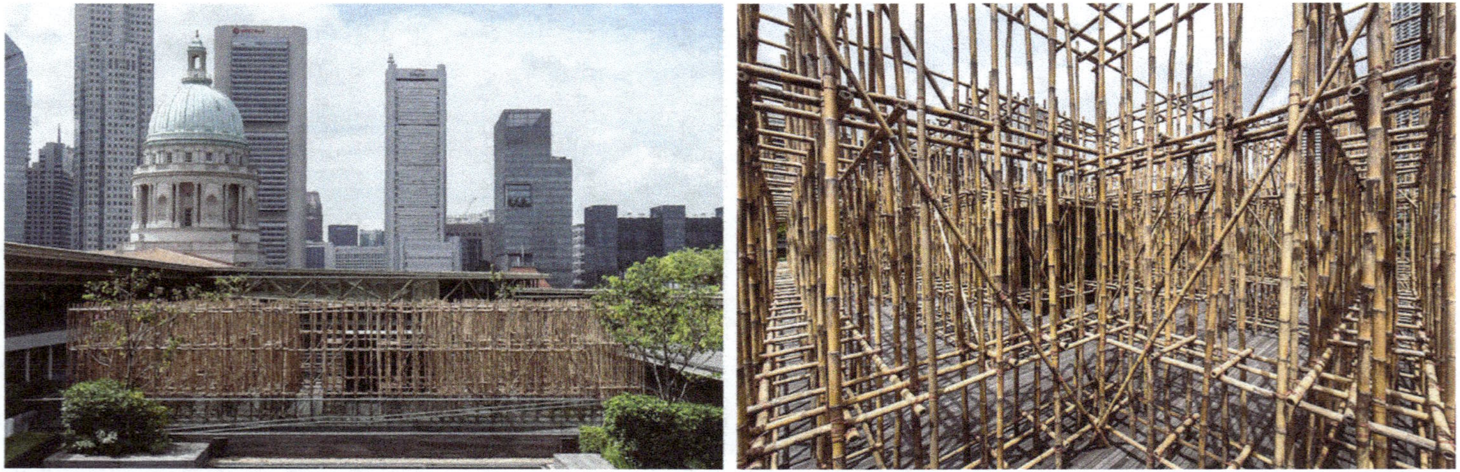

FIGURE 3.7 (left) Rirkrit Tiravanija, *Bamboo Maze* (2017); (right) *Bamboo Maze* close-up

Source: courtesy of the artist, Gavin Brown's enterprise, New York/Rome and the National Gallery Singapore; © Rirkrit Tiravanija

As Nicolas Bourriaud suggests (2002:18), "all works of art produce a model of sociability, which transposes reality or might be conveyed by it". Bourriaud has invoked the term "relational aesthetics" to describe this approach. He seeks to identify "a set of artistic practices which take as their theoretical and practical point of departure the whole of human relations and their social context, rather than an independent and private space". The relational approach favours the exchanges of the "field" over the production of an autonomous artwork by an individual artist. Accordingly, it champions the site specificity and interactive character of the art process. This position is closely associated with artists like Tiravanija, who together with a number of other contemporary artists – such as Philippe Parreno, Liam Gillick, Gillian Wearing and Pierre Huyge, as well as curators such as Hans Ulbrich Obrist and Nicolas Bourriaud – seek to rethink the relationships between the artist and their audience and to re-configure the site within which the audience interacts with such work and on occasion participates in its production. As Obrist has argued (2001a:129),

> the truly contemporary exhibition should express connective possibilities and make propositions. And, perhaps surprisingly, such an exhibition should reconnect with the laboratory years of twentieth-century exhibition practice. . . . The truly contemporary exhibition with its striking quality of unfinishedness and incompleteness would trigger *pars pro toto* participation.

Contemporary art, argues Bourriaud, employs a dialogical mode of address with regard to its audience. Such art, through its practices, sustains a distinctive conception of the public sphere as a space of "conviviality". Moreover, this orientation entails certain ethical and political choices on the part of the artist. It also emphasizes the unfinished and processual element in all art practice. A number of the artists associated with this approach conceive of both the art process and the exhibition context as akin to a laboratory. As Bourriaud himself has stated in an interview in *Art Forum* (2001:48) with regard to his own gallery, "we want to be a sort of interdisciplinary *kunstverein* – more laboratory than museum". And as we have seen, this approach has been adopted by a range of galleries and museums across Europe, including the Baltic Gallery in Gateshead, England; the Irish Museum of Modern Art in Dublin; the Kunstverein in Munich; the Fruitmarket in Edinburgh; and the Whitechapel Gallery in London, to name but a few.

This relational focus serves to diminish the significance of the finished and exhibited art object as the focus of visual arts practice. All work is effectively "work in progress". The relational approach sees the foregrounding of research processes as central to the reconfiguration of the work of art and its reception. Work in progress is effectively the artwork as an integrated process of research and development, exhibition and reception. The research context that guides the artist's structured interactions with their audience becomes a key element of the work. The work of art in effect is constituted jointly by the artist and their "witnessing public" (Bishop 2004) in a site-specific encounter not unlike an ethnographic event (though often it would seem with a diminished sense of ethical responsibility in comparison with the concerns of a social anthropologist or documentary film-maker). These strategies of engagement often employ complex forms of documentation which are also now seen as intrinsic to the artwork.

The work of Tiravanija and the theorizations of Bourriaud can be seen as concerned with what anthropologists call the "gift relationship". Anthropology since the work of Marcel Mauss (1925/1990) has had an acute interest in the cultural meaning of making and receiving gifts (Strathearn 1988). Indeed, some critics have commented on the parallels between relational art and contemporary anthropological practice. In the 1920s, Mauss, as a comparative anthropologist with a particular interest in Polynesian cultures, explored how the obligation in various cultures to give gifts sustains forms of social reciprocity. That which is given in turn calls forth a response of either equal or greater value to that which was originally received. His account focuses on not only the obligation to reciprocate but also the beneficial effects of this response in building forms of social solidarity within a society. Through such exchanges, the donator gives of themselves and their talents and the receiver acknowledges the gift as freely given and as confirming, or configuring, a social bond. Can such bonds be generated by contemporary art practice?

One of the features of the relational aesthetic is an acknowledgement of the interpretative power ascribed to the reader/viewer, something which is more generally a feature of poststructuralist accounts of the text/artwork. The role of the author/artist in the production of cultural meaning is correspondingly seen as diminished. Instead, it becomes supplemented by the activities of the curator, who is now both the arbiter of the significance of the work and defender of the authority of the artist as producer. As Clare Bishop (op. cit.:52) persuasively argues in her critical review of this art approach,[12]

> such work seems to derive from a creative misreading of poststructuralist theory: rather than the *interpretations* of a work of art being open to continual reassessment, the work of art *itself* is argued to be in perpetual flux. There are many problems with this idea, not least of which is the difficulty of discerning a work whose identity is wilfully unstable. Another problem is the ease with which the "laboratory" becomes marketable as a space of leisure and entertainment.

The Disneyfication of the gallery space as a site of recreational pleasure is nowhere more apparent than in Philippe Parreno's 2013 exhibition *Anywhere, Anywhere Out of The World*. Parreno radically transformed the monumental space of the Palais de Tokyo in Paris, the largest and most imposing museum in France dedicated to temporary exhibitions of contemporary art. With the support of Bourriaud (the gallery's director), he turned the building itself into the focus of the "visitor experience". Parreno uses sound, image and performance to guide the viewer on a journey through a range of his works, old (he includes for instance a multiscreen installation presentation of his and Douglas Gordon's *Zidane* film) as well as pieces specifically produced for the show. The exhibition was staged as a dramatic composition. Its use of music, lights, gigantic projected images and assembled objects is designed to shape the visitor's perception of the work in an immersive and polyphonic experience, which we get a sense of in Figure 3.8.[13]

Rather than seeing exhibition simply as a collection of individual works presented in an abstract space to a disaggregated and disembodied set of spectators, Parreno has sought to explore the exhibition event as a unified social experience in which the audience can be invited to participate. In a response to an interviewerwho asks whether his work was "an exhibition or a creation", Parreno answers

> Both. The exhibition becomes a strange premiere. I'm less an object maker than an exhibitions producer.
>
> Contrary to the cinema or the theatre, here the attention is floating (the viewing) is not mandatory. It's more like a garden where one walks, you find a silhouette against the light . . . its more about those things than the classical process of presenting the audience with an authoritative vision.[14]

Other artists take a withering view of this sort of "exhibitionism".[15] The English artists working collectively under the moniker Jeffrey Charles Henry Peacock (JCHP) have sought to critique contemporary exhibition practice. They ask (2014)

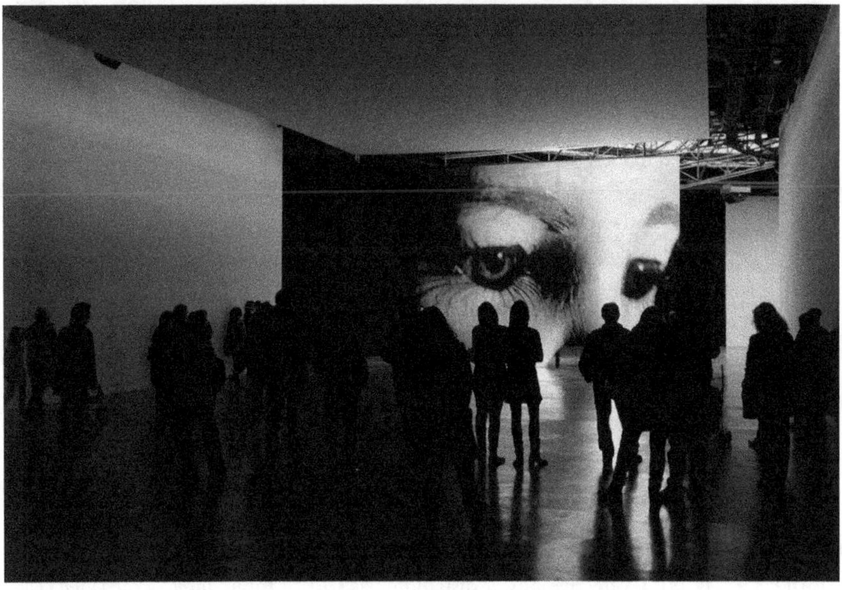

FIGURE 3.8 Exhibition view: Philippe Parreno, "Anywhere, Anywhere, Out of the World", 2011, Palais de Tokyo, Paris, 2 *TV Channel*, 2013, LED panel screen 768 × 576 cm – on screen: *The Writer*, 2007

Source: courtesy of the artist and Pilar Corrias, London; photo credit: © Andrea Rossetti

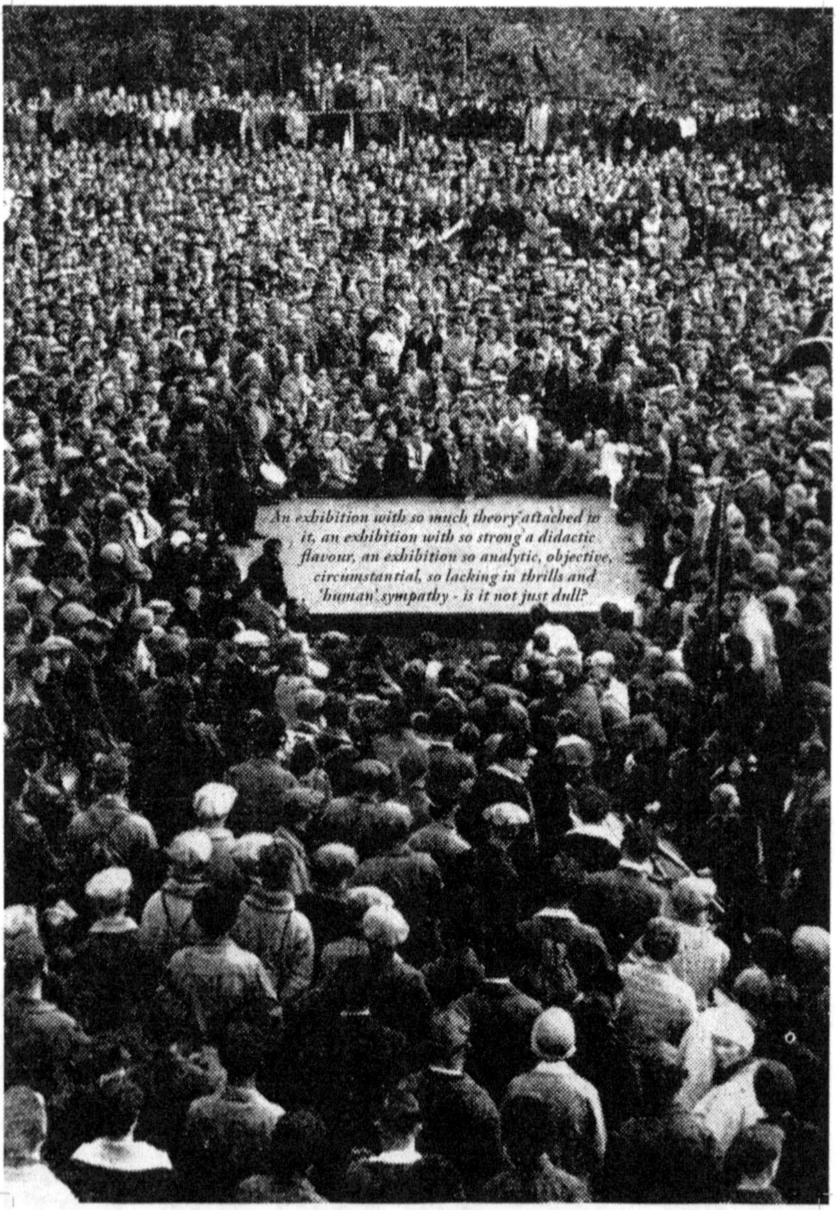

FIGURE 3.9 JCHP Collective poster distributed as part of the *Critical Décor: What Works* Exhibition Lanchester Galleries, Coventry, 2014

Source: courtesy of the artists

if there is "a way to engage in the distribution of our works in the context of an exhibition that remains critical and resistant to exhibitionism as it prevails in current art-hood – a self-certifying, self-aggrandizing celebration?"

Their Critical Décor exhibition, with its complex layering of images and propositions and allusions to historical determinations and critical anchors, makes considerable demands of its audience. But its underlying ethos is not didactic. The JCHP collective also draws on Marcel Mauss's exploration of the "gift relationship", but for different purposes. Indeed, they have provided perhaps the most consistent and radical application of this notion to the ethics and politics of social engagement within art practice. JCHP as a collective began producing and distributing its work in March 2005, posting out copies of work to those requesting it, thereby abolishing the need for a gallery space to exhibit work and completely bypassing the commercial process. As they announced in 2010, "by a significantly social reciprocity an audience will make contact with a work simply by being given it" (2014). Thus, JCHP founder Thom Winterburn, in September 2008, mailed out copies of his art book *Redux: The Exhibition of Mussolini* (a work of archival photographic images and unattributed text which explores Italian Fascism and its bloody overthrow as forms of mass inspired visual display, a plate from the book is seen in Figure 3.9) to those who had requested the work. This action constituted the exhibition and sought to offer "a symbolic form of exchange based upon the reciprocity of gifts rather than the equivalent of commodity exchange" (ibid.). The collective argues that "distributing work by giving it away, is an attempt to afford it a true value and proportion, perceptibly rending it worthless and lower than even merchandise, capitalism's heaviest discharge" (ibid.). Unlike the work of relational artists, who promote a vacuous spectatorship in their audience, the JCHP strategy seeks to address the core problem of *comprehension* that bedevils contemporary art practice. Given the lack of any clear criteria for evaluating the artwork, the curator, the institution and the market come to enjoy an enhanced interpretative role. This, they believe, can only be contested by presenting the artwork directly to an audience as a gift.

For the JCHP artists, the forms of reciprocity and social solidarity engendered by the gift relation provide an alternative basis for establishing the meaning of an artwork through the interactions of artist and audience. For, as the JCHP authors convincingly argue (2011:vii),

> a work can only be thoroughly comprehended – fulfilling its author's intentions and fulfilling its spectator's expectations – by being the manifestation of a form of ownership that obliges recipient and donor as one.
>
> The obligation to reciprocate initiated by the gift outweighs the pedestrian nature of spectatorship.

So, with the adoption of a range of socially engaged perspectives within contemporary arts practice, a range of ethical and political issues come into focus. These are associated with the relation between the artist and the subjects they represent or interact with, and with the forms of production and distributions they employ. Other disciplines call this parameter "research ethics" and refer to the pressing need for reflexivity within their disciplines. Artistic research may have something to learn from the social science disciplines – particularly social anthropology – where issues of representation, participation, professional responsibility and public interest figure within methodological discussion, setting limits for the researcher in terms of what is ethically acceptable.

3.6 Performativity

Contemporary art practice has also responded to the increasingly performative character of contemporary culture. It has employed a range of creative strategies to explore the role of performance in the construction of social relations and in the projection and play of identities. This interest in performance extends well beyond dramaturgical work proper and the traditions of performance art. Artists across a range of disciplines employ performance approaches which aim to transfigure the real and engage with the transgressive. In the course of this "play acting", the boundaries between fact and fiction, document and fable are eroded in significant ways. Facticity and knowledge are turned upside down, authorship reinvented. This approach to photography was pioneered by artists like Dieter Appelt (whose 1980 image *Canyon at Oppedette* is used on the front cover of this book). Appelt was a former actor from the Deutsche Oper in Berlin and is regarded as one of the founding figures of performance art.

This performativity is also apparent in the masquerading, photographic self-portraiture of Cindy Sherman or the more politically engaged work of Jo Spence concerned with challenging gender representations (Figure 3.10).

It is also present in the playful investigations and reportage of Sophie Calle. Calle employs lens-based forms of "record" and a form of "life-writing" to engage us as a storyteller for our time, calling into question documentary certitudes. For example, in her 1979 project *Venetian Suite*, she puts herself in the shoes of a detective. She sets the scene (2015).

> At the end of January 1981, on the streets of Paris, I followed a man whom I lost sight of a few minutes later in the crowd. That very evening, quite by chance, he was introduced to me at an opening. During the course of our conversation, he told me he was planning an imminent trip to Venice. I decided to follow him.

She trails her subject around the Italian city, surreptitiously photographing him and later exhibiting images of the anonymous man (Figure 3.11), providing a diaristic account of her activities. We do not know how much of this imaging activity is staged and how much candid. Again, as she says in a 2009 interview,

> If I am asked, I say it is all true – I am not able to invent. Afterwards it is other people's problem, not mine, if truth or fiction is a necessary criteria for them. I called one of my works *Histoires Vraies* [*True Stories*] to try to get rid of that question, but then I was asked, "Are the true stories true?" So how can I answer?

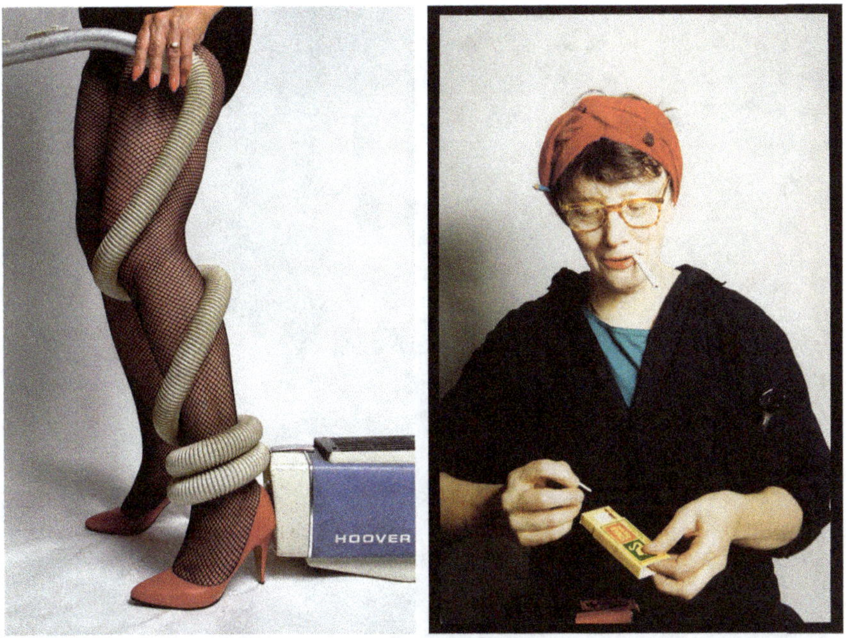

FIGURE 3.10 (left) Jo Spence, *Libido Uprising* (1989), collaboration with Rosy Martin, archival pigment print, 105 × 70 cm; (right) *Photo Therapy: My Mother as a War Worker* (1986–1988), colour photograph mounted on board, 25.3 × 16.5 cm

Source: © The Estate of the Artist; courtesy of the Richard Saltoun Gallery

In the stagings of Tracey Emin, her bed becomes a sordid *mise en scene* and the paraphernalia of her everyday life a series of documents charting a life in crisis (*My Bed*, 1999). Photographic artist Geoffrey Crewdson employs film production techniques and elaborate staging involving actors to produce his melancholy, cinematic style images of suburban America and its psychodramas in projects like *Beneath the Roses* (2008). Documentary film-maker Nick Broomfield – *The Leader, His Driver and The Driver's Wife* (1991), *Tracking Down Maggie* (1994) – makes himself the leading protagonist in his films, pursuing and interrogating his subjects, microphone in hand, from his chosen stance as the vigilant sound recordist on a quest. All in all, across a range of art forms, and against the prosaic and indexical dynamics of the photographic process, we witness a sustained attempt to move performative concerns into the core of art practice.

Significantly, those involved in the teaching of performance studies – now seen as an interdisciplinary field that has expanded beyond theatre and drama to include film, photography and the rest of the visual arts – have been at the forefront of

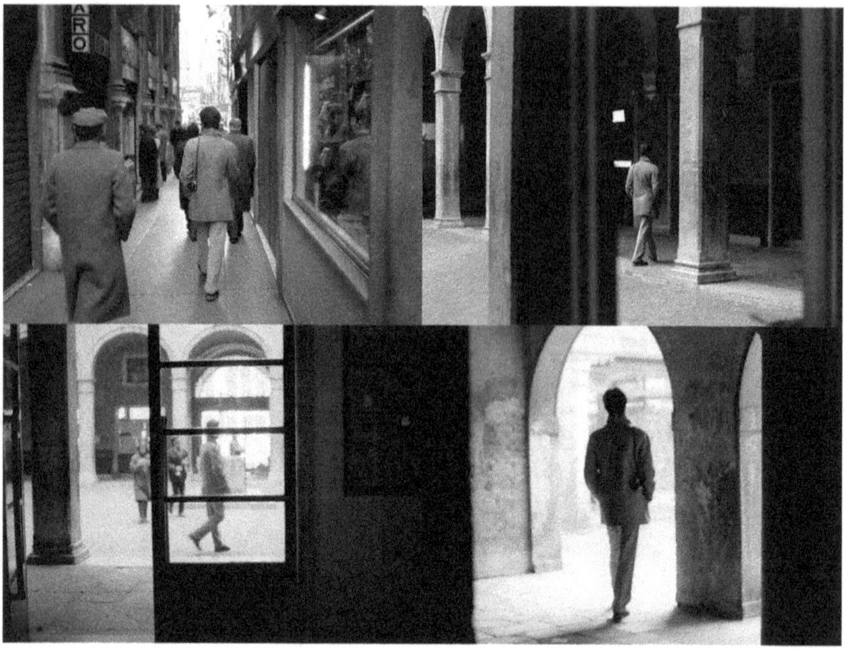

FIGURE 3.11 Sophie Calle, *Venetian Suite* (Detail) (1980)

Source: © Sophie Calle/ADAGP, Paris & DACS, London, 2017; courtesy of Galerie Perrotin

the movement to establish practice-based/led modalities of arts research. As Robin Nelson persuasively argues (2006:4),

> numerous instabilities in the diversity and ephemerality of performing arts practices pose particular challenges to ideas of fixed, measurable and recordable "knowledge". At the same time, however, the concept of "performance" has contributed a new conceptual map – and mode of knowing – to the academy and to research.

That said, the ephemeral nature of the performance event has led to a developed interest on the part of artist–researchers in capturing performance art activity via a range of forms of documentation, largely employing photography and video. Without such research capture, the ephemeral performance event has no ongoing status. How can you critically evaluate what you have not witnessed? On the other hand, the recording of performance using photographic means significantly transforms, and perhaps distorts, the original, embodied phenomenon.

To quote performance theorist Peggy Phelan's summative view (1993:146), "performance's only life is in the present. Performance cannot be saved, recorded, documented, or otherwise participate in the circulation of representations of representations: once it does so, it becomes other than performance". Yet, performance studies as an academic field has had to deal with the problem of the immateriality of its art object and the difficulties of calibrating its value in research terms. As Phelan admits, "the pressures brought to bear on performance to succumb to the laws of reproductive economy are enormous" (ibid.) – and nowhere more so, we might add, than within a corporate academy driven by research selectivity.

Moreover, while performance art, with its emphasis on process, event, embodiment and the problematizing of evidence and knowledge, advances a sophisticated postmodernist research practice, creative arts research more generally has been slow to address the broader implications of the notion of performance within the field of aesthetics. Indeed, it has been philosophers largely operating within a pragmatist tradition, rather than art theorists, that have made the most radical assertions about the performative character of all art.

According to the philosopher David Davies, what distinguishes art as a practice is that it is primarily a *performance*. And this is so not just in the sense that we refer to music, drama and dance as performance-led. Nor is it performative in the specific sense that Peggy Phelan intends when she observes, citing the work of Sophie Calle, that at present "the other arts, especially, painting, and photography, are drawn increasingly towards performance" (ibid.). Davies, in his densely argued book *Art as Performance*, asserts that art practice *in general*, including the visual arts of painting, film, photography and animation, should be viewed as a sort of "generative performance". For him, art practice needs to be explored in its own right *qua performance* and not merely viewed as a precursor to a finished artwork (1994:21):

> In the first place, our philosophical interest in "art" and in "artworks" is grounded precisely in such practice. It is because certain features of that practice puzzle us, or because the entities that enter into that practice fascinate us, that we are driven to philosophical reflection about art in the first place. (. . .) Put bluntly, there is no alternative but to start from critical reflection on our actual artistic practice because the very notions of "art" and "artwork" are parasitic upon that practice – artworks just are the things that play a particular kind of role in a particular kind of practice.

This pragmatist position asserts the "primacy of practice"[16] as the starting point for any theoretical account of art. For Davies, when we seek to provide a theory of art, we are subject to a "pragmatic constraint". A practice like art is, he believes, best justified not by referring it to an external tribunal of justification (such as a generalized conception of scientific procedure or aesthetic standard, for instance, or a general critical theory and interpretative mechanism such as deconstructionism) but rather through a patient codification of that practice and of innovations within it (such as those represented by the conceptual and performative strategies of contemporary art practice). The theorization of art as a field of activity is understood as the reflective process of bringing concepts to bear on art practice to more fully understand this. This can only be achieved through an attentive understanding of what artists actually do when they make work. This seems like a useful methodological tip for constructing an effective research paradigm for the creative arts.

For Davies and those who advance a pragmatist account of art, the central question in the epistemology of art is "the relationship between the generative act that brings a work into existence and the receptive act that is a proper appreciation of that work" (op. cit.:26). The great virtue of this way of viewing art is that it seeks to link the "focus of appreciation" of an artwork – how we make sense of and evaluate it – with an understanding of its auspices as a generative act. And in so doing, it abolishes the demarcation which has bedevilled the creative arts as a research domain, namely, the relationship between the act of aesthetic judgement of a manifest work and the act of understanding art as a practice and performance through which work is produced. Davies concludes that contemporary aesthetics must start from a serious consideration, or what I have called an immanent critique, of artistic practice.

3.7 The practice turn in contemporary theory

Some social theorists (Cetina, Savigny and Schatzki 2001) have detected a broader, more paradigmatic shift within the humanities and social sciences, making practice core to contemporary theory. They identify a "practice turn" across a range of theoretical disciplines, arguing (78) that

> it is through action and interaction within practices that mind, rationality and knowledge are constituted and social life is organized, reproduced and transformed. During the past two decades, practice theory has emerged as a potent challenger to prevalent ways of thinking about human life and sociality, which have until now focused either on individual minds and actions or social structures, systems and discourses.

The new focus on practice within contemporary theory seems to focus on at least four major tenets. Firstly, that social practices are constitutive of social reality. Society is socially constructed, as are the forms of subjectivity, object-hood and knowledge which appear within it. Secondly, practice involves non-propositional forms of knowledge, whether in the form of direct sensory and emotive experience or in mediated experience captured in image and sound. Thus, while language is a social practice, practice is not a text, or a bounded system, and cannot be interpreted as such. Thirdly, practice because it is embodied in social and cultural relations is inextricably bound up in ethical and political issues of conduct. Finally, the practice turn undermines the various forms of theoreticism – systems theory, semiotics, positivism, structuralism – which have been dominant within the humanities and social sciences. Such forms of theory tend to marginalize practice-based forms of knowing as a basis for research, including artistic activity. Moreover, these bodies of theory and analysis, in so far as they address practice at all, see it as implied by theoretical axioms and methodological procedures which are objective and value free.

As Theodore Schatzki summarizes it (ibid.:10), "practice theorists conceive(s) of practices as embodied, materially mediated arrays of human activity centrally organized around shared practical understanding". The theorists of the "practice turn" are responding to various intellectual and political currents present across a range of philosophical positions, all of which assert the primacy of practice.

This position is, of course, associated in philosophical terms with American pragmatism. It is also to be found in post-Althusserian versions of Marxism which take their inspiration from political realignments within the French left after 1968 (the basis of the work of Alain Badiou (2012) and Jacques Rancière (2011)). Badiou (2012:xii) argues that French philosophy, in particular, has sought to abolish the division of concept and existence and to challenge the separation of theory and practice, seeking "to abandon the opposition between philosophy of knowledge and philosophy of action, the Kantian division between theoretical and practical reason, and to demonstrate that knowledge, itself, even scientific knowledge, is actually a practice". As he notes (2012:iv), French philosophy attempts

> to wrest science from the exclusive domain of the philosophy of knowledge by demonstrating that, as a mode of productive or creative activity, and not merely an object of reflection or cognition, it went far beyond the realm of knowledge.

Badiou, in particular, acknowledges the contribution that Gilles Deleuze has made in rethinking our concept of science. In Proposition III of the "War Machine" chapter in the book *A Thousand Plateaus*, Deleuze, together with his collaborator Felix Guattari, identifies a type of scientific practice christened as "nomadic". This they see as a type of investigative practice rooted in artisanal activity which "follows matter's immanent traits". As we have seen, they distinguish this nomadic or "ambulant" science from the forms of systemizing "royal science" which enjoy state patronage and institutional validity. Against a theoretical science of matter and mathematically formalized laws which generate a technology of control, Deleuze and Guattari emphasize the role of practice within nomadic or "minor" science. Such minor science, with its origins in artisanal pursuits like metallurgy, building and design, military engineering and of course art practice, is essentially experimental and grounded in practical engagements with matter and materials. It searches for explanations of discrete phenomenon, often in response to real-world problems. As they conclude,

> what we have . . . are two formally different conceptions of science, and, ontologically, a single field of interaction in which royal science continually

> appropriates the contents of vague or nomad science while nomad science continually cuts the contents of royal science loose.
>
> *(Deleuze and Guattari 1987:367)*

Clearly, art follows the pathways of nomadic rather than royal science, often providing the "counter rules" for established science with its practices in counterpoint with scientific thinking.

Similarly, feminism, with its long-term concern with questioning the barrier between personal experience and larger social and political structures, has played a key part in this transformation, providing its own set of "counter-rules" to traditional epistemologies. Anglo-American ordinary language philosophy, with its historical antipathy to metaphysical discourse and concerns with the pragmatics of language use, has also created an anti-systemic dynamic within philosophy. The practice theme is also present in more recent studies of scientific activity which focus on what scientists actually do when they make science and not on what logicians claim they do (Latour 1988).

So, across a wide range of philosophical approaches to understanding art and science, knowledge and political conduct, a concern with practice is manifest, as is an interest in rethinking the theory–practice organization of knowledge.

Indeed, as critical theorist Alfred Sohn-Rethel[17] convincingly argued over 50 years ago, this hierarchy between theory and practice follows a rigid historic social division within capitalist society between intellectual and manual labour. Or, as Deleuze and Guattari put it (2013:429–430), "the way in which a science or, or a conception of science, participates in the organization of the social field, and in particular induces a division of labour, is part of that science itself". Royal science, with its hierarchical organization of theory and practice, form and matter, is inseparable from "a society divided into governors and governed, and later intellectuals and manual labourers".

Art activity has always rested uneasily on the cusp of this division. It aspires to a status higher than the craft activities of the artisan. However, it is often unclear where art's intellectual claims reside. Do these rest in its capacity for unique design, inspiration and innovation and its ability to address intellectual and cultural themes? Or do they lie in the artist's ability to mobilize research and other critical resources to finesse and contextualize their work and achieve a public mode of address?

And it is, I think, in relation to this "practice turn" and search for an appropriate mode of address that a range of new methodological possibilities seem to be emerging across the arts, humanities and social sciences. Art educators have eagerly seized upon these various developments, as they appear to intellectually underwrite an intrinsic research impetus within the creative arts *as a field of practice*. Research is now a core dynamic within art practice. Moreover, this development is occurring just as a range of humanities and social science disciplines seek to expand their methodologies to deal with the experiential, sensate and performative aspects of social reality by beginning a conversation with art practice.

For example, in a scholarly field like the history of art, we have recently witnessed a renewed focus on the detailed description of the *practices* of historic artists. Martin Kemp's pioneering work on Leonardo (2006a), discussed in Chapter 4, focuses on the experimental and design aspects of the artist's work, which can be conceived of as a type of minor or nomadic science. By focusing on the fine detail of that practice, he manages to illuminate a range of broader issues about the relationship between art, natural philosophy and technological endeavour in the high Renaissance. Ralph Steadman's (2001) studies of the painting practice of Vermeer explore the artist's understanding of optics, and, in particular, his possible use of the camera obscura, a proto photographic device which could project an accurate inverted image of a scene (not yet capable of being chemically fixed) for artists to trace and thereby make a realistic record for us in preparatory sketches for their paintings. From the available evidence, Steadman patiently "reconstructed" the room within which Vermeer executed some of his best-known paintings, placing both a camera obscura device and a contemporary camera within this modelled space to explore the possible construction of Vermeer's paintings (Vermeer may have painted as many as a dozen pictures in the original room employing a camera obscura). Steadman is thereby able to provide evidence to support the view that Vermeer did indeed use a camera in his work, whether directly tracing images produced by it or employing its projected image to assist him in matters of perspectival rendering and of capturing naturalistic detail.

Similarly, art historian Barbara Maria Stafford, in her polemical text *Good Looking: Essays on the Virtue of Images* (1996), identifies the historic role of the visual in the production of scientific knowledge. In her earlier text, *Body Criticism: Imaging the Unseen in Enlightenment Art and Medicine*, she argues that this is apparent in the work of artists and illustrators in the eighteenth century, for example, in that of the Roman printmaker Giambattista Piranesi and in the work of a range of medical illustrators.

Stafford's broader critical project has been to reject the over-reliance on linguistic models within the study of visual culture. This tendency she sees as a characteristic of poststructuralism. The "linguistic turn" within the arts and humanities has encouraged a distrust of sensory experience and led to a difficulty in engaging with the materiality of the visual in western culture. As she puts it (1996:5), "an educational economy materially based on language has either marginalized the study of images, reduced it to a subaltern position or appropriated it through colonization". Visual culture too often involves a reductive analysis of the visual field, which is treated as simply the play of signs which can be analysed as part of a language system, thereby collapsing the distinctive phenomenal and sensate character of the image and visual textures into a "homogenizing, language-driven interdisciplinarity". Her work seeks to celebrate the role of the visual image in human understanding, and she argues that "imaging, ranging from high art to popular illusions, remains the richest, most fascinating modality for configuring and conveying ideas" (1996:4).

In short, we seem to be witnessing a renewed interest on the part of art historians in mapping the distinctive visual practices of individual artists and cross relating these to broader intellectual currents, including the contribution of artists to science.

In the field of lens and screen studies, similar developments are apparent. Film theorist Laura Marks, in her pioneering work *The Skin of the Film* (2000), engages, in turn, with the practices of inter-cultural cinema (2004). She recognizes at the heart of a range of avant-garde films produced around historical themes in the 1990s a dialogue between critical theory, historiography and film practice. Her work has led to a refashioning of critical concerns within film studies around the haptic and mnemonic dimensions of the filmic image and its role as "prosthetic memory" – particularly for diasporic communities. Marks is one of the few contemporary film theorists prepared to admit that film practice – particularly in the avant-garde tradition – can involve at its core a dialogue between critical theory and creative strategy. Discussing the work of experimental film-maker Trinh T. Minh-ha (whose work I discuss in Chapter 9) in her seminal *The Skin of the Film* (2000:xiv), she argues that

> the works I examine in this book are themselves works of theory, many explicitly so. They are not waiting to have theory "done to them"; they are not illustrations of theory but theoretical essays in their own right. The works themselves have developed a sophisticated argument for how cinema can represent embodied experience and why it should do so.

Marks here seem to echo Sol LeWitt's argument (1967) made decades earlier to mark the birth of conceptualist art that "the philosophy of the work is implicit in the work and it is not an illustration of any system of philosophy".

3.8 Conclusion

Our focus in this chapter has been on identifying the elements that might make up a distinctive research culture for the creative arts. I suggest this is likely to involve:

1. A consideration and critique of current developments within contemporary arts practice which are encouraging the development of an artistic research culture
2. The dialogue between arts practice and critical cultural theory. This now involves two-way traffic – the artist–researcher seeks to critically contextualize their practice in conversation with a range of other disciplines, while a "practice turn" has occurred across the arts, humanities and social sciences, with the art process now acknowledged as an important methodological resource
3. A consideration of pedagogic approaches in the arts which can be identified as one of the consistent sources in the development of an artistic research culture as a form of "nomadic science"

4 The consideration of exemplars of artistic research, both historical and contemporary, from which a range of general methodological lessons might be drawn and from which a tentative body of middle-range theory[18] based on experience and structured reflection on this might emerge

These, then, are likely to be the sources we will need to draw upon to develop a research culture for the creative arts and media. This foregrounding of practice and experiential learning might, one hopes, help to arrest the current drift towards positivist method that sadly is already discernible in most Anglo-Saxon adumbrations of arts research.

So, our call in this prospectus for artistic research is "back to basics". This means an engagement with the best of contemporary arts and media practice and critical thinking and a move to a more philosophically informed methodological position, "the primacy of practice". This standpoint regards knowledge itself as a practice and subject primarily to practical considerations relating to human purpose, value and action. This position, in turn, opens up arts practice as research to a broader cross-disciplinary field of critical enquiry. So, my approach seeks to be attentive to the creative and critical dynamics at the core of all arts activity while at the same time being open to a broad range of intellectual considerations and methodological approaches shaped by these cross-disciplinary affiliations and dialogues.

Notes

1 As Buchloh argues (1990:126),

> in the absence of any specifically visual qualities and due to the manifest lack of any (artistic) manual competence as a criterion of distinction, all the traditional criteria of aesthetic judgment – of taste and of connoisseurship – have been programmatically voided. The result of this is that the definition of the aesthetic becomes on the one hand a matter of linguistic convention and on the other, the function of both a legal contract and an institutional discourse (a discourse of power rather than taste).

2 The notion of immanent critique is a critical method of dissecting a cultural or intellectual system. "Boring from within", it seeks to reveal the internal contradictions and limits of such systems of thought and theory and how, in particular, various ideologies may play a destabilizing role within them (Bhaskar 2008).

3 Peter Bürger is the critic usually identified as defending this claim, and certainly his key text, *The Theory of the Avant-Garde* (1976), identifies the notion of a "Duchamp effect" transforming art practice after 1917.

4 *Fountain* was submitted for the exhibition of the Society of Independent Artists and summarily rejected ("simply suppressed", as Duchamp himself says) by the exhibition committee. It was subsequently photographed by Alfred Stieglitz and gained its initial notoriety when the image of the urinal was published in the avant-garde art magazine *The Blind Man*.

5 Gustave Courbet (1819–1877) was the leader of the Realist movement in French painting, whose canvasses mobilize visual sensation to portray social and political narratives.

6 Duchamp insisted that his original intention was not to make a work of art out of such readymades. The early examples, such as his 1913 version of *Bicycle* (a bicycle fork with

front wheel mounted upside-down on a wooden stool), did not employ the term, which he seems to have adopted after he took up residence in New York in 1915.

7 Kossuth was one of the first artists to use the term "Conceptualism". See Buchloh (1990).

8 This strategy was taken up by the Art and Language Group in London and in the US by Lawrence Weiner, whose work also sought to explore the potential for language to serve directly as an art form – another challenge to the retinal within art.

9 That said, we can over-emphasize the rationalist and scientific character of the Conceptualist revolution in art. Few Conceptualists actually believed that art was becoming a science. LeWitt argues in his *Sentences on Conceptual Art* (1967) that artists plough a different furrow to scientists. In particular, he argues that "conceptual artists are mystics rather than rationalists", concluding that "rational judgements repeat rational judgements. Irrational judgements lead to new experience".

10 Despite this concern with filmic craft, Doherty, like a number of fine artists who employ filmic means in their work, has been reluctant to locate his work within any cinematic tradition or discourse – including avant garde film traditions – preferring to focus on the ideational and experimental element of the work.

11 Paul O'Neill and Mick Wilson (2015:14) invite their readers to "consider the relationship between curating, the curatorial and the practice of "research". For them, this is more of a case of advocating a particular model of curatorship as an expanded practice that involves "critique, editing, education, and fund-raising practice".

12 In the past, the production of such films would have been beyond the resources of galleries. Moreover, in Britain and Ireland there was a strong tradition of arts documentaries and arts magazine programming within public service broadcasting providing imaginative and well-crafted films on individual artists. With the decline in such programming and the reduction in the costs of documentary film production, many galleries have themselves taken on commissioning and editorial roles in this area.

13 Clare Bishop has provided a searing critique of relational aesthetics (2004), raising doubts about the self-serving character of the laboratory and processual approach embraced by Bourriaud. As she notes (53),

> an effect of this insistent promotion of these ideas of artist-as-designer, function over contemplation, and open-endedness over aesthetic resolution is often ultimately to enhance the status of the curator, who gains credit for stage-managing the overall laboratory experience.

14 A version of the interview with Parreno can be seen at: www.youtube.com/watch?v=SL-0pZZ9UmU (accessed 17/03/2016). For a sense of the work discussed, see: www.youtube.com/watch?v=M1RWxQaM5mc.

15 However, relational aesthetics is not just a theory with a wobbly conscience. Curators like Bourriaud in France and Obrist in the UK have access to significant institutional resources to give material shape to their theories and have, in effect, created a network of "franchised" artists operating within the "relational brand" – Phillipe Parreno, Douglas Gordon, Liam Gillick, Pierre Huyghe, Beatrix Ruf. This roll call is, for instance, reflected in the constitution of the group advising the Luma Foundation on the building and programming of a massive new gallery of contemporary art in Arles on the site of the former railway repair yards. This centre, when finished, will be the largest gallery of contemporary art in Europe outside London and Paris (and this in a city of just 50,000 inhabitants).

16 The immediate reference here is to Nicholas Rescher's 1973 book of essays of the same name. Rescher argues (xiii) for "the subordination in the sphere of factual knowledge of theory to practice" and stresses "the extent to which the justificatory mechanisms of our knowledge regarding the empirical features of nature are controlled and undergirded by considerations of practical reason relating to human purpose and human action". More generally, the notion of the primacy of practice is central to the philosophical

pragmatism of Charles Saunders Pierce and William James. Both argued that rather than trying to identify the essence of truth via prescribing a binding logic of scientific methodology, pragmatists should try to describe how our practical engagement with the world conditions the knowledge apparatus we bring to bear to understand that world.

17 Alfred Sohn-Rethel (1899–1990) was a French born, German critical theorist loosely associated with the Institute for Social Research in Frankfurt in the 1920s. He met and corresponded with Walter Benjamin and Theodor Adorno and later settled in England. He did not publish his major work, *Intellectual and Manual Labour: A Critique of Epistemology*, until 1977. Deleuze and Guattari seem unaware of his work.

18 This approach to sociological theorizing, associated with the US structural functionalist sociologist Robert Merton (1968), aims to integrate social science theory and empirical research. Middle-range theory usually starts with an empirical phenomenon or social problem rather than a broad theoretical formulation, such as the notion of a social system and its structural dynamics and abstracts from the particular case studied (e.g., the investigation of deviance in a particular social location), in order to create a number of general statements of limited applicability about the social phenomenon studied that can be verified by data and experience. A number of artistic researchers have identified the notion of middle-range theory as one which might be usefully explored with visual arts study, which departs from a specific field of investigation, usually constituted by the work of a single artist, and subsequently seeks to arrive at a more generalized understanding of art phenomena.

4

THE SPECTRE OF SCIENCE

4.1 Introduction

In this chapter, I explore the interactions between art and science. I sketch the positivist model of scientific explanation which has set the terms of debate about creative arts research as it has for other subjects. But is this "logic of scientific discovery" applicable to the arts and humanities and, in particular, to the creative arts? Certainly, within contemporary art, research practice is seen as being at the core of artistic innovation, and the smart talk is of experiment, enquiry and rigour. Moreover, with the critique of positivism and the recognition of the pluralistic character of scientific endeavour, a new dialogue between science and the creative arts now seems possible. But how serious is this postmodernist concern to link art and science? To address this question, I explore some contemporary examples of collaboration between science and the arts. Hopefully, these cases can help us to identify the points of commonality and difference between the two fields.

4.2 The methodological drive in artistic research

As artistic research within the university has expanded, creative arts academics have come under increasing pressure to clarify the research methods they employ in their work and to identify their "results". Indeed, a cadre of arts academics argue that if the status of practice-based research is to be advanced, then greater attention needs to be paid to articulating the methodological auspices of such research.[1] The argument goes that art researchers, like those in any other discipline, will need in any project to specify a research context for their investigations, identify a clear set of achievable research questions to be addressed and elaborate the methodological basis of the approach adopted. They must review the research already conducted in a field of study. Research, after all, is cumulative in character, and new lines of investigation are built on existing knowledge and on the application and refinement of established methods. Researchers are expected to identify the particular contribution their project might make to a field and to submit their research for peer review.

Others caution against any such methodological turn in creative arts research which might elevate the systematization of investigative process over understandings arrived at through studio-grounded activity. Indeed, some have raised the awkward question of why artists would wish to identify their working methods with scientific methodology rather than with long-established studio and field practices. They counsel that an emphasis on peer-reviewed research outputs might also distract artists from taking creative risks and from innovative work that does not easily fall within institutionally validated, methodological guidelines.

Stephen Scrivener, in one of the more insightful interventions into this debate (2002:1), attacks the notion that art making "yields knowledge that is independent of the actual art objects produced". Scrivener as a philosopher is in fact generously disposed towards visual arts practice. Indeed, his primary concern is that the creative arts might be diminished by the new-found enthusiasm of arts academics with "research". In his classic 2002 paper, he argues (ibid.:2) that the objectives of the visual arts are significantly different to those of the sciences. His concern is that

> art may lose sight of its own objectives by adopting predefined notions of research. Indeed, art needs to ensure that what it chooses to call research contributes to its interests first and foremost. Hence, my aim is not to diminish the significance of visual arts research as compared to scientific research, for example. Quite the contrary, I hope to show that art research performs an equally important but complementary function to that of the knowledge acquisition research domains.

Scrivener argues that the forms of understanding which attend art practice are experienced as forms of insight or "ways of seeing". They operate via the expressive power of the artwork and this is exercised through the senses as well as by the intellect. Accordingly, it is implausible to claim that "the primary function of an art object is to communicate knowledge". While an artwork may sustain a knowledge claim, this is clearly not its main aim. Arts practice, while it may encourage us to think in interesting ways about the world, does not seek to provide a systematic account of things as science does. Thus, artistic research is inextricably bound up in the mechanics of creative practice. Process trumps methodology.

Despite this cogent intervention and a range of other voices being raised that question the drift towards systemizing creative arts research, an unstable consensus has begun to emerge within the university-based creative arts community. The arts will have to sign up to the generic definitions of research prevalent in the academy if they wished to avail themselves of institutional support. Accepting the "King's shilling" means addressing the concerns central to the dominant model of scientific investigation: namely, the pursuit of methodological rigour, adequate evaluation of evidence, theory-based explanation and proof of the contribution of an enquiry to original knowledge. Art educators are being asked to acknowledge this model and to manage a transaction with scientific rationales. Most art educators find themselves struggling to preserve the character of creative arts education while adapting to the economic and bureaucratic logic of research performativity.

4.3 Understanding the logic of scientific discovery

This pressure has led some to attempt to delineate a model of artistic research heavily infused with a certain sort of scientific rhetoric. As one art educator has argued (Newbury 1996:25),

> at its simplest research can be understood as a process involving a period of information gathering, or in research terminology, data collection and a period of data analysis bounded on either side by some form of theoretical work. Generally, this takes the form of proposing a particular question about an object or group of objects (theory), observing the objects in question (data collection), reflecting on this observation (data analysis) and then answering the original question on the basis of this observation (theory).

Darren Newbury sketches here a loose version of what philosophers of science refer to as the *hypothetico-deductive* model. Scientific inquiry proceeds by formulating hypotheses that can conceivably be falsified by a test based on observable data. This model was initially advanced by logicians in the 1930s as an account of the advancement of scientific knowledge in that most regal of subjects – physics. Later, it became applied to all research domains. The model can be represented schematically (Figure 4.1).

Observations of the physical world establish patterns of occurrence. These lead scientists to pose various tentative hypotheses to account for particular phenomena by identifying their causes. Hypotheses are framed to guide an investigation and data is collected to confirm or refute these. Hypotheses are knowledge claims which, if empirically corroborated, become theories cohering with other bodies of theory. Early forms of the model required hypotheses to be *verified* by available evidence or experiment. Later refinements of the model, associated with the philosopher Karl Popper (2002), placed more emphasis on the *refutation* of hypotheses than on their confirmation.[2]

Scientific method is a powerful theory of knowledge which draws absolute boundaries between rationally justified forms of knowledge and mere belief. Hypotheses which do not present themselves to the tribunal of experience in a form capable of corroboration, or refutation, are deemed to be either meaningless or metaphysical (the two being held to be more or less identical by positivists).

If we get a significant degree of confirmation of a hypothesis over a range of cases, with no disconfirming instances, we are inclined to accept the hypothesis as a generalized proposition with explanatory power. This can take the form of an

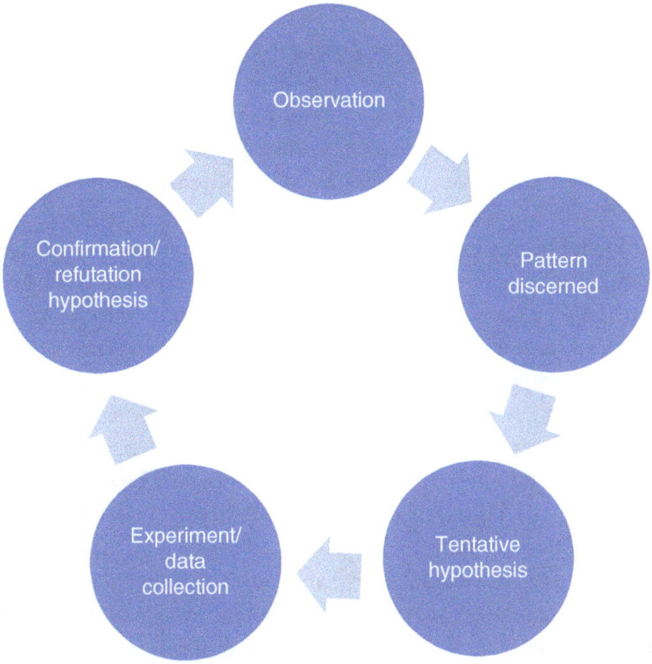

FIGURE 4.1 The hypothetical–deductive model

imputed *causal* relation between phenomena linked to a general law. Or, it can be a *probabilistic* statement providing inductive support for a statistically based prediction. Statistical laws state that under certain conditions we can expect certain outcomes with a calculable frequency. Causal laws, on the other hand, state that given certain initial conditions, certain outcomes will invariably follow. In each type of explanation, we compare the explanatory value of competing hypotheses in a field by testing how stringently they are corroborated by the available data.

Let me give an example. Anxious to chill a bottle of white wine for dinner and with guests shortly to arrive, I realize I have forgotten to place the wine in the fridge. In a panic, I deposit the bottle in my freezer for a rapid chilling. The guests arrive bearing their own bottle of wine, which we drink with the meal, and I forget about my bottle in the freezer. Hours later, I remember my wine and open the freezer; the wine is now frozen in the bottle, which has cracked. How do I explain this phenomenon? Well, water below 4 degrees Celsius changes its density and expands until it freezes at 0 degrees. Wine as a liquid is more than 90 percent comprised of water. Blown glass in certain structures shatters under specified degrees of pressure. Observing the conditions in the immediate environment and relating these to the incident of the cracked bottle, we conclude that the action of the freezer "causes" the liquid-filled bottle to crack. And, if we were to repeat the experiment with the same initial conditions, we might expect the same outcome. Explanation and prediction are in symmetry, linked by causal laws.

Generally speaking, in natural science explanation is held to occur when a particular physical event can be subsumed under a *covering law*, so that given a particular set of initial conditions, say C1, C2, . . . Cn (in my example the temperature in the fridge, the time period the bottle is exposed to this condition, the chemical constituency of the liquid in the bottle etc.), and with reference to a set of general laws L1, L2 . . . Ln (governing, for instance, the effect of temperature on the expansion and contraction of liquids etc.), an adequate explanation of a specific occurrence can be provided. There is in effect a deductive relation between an *explanans* and its *explanandum*, traced to a covering law. Hence the name for this model of scientific explanation, the *deductive–nomological* (DN). An explanation of a specific situation or event is deduced from a general law. Such a law not only explains a physical event but indeed can predict such an event if certain conditions obtain (Figure 4.2).

So, in my example, I have come to realize that if I am foolish enough to place a bottle of wine in my freezer for a prolonged period, it will unfortunately shatter. In other words, cause and effect are linked via reference to a binding law of universal application which has both explanatory and predictive power. If we can achieve one, we can ensure the other.

This is a very powerful model of explanation. Given the *symmetry of explanation and prediction*, the natural sciences have a very potent set of tools for handling man's interactions with the natural world via technological intervention. Moreover, the very notion that we can establish technical control over nature by drawing on the predictive power of science constitutes the basis of modern scientific activity and its

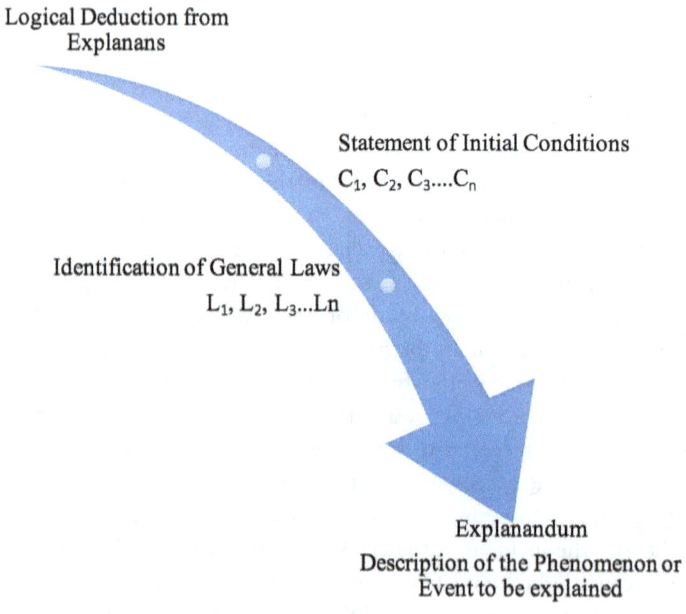

FIGURE 4.2 The deductive–nomological model

centrality to processes of societal modernization. Moreover, the hypotheses scientists employ to explain events in the physical world do not stand alone as provisional propositions but cohere within theoretical positions or what are sometimes called *paradigms*. Scientific hypotheses are not only elaborated via experimental method but flow deductively from law-like generalizations about the natural world drawn from physical theory (such as laws governing the impact of temperature on the molecular structure of liquids and probabilistic accounts of the effect of pressure on certain materials and structures).

4.4 The specificity of artistic research

Artistic research is neither hypothetic–deductive nor deductive–nomological. It is discovery-led rather than hypothesis-driven. It is experiential and reflective rather than data-driven or law-like. The investigative process is focused primarily on the making of an artwork and is often driven by curiosity and hunch rather than by methodological formula. Sean Hillen, one of the artists I interviewed for this book (whose work is discussed in Chapter 9), talks about any new art enquiry as involving "an itch that it's addictive to scratch". Actually, scientists also often speak in these terms with regard to a research problem that initially manifests itself as an irritation – something that cannot be easily be understood and just won't go away.

Nor is artistic research theory-led. It has a very loose relationship – if any – with that rag bag of critical, historical and speculative discourses known as "art theory". Generally speaking, it gets by with very little abstract theory, and such theorization that takes place does so *post facto* through structured reflection or via dialogue with other critical interlocutors.

Artists experimenting in their studios or in the field do not by and large naturally articulate their development work in the form of distinct research questions or hypotheses to be corroborated against specific empirical "data sets". Nor do they see themselves as testing the propositions of art and cultural theory. Their research, as we have seen, is focused on the task of making work, and the forms of documentation and analysis they employ are also harnessed to this task. What writing they do seeks to illuminate their practice, not stand in for it. Artists demand that you engage with their work in the form in which it is presented, whether as painting, film, installation, composition, performance or art proposition. Very often they have a greater interest in your reaction to their work as an audience than they have to its assessment by a committee of their academic peers.

Artistic research, because it is focused on the production and comprehension of specific artworks rather than on the testing of generalized hypotheses, is not predominantly concerned with general methodological precepts or conclusions. Still less has it any interest in divining law-like regularities in an artist's behaviour, or with arriving at binding judgements of taste in the aesthetic sphere. Hence, attempts to actually model artistic research on the natural or social scientific method, or indeed on the traditional humanities, seem doomed to failure. Arts research is not so much about producing well-defined and empirically answerable research questions as being open to various possibilities of discovery and insight that arise in and through the processes of art making. Such research is often more responsive to a range of professional concerns relating to craft, effective communication with an identified audience and forms of cultural engagement than it is to confirming or refuting academically framed hypotheses or elaborating general research methodologies. Such "approximate knowledge" remains, in Deleuze and Guattari's terms (2013:435), "dependent upon sensitive and sensible evaluations that pose more problems than they solve". This experiential openness contrasts with the methodologically driven character of institutionalized science which seeks to "isolate all operations from the conditions of intuition".

Some have tried to present artistic research as inductive in form. The artist–researcher is seen as proceeding in their work on a trial and error basis through various iterations of a project until a final design takes shape in the cutting room or studio or on a digital workstation. As we have seen, this iterative model has had particular purchase within design research. However, the focus remains the single artwork (or design) rather than arriving at a statistical generalization or probabilistic calculation. Where more general inferences are made by an artist in reviewing their work, the reasoning involved often seems of be *abductive* in character rather than inductive. This is to say the artist is not seeking to corroborate a hypothesis by assembling evidence of regularities but is rather sharing a particular insight at which they have arrived through experiential learning and reflection. The artwork produced is neither a data set confirming or refuting a hypothesis nor a knowledge statement, despite the current popularity within conceptualism of conceiving of the art process as a "proposition".

Others have been attracted to the conjectural side of Karl Popper's model of science. This seems to encourage the speculative and creative side of human enquiry valued by the creative arts. But Popper was a positivist through and through and expected conjectures to be expressed in the form of empirical claims that can be systematically tested against evidence. The artist, as we keep stressing, is primarily concerned with the progress of a specific work or project and with a grounded engagement with materials, media and voice. On the basis of their own practice, they do not normally seek to establish any series of law-like generalizations under which other cases of artistic activity could be subsumed and explained. The artwork aspires to being unique (although artists often acknowledge influences and commonalities of interest and style with other artists). Of course, some artists talk about their work involving "inspired guesswork tested by experience". Certainly, there is evidence that many artists engage in experimental strategies to advance their work. They employ models, maquettes and prototypes. They engage in work-shopping, scripting and rehearsal processes to explore ideas, test materials and to advance design approaches and creative strategies. That said, such background research activity is, as we have seen, primarily concerned with the development and delivery of a unique artwork. Artistic research is about the making and explication of the meaning of a distinct cultural product rather than the causal or probabilistic explanation of "art phenomena".

So, while science, or certainly the "royal" sciences of physics and chemistry, are systematic affairs, art very often is not. This is not to say that artists are not often methodical in their work. However, the methods they employ are often rooted in craft and studio practices and in professional protocols, habits and processes. And nowhere is this more so than in film-making. All film-making is a highly ordered activity resting on a bedrock of scripting, cinematographic, dramaturgical and design practices which structure and indeed constrain the creative and discovery processes at play in the development and production of a film work.

This discussion of the "logic of scientific discovery" might seem a million miles away from what we as arts practitioners understand as artistic research. However, these models of scientific process, of data collection, hypothesis testing and causal explanation, continue to exercise a considerable influence within the academy and among research policy makers. And the scientific mind-set tends to be all encompassing. Often it seems to assume that the world, including the humans which populate it, is constituted of physical matter and its dynamics explicable as such by reference to universally binding laws. Among the chaos of the universe, a certain order can be divined through scientific theorizing.

Artists are less inclined to accept this materialist premise with its reductionist logic. They may regard art as a humanistic activity primarily addressing aesthetic rather than scientific matters. Many believe that its research methods are closer to those of the cultural sciences than to the physical sciences. Others have a direct interest in matters spiritual, and in cultural meanings and sensibilities that lie beyond the purview of science. Many see art as being about subjective expression rather than objective calibration of the world. Still others see art as inextricably bound up in ideological and political dynamics and requiring a distinctive, critical apparatus to assist artists in negotiating this territory.

However, as researchers we have to acknowledge that science demonstrates great potency in its dealing with the natural world through its capacity to develop dependable methods for both predicting and controlling events. Scientific knowledge facilitates technological intervention and contributes to industry and the economy. This further establishes the authority of the natural sciences within society. And this is one of the reasons why every field of academic study will be required to justify its methods in relationship to this "deep structure" of scientific rationality and instrumental reason. The language of research embodies a scientific epistemology. The key terms in research discourse – objectivity, evidence, metrics, hypothesis, explanation – derive from this source. It is a deep structure in so far as it rests on the fundamental principle of the *methodological unity of science*. So, for positivists, while there may be a plethora of different disciplines, each with their own perspectives, theoretical architectures and methodological preoccupations, in so far as these fields are, or aspire to be, scientific in character, they rest on a common logic of investigation. Artistic research is expected to fit into this broader model.

Yet, as we have seen, artworks seem to have a different character to the generalized character of scientific propositions. Of course, art historians and critics often seek to detect within an artist's oeuvre adherence to a distinctive stylistic approach which unifies their work or to trace a general set of preoccupations that occupy them. They identify in an artist's membership of a "movement" a congruence which binds together a cohort of artists around shared approaches, concerns and methods. As we shall see in Chapters 6 and 7, artists have at various times sought to codify artistic practices and languages, particularly when seeking to communicate these to their students via a transformed art educational practice. The modernist avant-garde even chose to invoke the idea of a "science of painting" and of visual art. More recently, those critics who employ semiotic tools to dissect cultural formations see the artwork as a complex of signs which can be objectively described with some degree of analytical purchase. And, as we have seen, both contemporary curators and some artists have been attracted to a laboratory model of practice.

Many artists, on the other hand, wish to start each work afresh, without established preoccupations. They often try to set aside approaches employed in earlier work in favour of creative experiment and ludic possibility. Some scientists view their disciplines in a similar light and see both art and science as based in a shared concern with invention. They argue that the emphasis on the conjectural character of scientific methodology, and on simple curiosity and problem solving, facilitates these commonalities between art and science.

However, scientists usually seek to delimit an investigation by formalizing an initial set of hypotheses and a methodology to guide their laboratory and field work. Data is collected and calibrated mathematically in a methodical manner. Corroboration of hypotheses allows more general theories to be defended and new frontiers of knowledge to be explored. The research questions posed in a scientific enquiry are developed within existing lines of investigation and in relation to the current state of the field and its conceptualization – or as a small extension of existing knowledge. This is not to say that artists do not employ delimiting and framing devices to achieve focus in their work, whether in terms of style and theme, signature and mode of address. However, art practice is by nature a rather open-ended mode of discovery largely shaped by the "down and dirty" creative demands of the specific project being tacked.

Moreover, scientific practice, whether conducted in university labs, research institutes or industrial settings, is, as Bruno Latour (1979, 1988) and Barnes, Bloor, David and Henry (1996) observe a highly institutionalized set of social activities subject to binding organizational forms, corporate imperatives and class and gendered work cultures. On the other hand, arts research is largely conducted by individuals, or in the case of design and media activity, in relatively small creative teams. Employing a discovery approach within an art project can lead you anywhere (although in many art and media disciplines, professional and commercial considerations may constrain and structure the urge to discover). In addition, while the exercise of the creative imagination is for scientists an attractive feature of the arts and one they would be happy to see enshrined in scientific practice, it is generally not how they themselves proceed in their own field of study, which like any profession has its own structures of acceptable practice and disciplinary accountability.

Yet many artists have a fascination with science. They have, for example, displayed an intense interest in optical matters, drawing on a rich vein of art historical writing on perspective and vision by figures like Erwin Panofosky (1991), Sam Edgerton (1975), Hubert Damisch (1995), Martin Kemp (2006c) and Jonathan Crary (1990), to name but a few. In a digital age, many artists concerned with questions of perception and memory have experimented with now technologically redundant imaging processes (for example, Tacita Dean's fascination with using 16mm and 35mm celluloid film stock in her work, or Grace Weir's use of early photographic processes, discussed below) or sought to appropriate new visual technologies such as digital scanning (Katherine Jebb) or MRI technologies (Marc Didou, discussed below).

In general, I would argue that there is no necessary intellectual conflict between the advancement of art practice and artists being attentive to scientific matters. Indeed, a broadly scientific disposition on the part of artists – albeit one driven by curiosity and puzzlement rather than system and measurement – may be a condition for such advancement. I devote my next chapter to exploring the successful science–art collaborations that took place in the Renaissance. Later, I look at the intersection between scientific method and art practice that characterized the work of the modernist avant-garde in Russia and Germany, where there are also lessons to be learnt for current arts research. A genealogy of art–science collaboration can be traced to inform current research.

4.5 Against method (and for art)

However, a positivist modelling of scientific practice remains the elephant in the room. The German social philosopher Jürgen Habermas has drawn our attention to the intimate relation between the forms of technical control over nature inscribed in modern science and the structures of social control entailed in the rationalized decision-making processes of contemporary bureaucracy and the tutelary state. Habermas recognizes both modalities of instrumental control as core elements of an advanced industrial society (1974:254). Capitalism seeks to

> look after its survival on the escalating scale of a continually expanded technical control over nature and a continually refined administration of human beings and their relations to each other by means of social organization. In this system, science, technology, industry and administration interlock in a circular process.

Habermas operates within a tradition of critical theory that seeks to locate science as a social practice within advanced capitalist society. He treats positivism as a particular reading of the empirico-analytical sciences and one which erects a restrictive and ultimately erroneous theory of knowledge. From the seventeenth century on, the natural sciences began to employ the rigour of mathematical form in its theorizations together with a range of experimental methods that probed cause and effect within law-like structures. Institutionalised science came to provide a privileged model of "well-founded" knowledge. This in turn marginalized traditions of "nomadic science" focused on practical problem-solving employing approximate knowledge often grounded in artisanal experience. From the Enlightenment on, the primacy of the scientific method was generally accepted by philosophers – whether of an empiricist or rationalist bent. However, it was not until the twentieth century that modern, theoretically driven science came to "coincide with knowledge as such" (ibid.:3). Habermas argues that up until then, knowledge was understood as a totality comprising not only scientific thinking but also practical and reflective reasoning, concerned with making informed moral choices and political decisions within society. In addition, the realm of art and of aesthetic experience was also acknowledged as contributing to wisdom and well-being, with beauty and truth seen as interwoven. With the emergence of positivism, this changed.[3] Both practical reasoning (concerned with moral and political issues) and aesthetic experience (concerned with the sensate self and with judgements about beauty and well-being) became marginal within the sphere of knowledge.

However, positivism has been subjected to sustained criticism from philosophers of science. Many have questioned the idea that science progresses via a steady growth of knowledge with new theories possessing superior explanatory power, replacing older, less extensive ones. This evolutionary idea has been gradually abandoned in favour of an account which sees the development of science as discontinuous and subject to radical *paradigm shifts*. This historicist position requires us to rethink our notions of scientific truth and method. As a result, art and science seem to realign in a more positive manner.

Thomas Kuhn, who trained as physicist before becoming a historian and philosopher of science, was to argue that these paradigm shifts – such as the transition from Newtonian to quantum physics – open up new perspectives and methods (including what might constitute notions of evidence). As one paradigm is abandoned and another adopted, our scientific world view is transformed. Kuhn argued that the notion of scientific truth operative at any given moment cannot be established solely by any objective criteria, such as the sort of protocols of verification favoured by positivism. Indeed, he observed that the failure of an experimental result to conform to the dominant scientific paradigm is very often not likely to be seen as refuting a theory within that paradigm. Rather, it will often be treated as a procedural mistake committed by the researcher.

For Kuhn (1962), notions of scientific truth are primarily arrived at through forging a professional consensus within a scientific community. As anomalous experimental results build up and cannot be explained away, a once-dominant scientific paradigm reaches a point of crisis. At this point, a new "revolutionary" paradigm may emerge (as with the Copernican revolution in cosmology or, in the early twentieth century, with the general theory of relativity). The new paradigm subsumes the results of previously held theories along with the anomalous results into one explanatory framework, leading to the acceptance of the new theory. As we shall see in Chapters 6 and 7, this model, or something like it, has been employed to explain rapid transitions in visual arts practice such as those characterized by the emergence of a modernist aesthetic with its drive towards abstraction and new forms of social engagement.

Science (and indeed art) is characterized by competing paradigms, and the theoretical and evidential claims and sensibilities of these are often *incommensurable*. That is to say, they represent irreconcilable accounts of reality which cannot be objectively evaluated by reference to a neutral framework of assessment (scientific or aesthetic) or to any tribunal of experience. This seems particularly the case in the creative arts, where the modernist break seemed to open up a chasm within visual arts and design, musical, theatrical and dance practice separating nineteenth-century "traditional" art from avant-garde practice.

Thus, our comprehension of science can never rely wholly upon "objectivity". Science, like art, must account for subjective and ideological points of view as well. All objective conclusions are ultimately founded upon the subject position and worldview of its researchers and on the machineries of observation and measurement operationalized by scientists to conduct their research (something that became very apparent in quantum and particle physics with its debates about the duality of matter but which has also manifested itself more recently within critical studies of visual culture, where the theme of the technological and discursive structuring of vision has become a core one).[4]

Paul Feyerabend, another physicist–cosmologist with developed philosophical interests and a taste for the creative arts,[5] was to push the critique of positivism further and in a direction that provides a favourable basis for exploring the linkages between science and the creative arts. Feyerabend started his intellectual career closely associated with the Vienna circle of logical positivists. Indeed, he freely admitted (Feyerabend 1995b) to have been a "raving positivist" during his period of graduate study at the University of Vienna in the early 1940s. After World War II, in which he served in the Wehrmacht on the Russian front, he was to study under Karl Popper at the London School of Economics. He soon began to question the central tenets of neo-positivism, including the rigid falsificationism of his mentor. As we have seen, Popper had adopted a strongly normative epistemology, seeking to lay down optimum methodological rules for scientists to follow.

Feyerabend gradually moved to a stance of *theoretical pluralism*, proposing that scientists should exercise their creativity by constructing and defending as many alternative theories as possible – even if these were initially purely thought experiments not immediately capable of being empirically confirmed. He may have been thinking of the *Gedankenexperimenten* (thought experiments) that Albert Einstein employed to develop the theory of relativity. Feyerabend, like Einstein, felt science and art shared many features in common, including the exercise of conceptual creativity and a commitment to formal rigour – qualities which marked both modernist art and quantum physics.

Feyerabend found support for this theoretical pluralism within the history of science, particularly in that of the seventeenth-century scientific enlightenment. His work explores the transition from geocentric to heliocentric astronomy, that is to say the scientific revolution associated with Copernicus (1473–1543) and Galileo (1564–1642), both of whom challenged the prevailing, religiously endorsed orthodoxy that the earth was the centre of the universe. Their cosmological data and calculations, based on telescopic observation of the solar system, indicated that the earth was in fact in orbit around the sun as the centre of our solar system. However, Feyerabend revealed that Galileo, the so-called father of scientific method, was not always as scrupulous in his scientific methods as he was later represented to be in positivist accounts of scientific progress.[6] The appeal of the new cosmology to Galileo or Kepler was not purely its empirical character, which in terms of observational data about planetary movement remained incomplete, but rather the coherence, simplicity and indeed elegance of their theory. Science had always had an aesthetic dimension.

As his historical studies of science proceeded, Feyerabend sought to downgrade the importance of empirical arguments in processes of scientific proof. He began to admit that aesthetic criteria, personal whims and social factors may have a far more decisive role in the history of science than we might think. Accordingly, science and art might be closer bedfellows than positivism acknowledges. Feyerabend observes how a range of "non-scientific" procedures and forms of knowledge, some of them derived from art and artisanal practice, were employed by scientists like Galileo (as they had been by Leonardo da Vinci) alongside experimental methods. Galileo, of course, was heir to the rich intellectual and artistic traditions of the Florentine Renaissance (a tradition of vernacular science we discuss in the next chapter). Renaissance artists and architects had succeeded in translating physical space into mathematical terms of proportion and perspective in order to produce artistic works that enabled the accurate, if illusionistic, portrayal of a three-dimensional world in the two-dimensional forms of painting. As a would-be painter, Galileo would have been aware of Filippo Brunelleschi's demonstration of linear perspective in Florence c.1420, almost two centuries earlier. He would have appreciated the mathematical spirit and empirical endeavour embodied in this essentially artistic experiment. Moreover, he himself had mastered perspectival drawing, producing accurate, realistic portrayals of physical space for the purposes of both scientific illustration and theoretical projection.

In his 1975 book *Against Method*, Feyerabend (fourth edition, 2010) sets out a radical epistemological position that totally breaks from positivism. His stance has been described as "epistemological anarchism". He argues that there is no such thing as *the* scientific method as portrayed by the logical positivists. As he mused, if we insist on a general methodology which will not inhibit scientific progress, then "the only 'rule' it will contain will be the useless suggestion: 'anything goes'" (p. vii). His study of the history of science taught him that successful scientists are often *methodological opportunists* and that science is characterized by a plurality of methods. Feyerabend is "against method" in the sense that he is completely opposed to the idea that a normatively binding universal logic of scientific discovery can in any way be binding on scientists or indeed any group of researchers seeking to advance our understanding of the world. Of course, all artistic research is in this sense against method. It proposes its own forms of investigation which, while mindful of scientific norms and standards of proof, primarily employ artistic means to explore creative art questions.

In *Against Method*, Feyerabend does not explicitly discuss the relationship between science and art – except in one footnote, where he notes that if we assume that science and art share a problem-solving attitude, then we are forced to admit that there are in in fact fewer differences between the two than have been supposed.[7] While the pluralism Feyerabend advances opens up a possible dialogue between the sciences and the arts, he was forced to conclude (ibid.:6) that

> the remaining differences are enormous. Scientists might rhapsodize about the unity of all human efforts; they may redden with excitement when speaking about the artistic aspects of scientific research: but their tolerance vanishes when the aspects become real, enter their laboratories and wish to be heard.

He recognized the role of the creative and indeed fictive in scientific theorizing activity, but also reluctantly accepted that

> a large amount of philosophical rhetoric is devoted to showing that philosophical (scientific) arguments are NOT a special kind of fiction. We are not too far from the truth when asserting that the beautiful arias that are being sung about the unity of the arts and sciences are nothing but hot air designed to conceal and to protect the strong antagonisms that still exist.
>
> *(ibid.:7)*

However, Feyerabend saw himself as having undermined the logical arguments for science's privileged position within culture, and much of his later work was a critique of the hieratic position of science within western societies. Because there is no universal scientific method, he concluded, we can't easily justify science as the optimum route to acquiring well-founded knowledge. Nor should we denigrate art for not following the protocols of scientific investigation. Feyerabend concludes – in a critical position which takes him close to Habermas and Deleuze's accounts of science – that scientific logic in effect prevails because of the authority it enjoys in advanced industrial society. This resides in its technological application in the industrial economy. Other traditions of thought, such as artistic ones, despite their concrete cultural achievements, have never been given a chance. The truth, he suggests, is that science is much closer to myth and to the creative arts than any scientific philosophy is prepared to admit. It is one of the many forms of thought that have been developed by man, and not necessarily the best. It is, as he says, "conspicuous, noisy, and impudent, but it is inherently superior only for those who have already decided in favour of a certain ideology, or who have accepted it without ever having examined its advantages and its limits" (2010:295).[8]

4.6 Artists and science

But how has this relationship between art and science played out at the level of practical collaboration between the two cultures? Over the last years, we have seen the re-emergence of the modernist view that artistic practice can be modelled on the laboratory and artistic invention seen as akin to scientific experimentation. The most enthusiastic exponents of this position today are of course art curators. They are the brokers of contemporary art product. They are also the key gate-keepers facilitating or denying artistic access to the contemporary art gallery and mediating between the artist and their audience. However, despite their professions about the links between science and art, many of their interventions seem to lack either intellectual engagement or creative risk taking. Nor is there much evidence that contemporary curators have any familiarity with either current scientific practice or analytical discussions about artistic research methodologies that might form the basis for plotting the science–art nexus.

Instead, we have postmodernist gestures. So, Hans Obrist, the artistic director of the Serpentine Gallery in London, together with a number of other high-profile curators, advocates a model of "laboratory" research to guide both art making and the practices of curation/exhibition. But this "enabling rhetoric" seems primarily about facilitating a range of promotional strategies. These, as we have seen, are partly concerned with offering a participatory model for the visual arts which in many ways is knowledge-led. Obrist embraces an "open exhibition" model treating all art as work in progress which can be shared with an audience. Science, despite the best efforts of advocates for the public communication of science, does not operate with a participatory model. Its key elements are specialist technical expertise, public and corporate funding, the patient pursuit of experimental outcomes and the generation of data-based knowledge supporting theoretical advances. Ideally, these methods lead to favourable technological outcomes.

We get a sense of the self-serving flippancy that can attend discussions of the art–science nexus in an interview that Obrist conducted with the English artist Damien Hirst in 2008. This was associated with Hirst's exhibition at the Serpentine Gallery, *In the Darkest Hour There May Be Light*. Obrist asks Hirst about his fascination with scientific matters (present, for example, in Hirst's trademark display of animal bodies in formaldehyde-filled glass containers in a manner akin to the preservation of scientific specimens as seen in Figure 4.3 below).

HUO: Some of your work links to display features in science museums, and other works have more to do with scientific formulas. I'm interested in this relationship to science. Can you talk about that?

DH: I just hitched a ride on science – or not really science – it was medicine. It's just collage, isn't it? Art is always very simple, or good art is always very simple. I took science in the way that Picasso took the bike seat and the handlebars and made the bull's head. I mean, there's nothing complicated about it. Science seemed to be getting people's attention and art didn't, so I hitched a ride on that. Or people were believing in science and

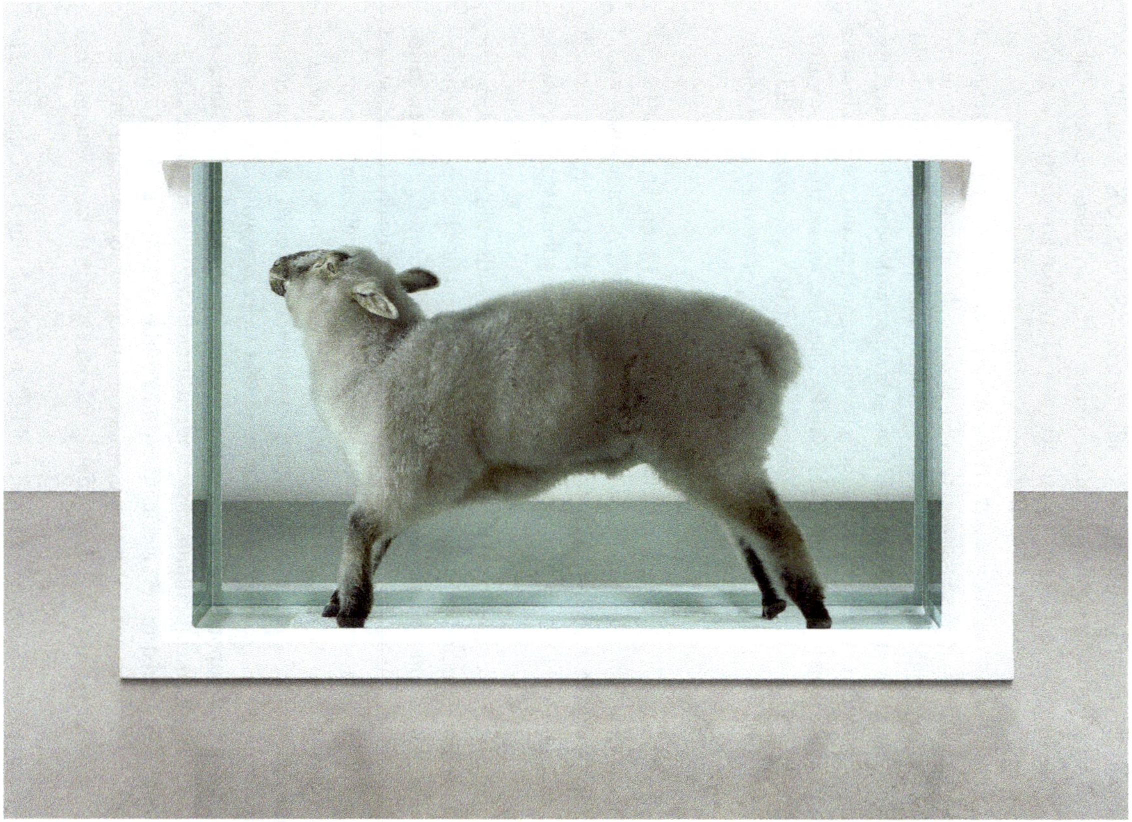

FIGURE 4.3 Damien Hirst, *Away from the Flock* (1997)

Source: © Damien Hirst and Science Ltd. All rights reserved, DACS/Artimage 2018; photo by Prudence Cuming Associates Ltd.

> questioning art, so I just took it very directly and used the science. It's been a very rich vein for me.

Hirst's tongue in cheek enthusiasm for things scientific extends to naming his purpose-built studio complex in Stroud in the south west of England "Science studio" and naming the company which deals with his work "Science UK Limited". This essentially is a figurative appropriation of science within art discourse. Really, it is a bit of codology. Of course, artists can get their ideas from anywhere – and why not science? However, such posturing does not make scientists out of artists! Nor does it introduce into contemporary art the investigative and conceptual rigour of the sciences, nor sensitize scientists to the contribution that artistic imagination and creative strategy might make to a scientific picturing and questioning of the world.

Of course, many artists are critical of the relativism and pseudo-intellectualism that bedevils much contemporary art, and scientific rigour remains a sort of lodestar for them. For example, the artists of the JCHP collective, whose work, as we have seen, foregrounds the investigative dimension of art practice, recognize the difficulties which arise when we seek to compare the patterning of scientific development to that of art (2017).

> The history of science is punctuated by revolutions brought about by circumstances that allowed its practitioners to be puzzled. A comparison would be difficult to make in art's current conditions. In contrast to art, notions such as critical and theoretical mean something in an operational sense for science.

Yet, interestingly, contemporary art curators, together with exhibition designers, have had a not insignificant impact on one aspect of scientific culture – the science museum. This essentially nineteenth-century European institution has been structured around the public communication of science and traditionally largely focused on the history of invention. However, of late the science museum has been given a substantial makeover to refit it for the demands of the culture industries. Alongside the introduction of a range of interactive experiences designed to enhance the "visitor experience", some museums have sought to engage with the ethos and aesthetics of contemporary art as part of this transformation of the science museum from fusty Victorian emporium to postmodern attraction. Visual artists and designers have been invited to recast scientific issues within broader narratives such as ecological threats, the moral panic around the impact of robotics and information technology on society and anxieties about our capacity to transform man and nature via bio-engineering.

For example, the Science Gallery in Dublin[9] in 2017 mounted an exhibition, *In Case of Emergency*, publicized with the strapline,

> What's the difference between a collapse, a downfall, and a downright apocalypse? How will it all end, and why do we love to wonder? Ice or fire,

> zombies or bombs? Out with a bang or a whimper? And can we do anything to stop the decline?

A number of contemporary artists were invited to muse on various manifestations on our ecological and economic crises and on the scientific response to these. Robbie Collins, in his piece *64 Things to Worry About* (Figure 4.4), presents us with an installation composed of an 8 × 8 grid of diffused lights, "designed to be hung on a wall in place of a clock, providing a heat map of the state of the world". Each light, we are told, is designed to represent "a specific danger indicator", signalling radiation levels, seismic activity, global warming and "the use of certain critical words on social media"(Science Gallery 2017).

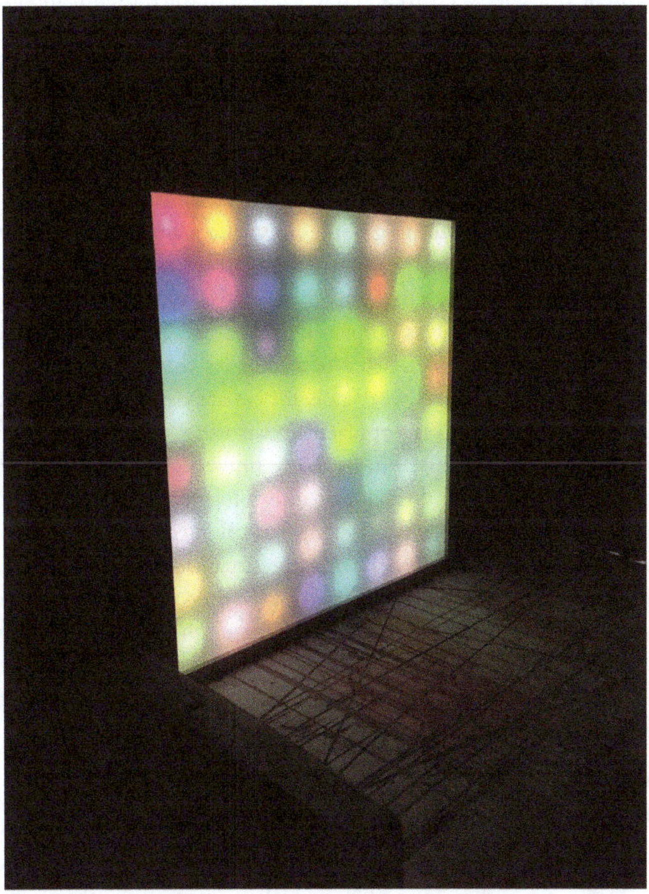

FIGURE 4.4 Robbie Collins, *64 Things to Worry About*, Science Gallery Dublin, 2017

Source: courtesy of the artist

Catherine Sarah Young's installation, *An Olfactory Portrait of the Amazon Rainforest* (Figure 4.5), explores the issue of the vanishing rainforest and its ecosystem of flora, fauna and indigenous peoples via scent, in this case by presenting us with a collection of perfume bottles with labels such as *Peanuts* or *Eucalyptus*. As the gallery website announces, "smell is closely related to memory, and so in smelling the scents, perhaps you can recall some of your memories and create a personal relationship with the rainforest". The use of this promotional language, pitched somewhere between scientific comment and the ironic and knowing discourse of contemporary art, is certainly unsettling. But the flippancy can jar with the seriousness of the topics addressed, and this is recognized particularly by the artists concerned with our ecological crises. We come away from the exhibition with the impression that artistic panache has been favoured over scientific explanation or indeed coherent political critique.[10]

These interactions between art and science within the museum and gallery seem increasingly to be characterized by a little too much art and perhaps not enough science. Despite that, there has been a largely positive response from the scientific community to these faux gestures made by sections of the contemporary art community towards scientific concerns. Over the last decades, a range of science institutes and university science departments have hosted artist-in-residence programmes, with artists invited to work alongside scientists but following their own lights. A number of research funding bodies now also support collaborative research endeavours involving artists working with teams of scientists.

However, it has to be said that much of this activity has been largely window dressing. It mostly involves employing the artist, their craft and their communicational reach to engage the sort of audiences that scientific institutions do not easily reach. It has, in other words, been primarily concerned with the public communication of science. This popularizing thrust is, as we have seen, present in the contemporary science museum. In many university science departments and institutes, artists are allocated a predominantly illustrative and educational role in facilitating "engagement" with non-scientific audiences. As anthropologist Marcus Banks caustically notes (2016:11), in these situations, "the artist appears to add little if anything to the scientific research itself other than to make it more cuddly". To use a historical analogy, the artist in these schemes often appears like the eighteenth-century Irish itinerant harper summoned to the big house to provide entertainment for his aristocratic patrons but with no hope of real participation in their ascendancy culture, nor of achieving any security of tenure.

4.7 Case studies of art–science collaboration

To conclude this chapter, I consider two specific art projects which explore the relation between art and science in a critical manner rather than simply popularizing science. The first of these is by an established Irish artist with extensive experience of collaborating with groups of scientists, the second by a former PhD student, now film and visual culture studies academic, who as part of her doctoral

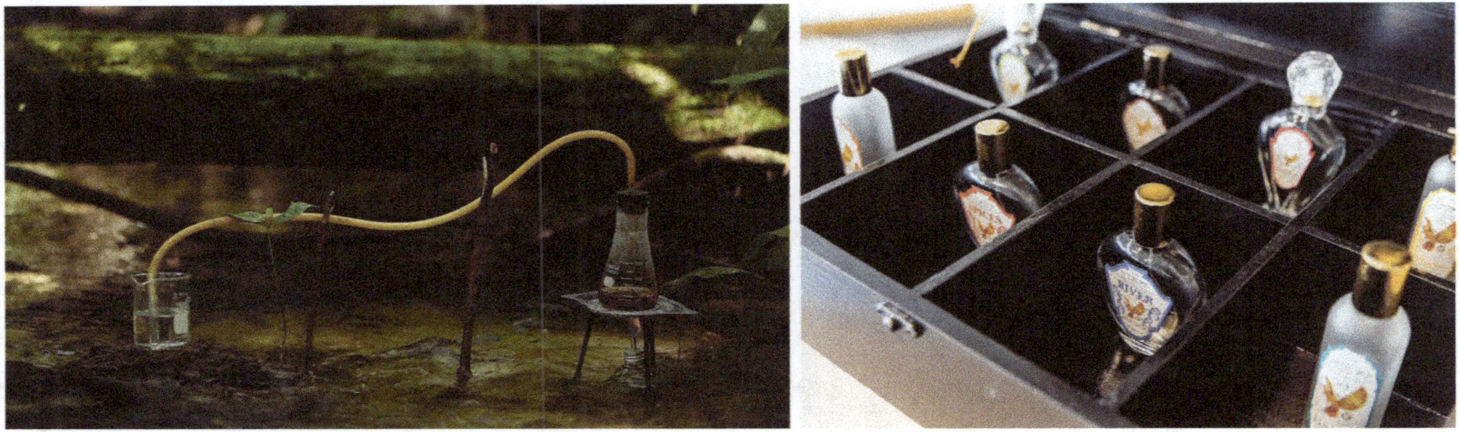

FIGURE 4.5 Catherine Sarah Young, *An Olfactory Portrait of the Amazon Rainforest*, Science Gallery Dublin, 2017

Source: courtesy of the artist

research reflected on the disparate discourses and practices of medical science and the creative arts. The former example gives us insight into the working research practices of an independent artist who often works within an academic science environment but who is not an academic, either by profession or temperament. The latter involves a set of philosophically informed reflections on art and science arising out of a programme of doctoral artistic research. Hopefully, a discussion of these projects reveals how artistic research can have a complementary function to scientific investigation.

Grace Weir is a leading Irish artist working primarily in the moving image and in installation practice. She has held a number of residencies in university schools of science in Ireland and the UK, working among various groupings of scientists over prolonged periods of time. Grace represented Ireland at the 49th International Venice Biennale in 2001 and has exhibited widely internationally. She was recently Artist-in-Residence in the School of Physics, Trinity College Dublin, and showed a range of her work emerging out of her engagements with scientists in a major solo exhibition at the Irish Museum of Modern Art (IMMA) in 2016.

Weir is concerned with aligning the sorts of objective knowledge generated by science with a lived experience of the world – in her case, usually filtered through her camera. She has developed a fairly unique approach to researching her projects based on a series of open conversations and joint experiments with scientists. The IMMA exhibition, titled *3 Different Nights, Recurring*, presents a range of her work produced over a twenty-year period. This was shown together with three new video pieces developed during her period of residence at Trinity College (https://www.youtu.be/h39_U3jkCKE).

The exhibition's title alludes to an astronomical drawing of the Whirlpool Galaxy (also known in scientific terms as Messier 51) made in 1845 by William Parsons, the third Earl of Rosse, with the assistance of the 72-inch Leviathan telescope, one of the largest telescopes in the world at this time. Parsons repeated his drawing of the galaxy over three nights as a form of proof of his discovery of the spiral nature of galaxies (the Whirlpool galaxy being a favoured example used to explore this structure). The drawings pre-date photography.[11]

Birr Castle, the seat of the Rosse family, has proved a fertile site for Weir, as she has also documented the original darkroom/photographic laboratory of Mary Rosse, Parson's wife. She, like many early photographic pioneers, combined her interest in the new medium with an interest in optics and chemistry, collaborating with her husband in the design of the Leviathan telescope. Rosse's darkroom was built in 1854 following advice from Fox Talbot, the inventor of the calotype photographic process. It has survived to this day more or less intact as a time capsule of scientific life in Birr Castle. Weir was able to visit and photograph the room before it was carefully dismantled and relocated to a new Science Centre on the grounds of the castle for permanent exhibition.

The resulting film, *Darkroom*, provides an intriguing moving image documentation of the remaining traces of this early scientific endeavour, before its "makeover"

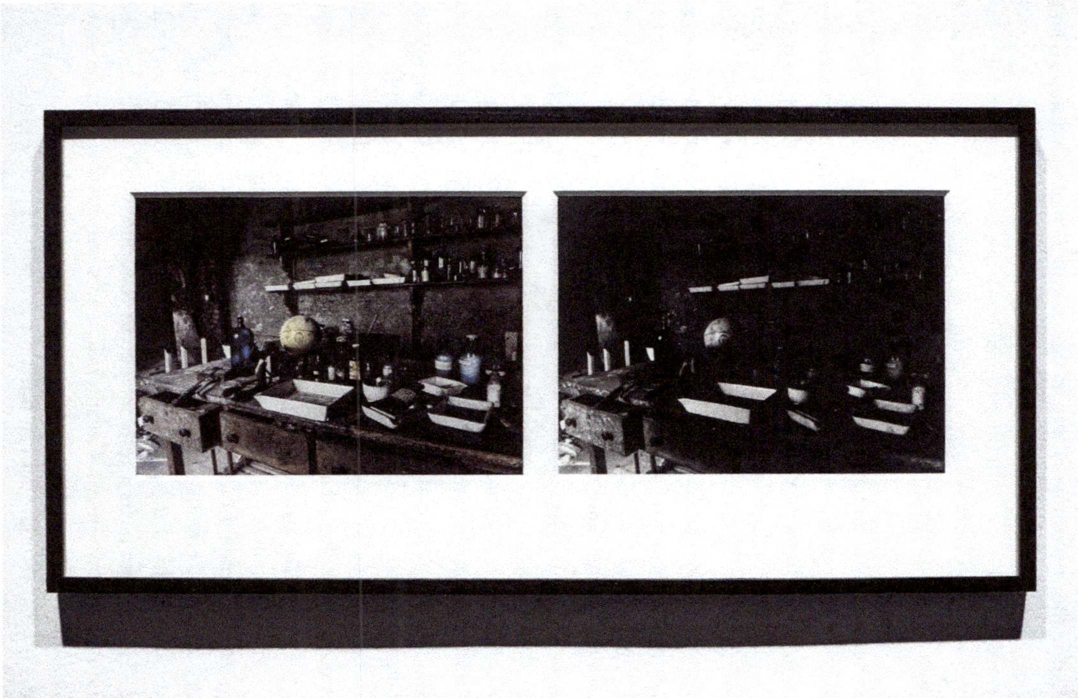

FIGURE 4.6 Grace Weir, *Dark Room* (2015), (left) digital photograph, (right) calotype copy of digital print, Irish Museum of Modern Art

Source: courtesy of the artist

as a tourist experience. However, it is also an artist's reflection on the role of time within photography. As the IMMA catalogue for the show tells us,

> Weir was interested in the relationship between reconstruction and photography – both attempt to fix a moment in time. Having visited and photographed Rosse's original darkroom prior to its dismantling, Weir returned to this space, now empty, to film a sequence of video shots. The two different projections, one filmed in the original but empty space and one in the reconstructed space with the original contents, mirror each other. Shown side by side, a strangeness forms in terms of location and time but a lucid connection is made through the activity of photography.

The work is neither historiographical in form nor scientific in character. It is not a documentary exercise concerned with the history of science. As a documentary film-maker,[12] Weir might have wished to pose the question of why the Rosses, like other members of the Anglo-Irish gentry who possessed photographic equipment, were not tempted to use the new medium with its power to document the distress of the famine which raged at this time.[13] A historian of science may have wished to explore the early photographic processes Rosse employed and to explore her use of lenses and knowledge of optics and chemistry. Certainly, Weir's work is alert to all these possibilities, but they are not her central focus.

In an adjacent room, Weir presents us with another take on the Birr project. A blown-up digital colour photographic image (Figure 4.6 left) portraying the Mary Rosse darkroom is mounted alongside a black and white photograph, a copy print of the digital image (Figure 4.6 right), produced by employing the very calotype photographic techniques that would have been used by Rosse in her darkroom. This diptych can be read as a further meditation on photography and the passage of time. Photography provides perhaps the most accurate documentation of the past and was initially hailed as a scientific instrument of record.[14] Photography creates a sort of simulacrum of the past (a "second degree original"). And this problematic relation of the photographic image to history is explored in Weir's work. Her subsequent videography of the reconstructed faux studio captures a site from which history has been effectively evacuated. Weir is conversant with the critical debates around the status of the photographic image and wants to alert us to these complexities. Her project bears a resemblance to Katerina Jebb's work, the *Balthus Inventory*, discussed in my previous chapter. It shares concerns with time and the photographic image and with documenting material traces of the past, such as those preserved in a working studio that can function as a sort of time capsule.

However, in employing early photographic techniques, Weir brings a formal reflexivity to bear on the issue of how we picture the past. Her strategy draws our attention to the means and mechanisms of representation and to questions of identity and change. And while Weir does not engage in a feminist heroicization of Mary Rosse (the installations give us little historical detail with regard to Mary's social situation), we sense her empathy with this female pioneer seeking to make her way in the male-dominated world of colonial science and photography. Indeed, this empathy might not be not unrelated to her own position as a woman operating within the male-dominated world of an academic physics department.

In her project *Black Square* (2015), Weir explores the difficult astronomical process of producing an image of the black hole that lies in the exact centre of the Milky Way Galaxy. With her film crew, she documents the work of an astronomy team from Munich working under astrophysicist Reinhard Genzel, based at the Radio Telescope facility at Cerro Paranal, in the Atacama Desert in Chile. The team has been seeking to prove the existence of this black hole by tracking the stars around it for the last sixteen years. Her project is not primarily an exercise in science education, but a transaction between the worlds of art and science. So, alongside the filmic documentation of the work of the Cerro Paranal team, Weir presents us with three photographic ink-on-paper works titled *3 Different Blacks* (2015). This is a series of ephemeral works (Figure 4.7) printed with non-lightfast inks which will fade in time. Weir thereby introduces a certain temporality into her images that parallels the play of time in the universe. The images allude to the black holes sought by the astronomers. Weir may also be referencing perhaps the most the iconic work of non-objective art – Kasimir Malevich's 1915 painting *Black Square*.

Grace Weir's work has involved a sustained conversation with the world of science, and in particular physics, conducted over several decades. But her research is clearly that of an artist, not a physicist. And while as a good ethnographer she has familiarized herself with the cultural habits, language and professional practices of her community of study, she clearly has not "gone native". She patiently observes her subjects and diligently attends to their culture and sacred texts. However, the work she produces is the independent voice of an artist.

The second project I want to discuss was conducted by the Italian scholar Silvia Casini as part of a PhD she pursued in film and visual arts at Queens University Belfast. Sylvia's background is in philosophy and the material and visual culture of science and medicine. She has a developed interest in art–science collaborative projects. Her research investigated "and sought to make visible" the aesthetics of Magnetic Resonance Imaging (MRI), an imaging technology which normally remains within the confines of the clinical laboratory and is not easily available to artists for use in their work.

Her enquiry took its lead from philosopher Jacques Rancière's definition of aesthetics as a "mode of articulation between ways of doing and making" (2004:10). MRI seems to be the latest device in a long lineage of visualizing machines that artists historically have employed to open up new visual possibilities and to explore novel creative strategies. Her goal, as she tells us, was to "break the frame into which MRI is usually situated". This frame is, of course, a scientific and medical one. Casini (ibid.:11) seeks to release this imaging technology from this purely scientific application, making MRI and its images "available for other uses and for other narratives rather than those authorized in the laboratory".

As well as undergoing a number of MRI scans herself, she was able to use the images generated to produce an experimental film, *285 Looping Snapshots.*

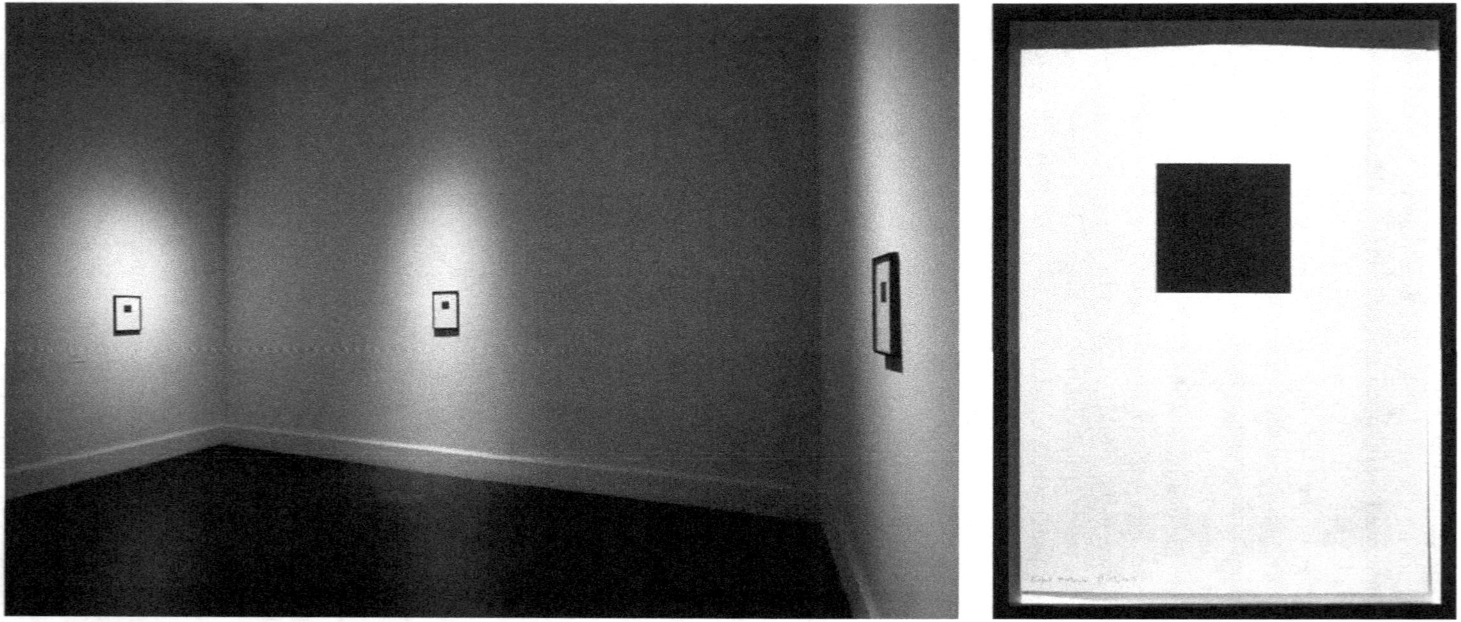

FIGURE 4.7 Grace Weir, *3 Different Blacks*, non-lightfast photographic ink on photo paper, 21 × 29.7 cm each, installation view and close-up, Irish Museum of Modern Art, 2015

Source: courtesy of the artist

Working with film-maker Eamonn Crudden, she animated her brain scans (which she describes as "a series of postmodern portraits") together with other video material recorded during MRI examination to fashion a moving image piece that explores issues of personal identity. As she notes, the MRI scan is now an established marker of individual distinctiveness and has replaced the fingerprint as an identification trace and as a forensic tool. The film in effect contrasts a first-person, "phenomenological" experience of undertaking the scan and the resulting sense of disembodiment caused by being encased in the MRI armature, subject to the objectivizing medical discourse within which the technology is normally mobilized. Patients often report a shock of recognition when presented with their MRI scans. Sylvia had also experienced this. In her dissertation, she describes the unease experienced in confronting her scan, an aggregate picturing produced from the sliced, cross-section images of her brain mapped by the MRI technology. She began to explore how MRI images might function as the raw material of an art project once freed from the functionality of medical diagnosis. Her research explores the use of MRI as a form of portraiture still bearing the mark of its forensic origins.

She also began working in a curatorial capacity with a number of artists who were using MRI in their work, including the French sculptor Marc Didou. Casini curated an exhibition of Didou's work for the Naughton Gallery in Belfast. This provided further opportunities for dialogue and propelled MRI as an imaging practice onwards on its journey from the medical laboratory to the artist's studio (Casini 2010, 2011).

As Marc Didou later reported to me in a 2018 email exchange, his use of MRI represented a sort of "methodological breakthrough" for him as a sculptor.

> I first started using MRI in 1996. I was looking for a way to get beyond traditional techniques of sculpture, whether modelling or physical carving. I felt the need for another approach which might re-engage with the human form and the artist's role. That's how I started using medical imaging – initially as a gesture of deconstruction of the figurative forms through which the body has been represented. I was now employing an abstract vocabulary drawn from another domain. Indeed, I really liked the idea that the image comes from an objective, scientific, extra-artistic source.

Both Didou and Casini seemed to find in MRI a liberation from traditional artistic and philosophical concerns. As Didou told me,

> I wanted to use a novel tool to look for, and work on, new forms of knowledge of the human body. I had the opportunity to witness the most up to date scientific research on medical imaging while employing my artistic intelligence to respond to the images that presented themselves to me. I had access to an extraordinary machine without really knowing how to use it.

> I was forced to invent a new and personal way of working. Where, I think, we could speak of a transformation occurring in my practice, is that the source of my new artistic work as a sculptor became the image and not the material, volume or genesis of the form.

Didou uses MRI as a tool of portraiture, mindful of the fundamental linkage between photography and death. The most direct allusion in his work, however, is to the historical use by artists of anamorphic lenses. Casini notes (2016:148) that

> in Didou's work . . . the mirror inside the MRI scanner is transformed into an anamorphic mirror; the sound of MRI becomes visible in the rhythms of the layers constituting the sculpture; the magnetic force is transformed into metal or other materials which attract or repulse the viewer; the immobility inside the scanner is created again through anamorphosis which obliges the viewer to assume a stationary viewing position and then to move again.

Didou plays with notions of lens manipulation and spectatorship in his use of an anamorphic mirror in his sculptural piece *Skull 1* (Figure 4.8). This piece is textured in layers like the assembled slices of an MRI scan. The mirror incorporated in the piece reveals in its reflection Didou's sculptured object, and in certain viewing positions the iconic image of a skull becomes visible. The visual output of the MRI scan, in the hands of the artist, retains its dual status as scientific trace and *memento mori*. This figure is the universal reduced face found in art depictions and in anatomical illustrations. In another sculpture produced for the Naughton Gallery, he gives this figure a name – *Eco* (Figure 4.9). As he told me,

> for the realization of *Eco* and that of *Skull* [4.8], I used the same images – exactly the same sequence of MRI scans. This may seem strange as they are very different pieces. However, after having observed MRI images of the head, I thought that it would be possible to realize one piece (*Eco*) by retaining the scan contours and, another (*Skull*) by exploring the images which correspond to the internal skull bone material. In this last piece, I falsified the thickness of the MRI slices so as to lengthen the sculptural form, making it abstract so that its true identity as a skull is almost hidden, only becoming fully perceptible by the anamorphosis of a mirror.

Didou uses an ancient visualization technique employed by painters to achieve new and indeed uncanny forms of legibility in his work (Holbein's painting *The Ambassadors* is only the most famous example of this technique). In giving three-dimensional form to the MRI portrait, Didou suggests the artistic lineage of different optical devices used by generations of artists to depict human identity and finitude. MRI is recognized as fitting into this line of descent.

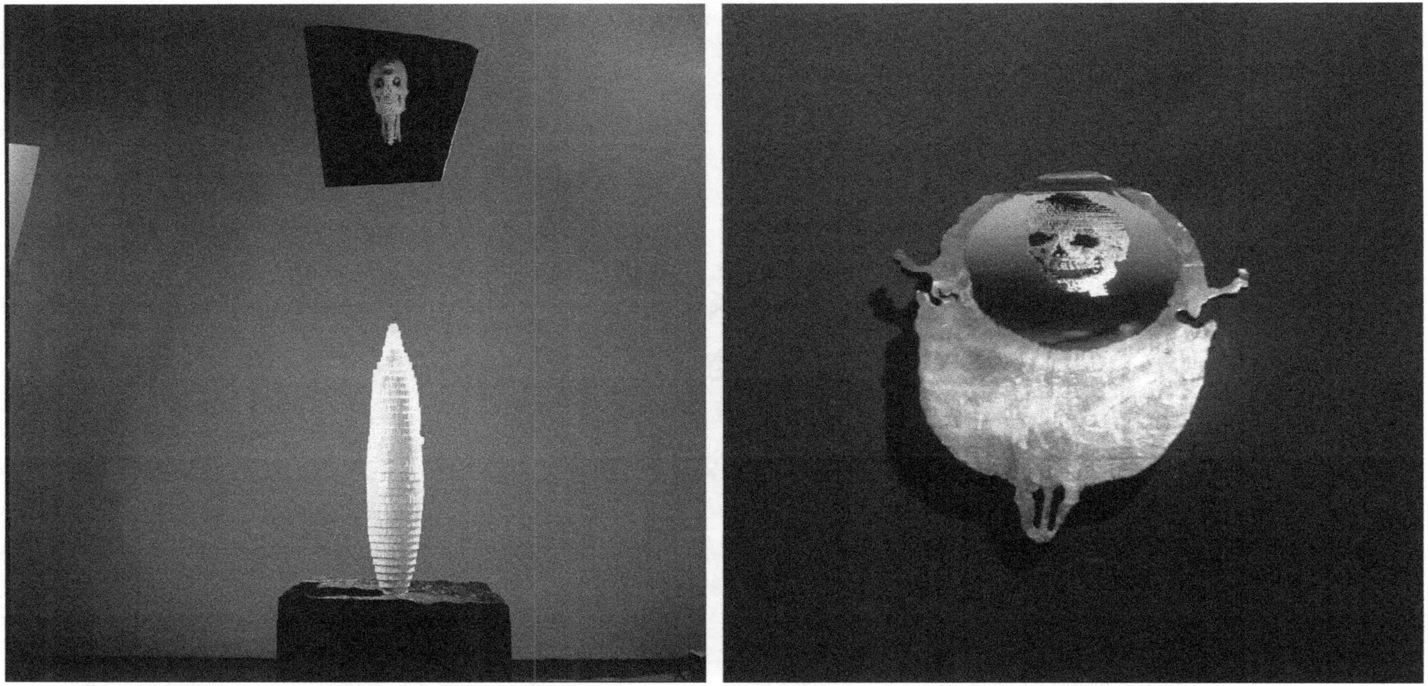

FIGURE 4.8 Marc Didou, *Skull 1* (2007), black and white marble with anamorphic mirror (and close-up), Naughton Gallery, Belfast

Source: photographs courtesy of the artist

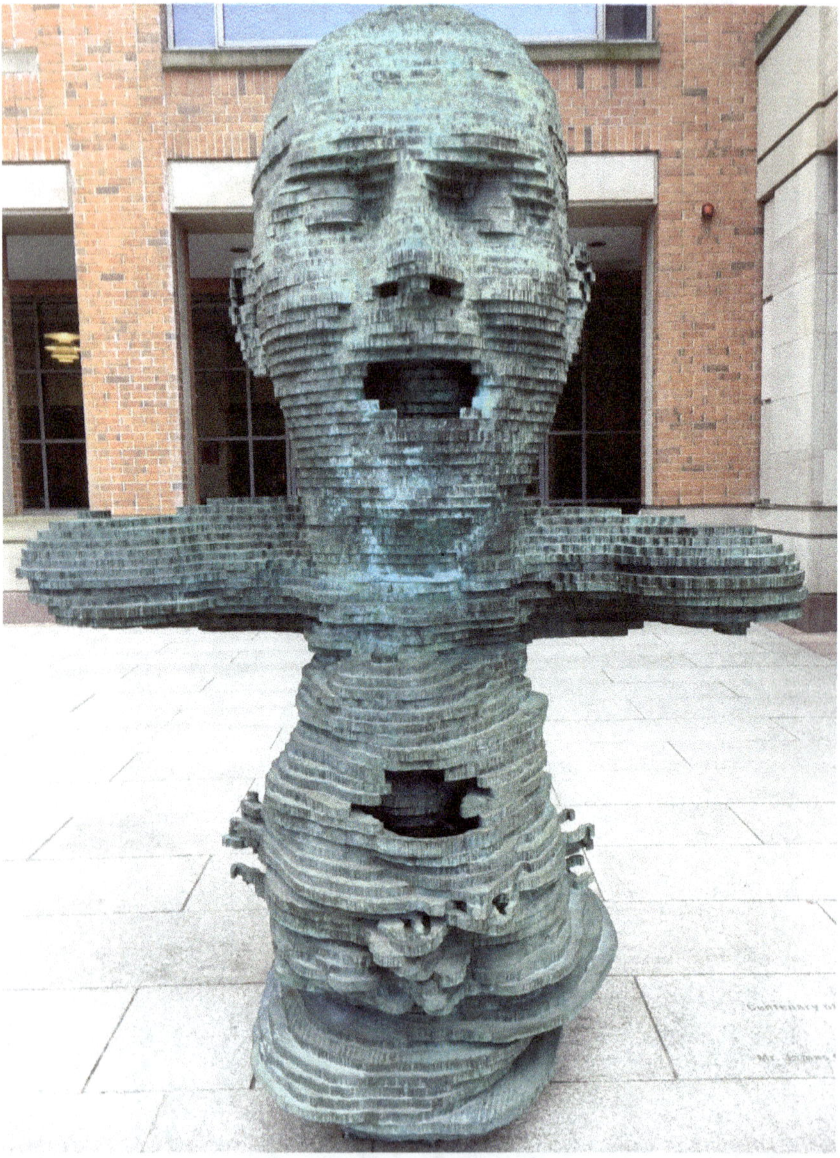

FIGURE 4.9 Marc Didou, *Eco* (2007), statue in bronze located outside the library of Queens University Belfast

Source: courtesy of the artist; photo by Hannah Healy Bell, 2018

4.8 Conclusion

Artists, then, can productively engage with science. Casini, working closely with sculptor Marc Didou, traces how MRI became a projective model for the making of original artworks, an appropriation which leads to a new configuration of this imaging technology beyond its purely scientific and instrumental applications. MRI is transformed into a visualizing technology, now mobilized within aesthetic practice. This sort of displacement of a scientific apparatus and practice and its artistic reconfiguring represents, I would argue, an important model of how science and art might interact. This synergy at the level of invention is, I think, the dynamic also apparent in Grace Weir's work. She closely observes a range of scientific practices, both historic and current, then re-appropriates this material to achieve distinctive artistic effects, whether by employing early photographic processes or by responding to the visual potency of contemporary astronomical observation. Art is quite capable of establishing a productive relationship with science – as it has over the centuries – but not by being subsumed under a universal method which sets limits to our modes of knowing.

Notes

1 Indeed, the proliferating discourse around practice-based research can often seem little more than a rhetorical ploy to establish arts practice as a respectable academic field of research on an equal footing with the established sciences and humanities subject areas.
2 However, as logicians point out, no number of confirming incidences linking two discrete events can allow us to conclude that they are linked by causal necessity. For Popper, it is not the accumulation of confirming incidences that is the essence of scientific testing. Rather, it is the elimination of hypotheses refuted by a single instance of unfavourable empirical evidence. For Popper, all knowledge is provisional. A scientific hypothesis has a fragile status as a knowledge claim and is afforded a truth status for only as long as it has not been falsified.
3 As Habermas observes (1972:67), positivism is a philosophical movement, but one which ironically denies its philosophical character.
4 Feyerabend was an accomplished singer and musician and in his teens studied singing in Vienna with one of the city's leading opera singers while developing his interest in science.
5 Feyerabend's study of the life and work of Galileo revealed the diversity of research methods employed by early scientists. Much of this work incorporated approaches that were in a strict sense not scientific at all – the use of abstract theories of cosmology, theological assumptions, residual neo-platonist views, rhetorical and figurative elements and, indeed, aesthetic considerations. As with Copernicus's argument for the heliocentric system, Galileo's position did not primarily rest on observational or other empirical data, even though he had the telescope at his disposal, but on mathematical and indeed aesthetic considerations.
6 He was to build on this supposition in his 1995 article "Science as Art". In this, he explored the parallels between science and art, which become apparent once we abandon the idea of science as a field rigidly regulated by methodological principles (1995a:1).
7 French philosophers have carried forward Feyerabend's attempt to plot the commonalities of science and art as modes of invention seeking, In Badiou's words (2012:iv), to "interrogate science for models of invention and transformation that would inscribe it as a practice of creative thought comparable to artistic activity".

8 The Science Gallery is part of Trinity College Dublin, a university with a long scientific tradition. Since its foundation, the Gallery has sought to interpret its brief in the widest possible way, engaging with a broad range of science, technological and design issues. This exhibition was funded by the Science Foundation of Ireland.
9 Some of the work in the exhibition is explicitly satirical – such as Mike Bianco's *Bush Blaster*, a phallically inspired, "strap-on" pollination cannon for dealing with the current honeybee crisis, which pokes fun at the human proclivity to try and "engineer technological solutions rather than make serious social changes".
10 Or to be more accurate, pre-dates any camera that could photograph at night (the earliest photographs in Ireland that have survived are the calotype photographs taken in Dublin, probably in 1843 and usually attributed to Fox Talbot, though probably by Calvert Jones).
11 In fact, Weir is an accomplished documentarist, as her film on the Irish modernist painter Manie Jellet, also displayed in the IMMA show, indicates.
12 The famine significantly affected County Offaly, with the workhouse in Birr just outside the castle walls overrun with famine victims during the period that Rosse and her husband were building their telescopes.
13 Photography fitted easily into a scientific frame of reference, with the indexical quality of the camera image chiming well with the positivist spirit of an industrial epoch. Indeed, the advent of photography may have led to a rebirth of Renaissance ideas about a fusion of art and science.
14 In line with current gallery strategies, Weir's exhibition is presented as an "activated project", and the curators encourage the artist to acknowledge the research process which informed this work and ongoing new work being developed by the artist and to share her insights with the audience through a series of lectures and experiments.

PART II

Antecedents and foundations

5

THE HISTORICAL ANTECEDENTS OF ARTISTIC RESEARCH

The Renaissance

5.1 Introduction

Artists have, then, flirted with science as a model for their practice. Scientists, in turn, have entertained dreams of artistic creativity. In the Renaissance and early modern period, artists *made* science. Scientists, for their part, often employed a range of artistic methods and visualization strategies. On the threshold of a new media age, we look back almost five hundred years to a historical period where art and science enter into an alignment of great productivity. In this chapter, I examine the intersection of artistic activity and science practice in the Renaissance period, particularly as manifested around the issues of perspective and realistic portrayal – topics that continue to fascinate us in a digital age.

Leonardo da Vinci (1452–1519) and his Florentine predecessors, the architect and artist Fillipo Brunelleschi (1377–1446) and Tommaso Masaccio (1401–1428), provide important historical exemplars of artistic research that we can still learn from today, while scientists like Galileo Galilei employed visual representation as a key element of their experimental and theoretical methods. We delineate the models of artistic research apparent in the work of artists such as Brunelleschi, Masaccio and Leonardo, which, I will seek to show, still inform the creative arts today. By and large artists in this 200-year period (1410–1609) had the methodological and technical expertise to participate in an emerging field of intellectual endeavour that straddled art, architecture, design and engineering, philosophy and experimental science. Sadly, that is no longer the case. In the contemporary period, the specialist mathematical and computational tools employed by scientists in their modelling and theory building activities are no longer generally available to artists.[1]

That said, contemporary visual artists and, in particular, photographers and filmmakers have shown intense, and on occasion faddish, interest in the historical study of visual culture and in what Jonathan Crary (1990:1) has called "techniques of observation and measurement". At the precise moment when digitalization is transforming photography, film and art practice, lens-based artists and students of the media have developed a fascination with how optical instruments and practices have interacted with the visual arts and science down through the centuries. For, as the art historian Barbara Maria Stafford perceptively notes (2006:1), "newer, more glamorous technologies still glimmer with mysterious and sensual ancestors".

In this fascination with mirrors, lenses, the camera obscura, magic lanterns, optical games and a host of other historical visual machines, art history has blended, on the one hand, with the history of science and, on the other, with that of popular entertainment. A hybrid analysis has sought to trace the interactions between art and visual technologies and between scientific culture and popular distractions. At the same time, this fascination with optical instruments and the attempt to detail their historical use within art practice has fed into contemporary concerns about how our contemporary media apparatus is shaping artistic process and purpose (a somewhat paradoxical situation in a conceptualist era which decries medium specificity). It is as if with digitalization we sense that something is being lost in contemporary visual practice and that the pace of technological change can only be properly evaluated by the historical retrieval of long-lost technologies and practices.

This concern with the history of vision machines and systems has, in turn, provided a focus for an important strand of practice-based arts research. A generation of photography students have built and experimented with pin-hole cameras and camera obscuras[2] (Figure 5.1). They have explored Victorian stereoscopic images, revisited abandoned technologies like the instamatic camera and, in so doing, have, somewhat ironically, developed some of the critical tools to understand and evaluate the digital revolution. Film students have eagerly traced the history of the magic lantern, phantasmagoria, zoetrope and flick book (Figure 5.2), exploring what Laurent Mannoni (2000) has called the "archeology of cinema", a research exercise which has fortified film as an art practice. Indeed, as film historian Tom Gunning has convincingly argued (1990a), this fascination with proto-cinema and attention to the magic of early or what is sometimes called "primitive cinema" or "the cinema of attractions" has been an important source of inspiration for recent avant-garde film-making.[3]

Meanwhile, artists like the optical toys collector and recently deceased film-maker Werner Nekes (1986) have been instrumental in bringing an awareness of the rich pre-history of cinema to a new generation of film-makers. Certainly, his work has helped to renew our capacity for wonderment at the magic of the moving image. Similarly, David Hockney (2001, 2006) has employed the insights of a painter and the reach of television to popularize the debate about the working methods of Renaissance and Baroque artists and their use of optical instruments in their work. The title of his television series on the topic, *Secret Knowledge*, perfectly captures the occultist feel of a practice-led historical enquiry that confidently pushes beyond the boundaries of academic scholarship.

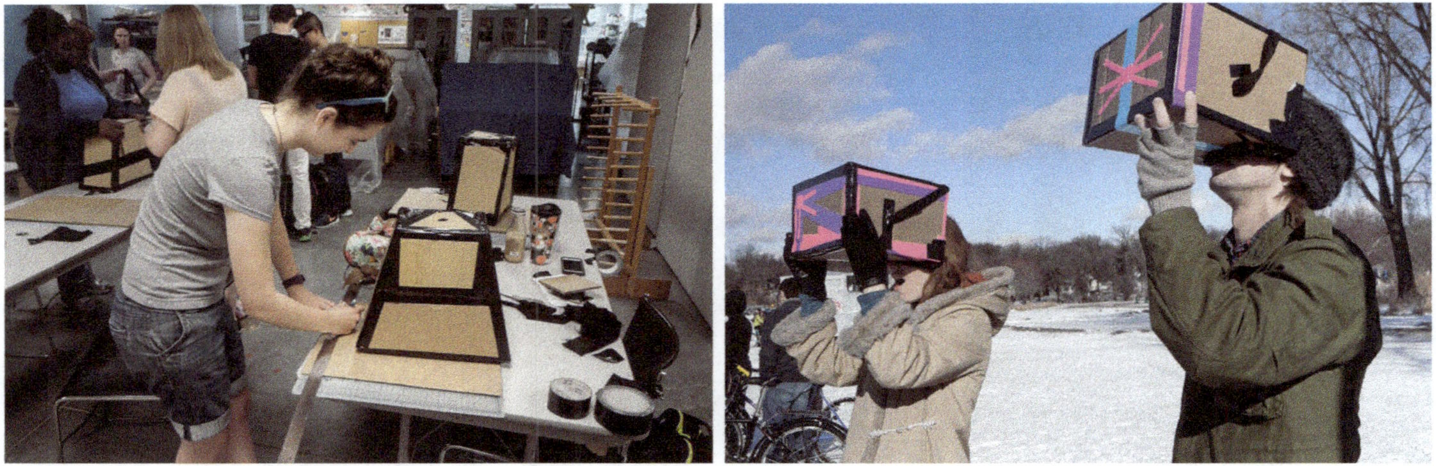

FIGURE 5.1 Students in Minnesota experiment with individual camera obscura

Source: photo courtesy of Andy Mattern

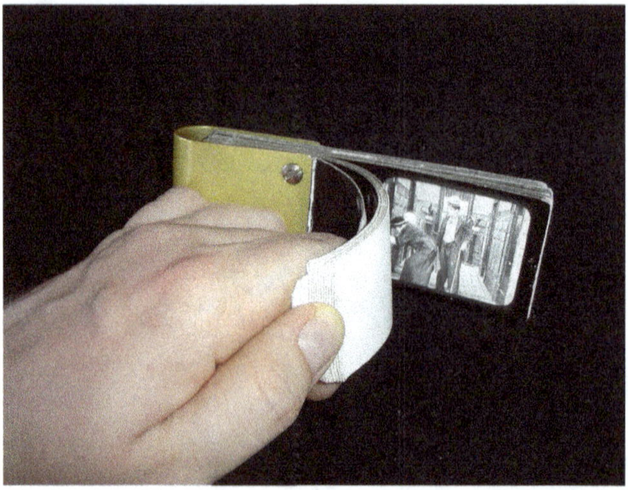

FIGURE 5.2 Flick book comprised of early cinema stills

Source: courtesy of the Museum of Technical Collections, Dresden

5.2 Renaissance vision

We now have detailed studies of how fifteenth-century visual arts practice experimented with reflecting mirrors (Warwick 2016) and the camera obscura (Hammond 1981; Kemp 2006c), and how artists engaged with planar geometry and optics (Edgerton 1975; Kemp 1990; Mark Smith 2014). We now understand much better how these optical tools assisted artists to perfect their command of perspective and naturalism. We have also begun to realize that, far from modernism having destroyed a perspectival ordering of the visual world, photography, film, television and now digital imaging continue to depend on and, in turn, reproduce these visual forms within contemporary mass culture.

Historians have also examined how in the sixteenth and seventeenth centuries artists responded to the visual worlds opened up by telescopy (Edgerton 2009; Knight Powell 2016) and, more generally, to the Galilean revolution in cosmology and scientific method (Kemp 2006c). Similarly, there has been an intense interest in how photographic capture transformed art practice, a focus which continues in our attempts to assess the impact of digitalization on the arts. In addition, speculations abound as to the impact of the quantum revolution in physics and post-Euclidian geometries on modernist visual practice after Cezanne (Damisch 1995). In other words, this fascination with the archaeology of visual media has directed our attention to a particular lineage of artistic practice that has been in constant synergy with scientific culture over the centuries around the topics of perspective, colour, shape and movement. Indeed, such a fascination has played no small part in sustaining contemporary ideas of artistic research and the aspiration to develop a "science of art".[4]

We also now have a series of detailed historical studies of Renaissance artist–theorists such as Brunelleschi, Ghiberti, Masaccio, Alberti and Leonardo. These reveal the overlapping worlds of art, science, invention, architecture and engineering in the Renaissance period and explore the processes of research and intellectual thought employed by artists in early modernity. For example, Martin Kemp has provided us (2006b, 2006a) with detailed studies of not only Leonardo's intellectual formation but also the diversity of his work as an artist–scientist. In particular, he has drawn our attention to the extent that Leonardo's practice – conducted across the disparate fields of drawing, painting, sculpture, architecture, engineering and science – was a unified one. The roots of this lie in the artist's belief in the divinely ordered character of the natural world and his capacity to explore the laws of nature through his visual practice and design investigations.

In particular, Kemp has examined Leonardo's modelling activity as an artist, designer and scientist, revealing a tradition of "vernacular science" (a concept which seems to bear a resemblance to Deleuze and Guattari's notion of "nomadic science") which prefigures mathematically formalized natural science. The science of art is vernacular not only in the literal sense that it is conducted on an oral basis with little or no reference to Latin written sources, but also in so far as it is draws upon the professional experience and insights of artists, designers, architects and engineers and atelier practices rather than the theoretical formulations of the scholastics. Kemp uses this term in relation to Leonardo's capacity (and that of later artisan researchers) to combine design practice with observational and experimental approaches to understanding the natural world.

Indeed, we should perhaps recognize Leonardo the artisan–engineer as the originator of a tradition of design research based on the building and testing of prototypes and the employment of an iterative process of investigation. In Leonardo's case, such research into mechanical and biological processes is primarily about achieving a pragmatic outcome to real-world problems by elaborating effective design solutions (whether in medicine, military engineering, hydraulics, aeronautics or urban design). However, his investigations also seek to contribute to our understanding of the natural laws governing mechanics and matter. Leonardo's "science" employs acute visual observation of the natural world, facilitated by his practices of sketching, painting and model making (neither telescopy nor microscopy were available to him). These traditional studio research tools are employed by the working artist in a manner not dissimilar to the experiments, mathematical formalizations and theoretical activities of the scientist, and indeed overlap in significant ways with the latter. For Kemp (2006a:3),

> Leonardo is now the very image of the *uomo universale* – the "universal man" of the Renaissance, somehow mastering a range of pursuits that would be unthinkable for a single individual today. His Florentine contemporaries, however, would not have been so surprised to find an 'artist' possessing such a range of skills in technology and even in science.

Leonardo was heir to a rich Tuscan tradition of mathematics, perspectival art and architecture, and science.[5] Moreover, his art activity functioned as a "well-founded" research practice, generating new forms of knowledge and insight. Leonardo's precursor, the Florentine sculptor Lorenzo Ghiberti (1378–1455), drew up the extensive list of knowledge competencies required by the practicing Renaissance artist: "grammar, geometry, philosophy, medicine, astrology, perspective, history anatomy, theory of design and arithmetic" (quoted in Kemp op. cit.:20). Ghiberti, the goldsmith and sculptor, found time to write about perspective and naturalistic depiction in his *Commentaries*, published in 1450, towards the end of his life. In this text, he annotated and commented on a range of scientific work on optics, geometry and perspective, indicating the synergies between the emergent vernacular science of art and the theory and natural philosophy of the scholastics.

Ghiberti's friend (and professional competitor) Filippo Brunelleschi was also an artisan–engineer who combined civil and military engineering work with architectural and other creative arts activity. He had, before commencing construction of the dome of the cathedral of Santa Marie del Fiori in Florence, built the austere, but beautifully proportionate, basilica of San Lorenzo in the city, a building which reveals his working grasp of the principles of perspectival rendering of architectural space. Ghiberti appears to have assisted him in his work on the Duomo as well as himself producing the decorative sculptural panels for the doors of the adjoining Baptistry of San Giovanni. The work of both artists, whether as sculptors, architects, engineers or inventors, is concerned with the intersection of geometry and optics, mechanics and invention, artisanal craft and scientific curiosity. As an architect tasked with the technical complexities of constructing the massive dome for Florence's cathedral, Brunelleschi seems to have had a firm grasp of Euclidian and planar geometry and of the metrics of projection as used at the time in building design. Like most artisan–engineers, he had an ability to draw and to paint, and it was usual for these master architects to keep notebooks of their sketches and drawings that might assist them in what we would now call their "R and D" activities.

These artists were trained in technical schools that focused on practical matters rather than in universities concerned with a scholastic curricula and abstract methods of reasoning. The *abacchi*, where Florentine professionals were educated, were institutions concerned with imparting practical, professional skills, such as the application of geometry to engineering, building and surveying activities, and of arithmetic to accounting tasks. Most of their graduates had highly developed drawing skills to assist them in their design activities. In particular, the artisan–engineers, as Samuel Edgerton notes (1975:38),

> had to have an unusual sense of what today is called "depth perception", for it was their job to imagine complex mechanical devices completely in the abstract, and then commit these designs to paper with a convincing illusion of three dimensions. It is scarcely surprising, in view of this, that it was an artisan–engineer who gave the Western world its first linear perspective system.

Brunelleschi, like many of his contemporary artisan–engineers, also appears to have had an intense interest in optics and to have employed a range of mirrors and lenses in his drafting work as an architect. Today he is largely known to art and media students for his exploration of the properties of linear perspective and his famous demonstration conducted around 1420 in front of the San Giovanni Baptistry (Figure 5.3). However, his scientific interests have to be seen as grounded in his experience as a working architect. Martin Kemp observes (2006c:19) that

> Brunelleschi designed buildings to be seen perspectivally – or perhaps it would be better to say that his insistence on spatial lucidity and proportional regularity ran through his invention of perspective, as a way to record the ordering of forms in space no less than through his production of an architecture which declared its regular spatial measure in both its whole and in its parts.

His biographer, Manetti, tells us that Brunelleschi, stationing himself several metres inside the main doors of the cathedral, executed a small-scale but naturalistically accurate painting of the Baptistry. This portrayed the front facade of the building, elements of its sides and the surrounding street scape as observed from his chosen standpoint (Figure 5.3 left).

Brunelleschi may or may not have employed a reflecting mirror to fix and render his image of the Baptistry (such as we see in Figure 5.3 right). He may or may not have had specialist advice from Tuscan intellectuals on the mysteries of optics. He certainly drew on his experience as an architectural draftsman to accurately portray buildings in a two-dimensional plan. As Kemp tells us (2006c:16),

> it is no coincidence that Brunelleschi should have invented this method of perspectival representation at a particular time and place in a society which was pioneering new systems for the design of urban environment as geometrical machines for the social life of modern cities.

The tablet image produced of the Baptistry was then treated, in effect, as a model or prototype. The reflecting mirror facilitates an exploration of the phenomenon of linear perspective and reveals the mode of spectatorship this visual system supports. Manetti tells us Brunelleschi drilled a small hole in his painting, "roughly the size of a lentil", at the vanishing point of the image (in this case at a central point of the door of the Baptistry). This perforation opened out to a wider aperture on the other side of the tablet, enabling the viewer to look through the painting. He then placed his eye to this small aperture, holding a mirror in one hand in front of himself while holding the painted tablet with his other hand (Figure 5.3 right). He squinted through the hole in the back of the painting to view its reflected image. The hole marks the point towards which the orthogonal lines in the painting converge at a central point on the doors of the Baptistry. The viewed mirror image captures his view of the Baptistry from the exact viewpoint where the painting was

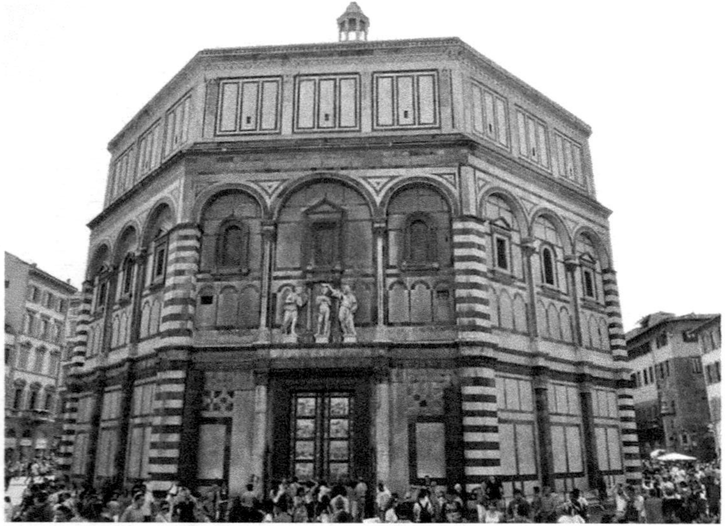

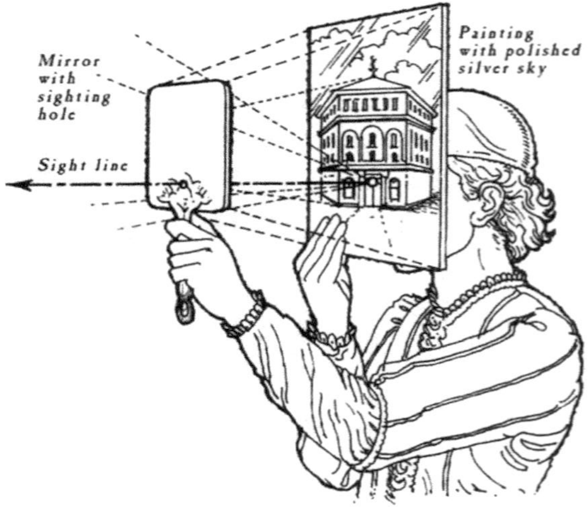

FIGURE 5.3 (left) The Baptistry of San Giovani, Florence today, seen from the Duomo opposite; (right) Brunelleschi's demonstration of perspective (c.1420), line drawing

Source: (left) courtesy of the author; (right) public domain

executed, in the darkened space just inside the front portal of the Duomo. Brunelleschi's demonstration employs a prototype and a design strategy to test a set of theoretical propositions.

Hubert Damisch,[6] who has provided us with the most penetrating reading of Brunelleschi's oft discussed demonstration, argues (1995:81) that

> what Brunelleschi's experiment demonstrates, in effect, is that the point we today call the "point of view" coincides, in terms of projection with the one we call the "vanishing point": both are situated at the intersection of the perpendicular sight line and the picture plane.

Brunelleschi was able to demonstrate how within a painting constructed according to the logic of this perspectival paradigm, the viewer is able to take over the painter's viewpoint. This subjectivity is effectively erased only to be replaced by a virtual subject constituted by the painting itself. A western visual tradition is born – but, also, a vernacular tradition of artistic research informed by, but distinctive from, the university science of the time.

The demonstration, like many of the Florentine paintings produced at this time, indicates the key properties of linear perspective: the systematic handling of perspectival space via creation of a checkerboard ground square perpendicular to the picture, capable of accurately handling the diminution of represented objects proportionate to their distance from a privileged observer; the convergence of the orthogonal line stretching as diagonals from the edge of the painting to its vanishing point, scaffolding this space into a rational order; the single monocular viewpoint arranged along the line of sight towards the vanishing point of the picture. These properties of perspectival painting would later be distilled and formalized by Alberti in his book *On Painting*, written in 1437 (Figure 5.4) and dedicated to Brunelleschi.

Alberti drew upon his detailed knowledge of Euclidian geometry and medieval astronomical graphics. He sought to combine the classical science of optics and its rich medieval development "perspectiva" with an understanding of the representational art he had encountered at first hand in Quattrocento Florence.

As a university educated, patrician intellectual, he was certainly more familiar with scholarly theories of optics, geometry and natural law than the artisan–engineers whose work he championed. But the initial intellectual impulse seems to have come from the practitioners.

As Jane Andrews Aiken argues (1980:70),

> if Alberti's approach to perspective emphasized mathematical purity because he had access to and depended on the speculative literature of classical and medieval astronomy, Brunelleschi's and Masaccio's approach was more pragmatic and closely tied to the symbolic affinities of medieval workshop geometry.

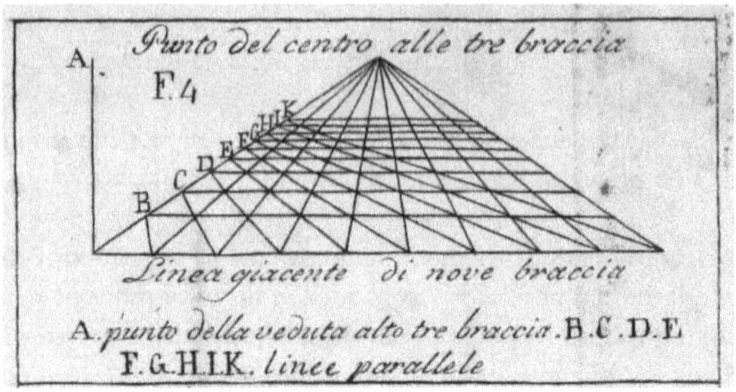

FIGURE 5.4 Diagram from Alberti's *Della Pictura* (1437)

Source: public domain

Alberti draws on Brunelleschi's insights, as he did on that of Florentine artists like Masaccio to later codify the visual system present in Renaissance art. By highlighting the reflective disposition within Renaissance art and giving this demonstrative form in his experiment with the panel and mirror, Brunelleschi provides us with an invaluable exemplar of an artistic research project in which creative practice and scientific aspiration chime. Surely this is the first example of a practice-based artistic project focused primarily on a knowledge outcome rather than simply the production of an artwork. It is also one which achieves its outcome via reflective practice (in this case, in a literal sense). The artist's practice is informed, in turn, by bodies of professional learning and scientific knowledge.

The use of perspectival techniques in painting to achieve an illusion of depth in a picture was, of course, already practiced in a highly refined manner by Renaissance artists from the time of Giotto (1266–1337), perhaps the first Italian artist to break free from artisanal social constraints and be afforded "celebrity" status. Accordingly, Brunelleschi cannot really be said to have "invented" linear perspective. However, he may have played a part in assisting his friend Masaccio in the design of what is perhaps the supreme example of perspectivally guided art in Florence at this time, the latter's painting *The Trinity with the Virgin, St John and Donors*, completed in the church of Santa Maria Novella in Florence in 1426 or 1427 (Figure 5.5).

Masaccio's work makes a different sort of contribution to this emergent science of painting, but one which is no less important. Again, the thrust of this iconic work is demonstrative. Samuel Edgerton and a number of other art historians have visually dissected this remarkable work to reveal the perspectival order present in it. However, Masaccio's fresco is not an abstract demonstration of perspectival rendering but part and parcel of a commissioned work with a range of other auspices and intended meanings.

The painting depicts a chapel whose depth seems to open up before the viewer behind the plane of the wall on which the fresco is painted. Inside this illusory space and framed by architectural columns and a vaulted ceiling, we see a crucified Christ

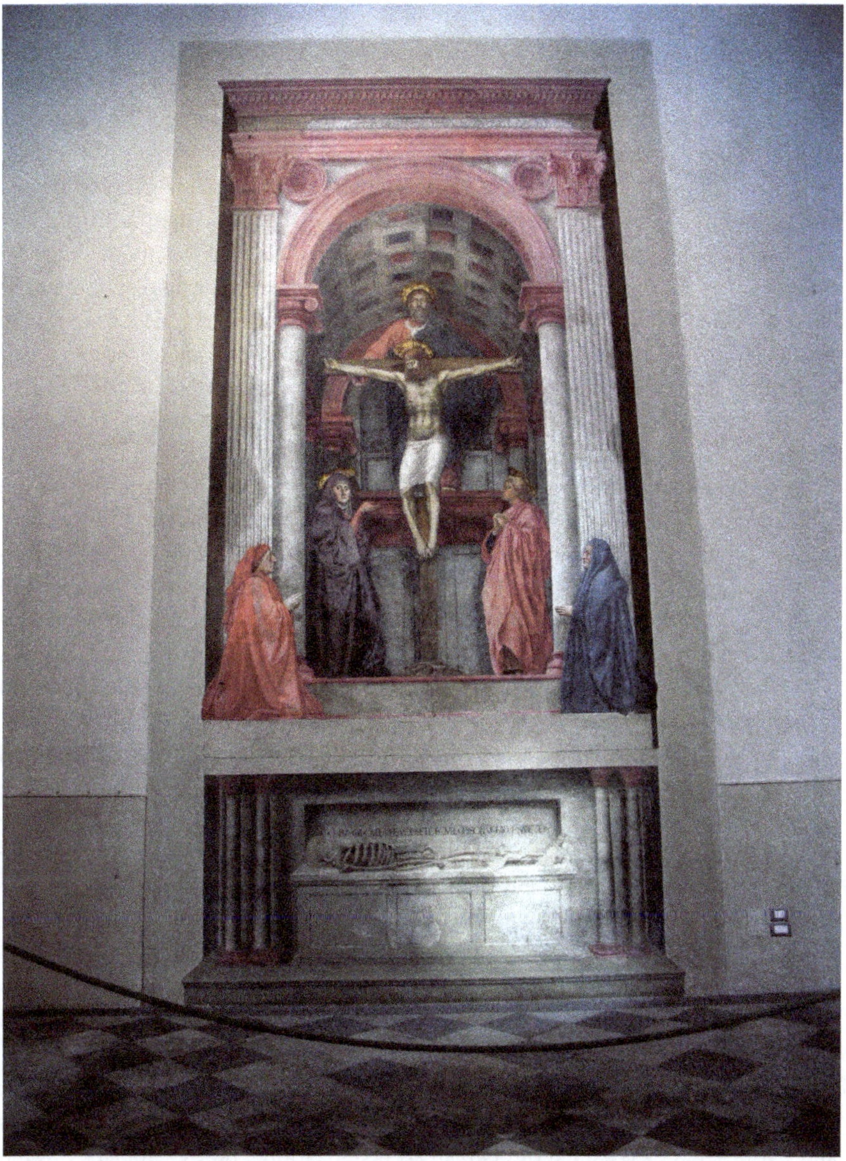

FIGURE 5.5 Masaccio, *The Trinity with the Virgin, St John and Donors*, Santa Maria Novella, Florence (c.1425)

Source: public domain

overlooked by a Godly figure. The figures of John the Apostle and the Virgin Mary are seen below the cross and are modelled with sculptural precision within the perspectivally ordered space. The bodily forms in this *mise en scene*, particularly the figure of Christ, seem to project forward into a common space with the spectator.

In front of the pillars which demarcate the entrance to the illusionist space of the chapel from the framing elements of the work, Masaccio portrayed in life-size detail Domenico Lenzi and his wife, the donors who funded this work. He paints them life-size and in realistic detail. These earthly creatures seem to occupy a different, quotidian space distinct from the divine realm staged within the inner painting.

It seems possible that Masaccio worked up his design for this painting from a three- dimensional model, or at least from the sort of a detailed architectural drawing that his colleague Brunelleschi could have provided. And this modelling enabled him to achieve the dramatic handling of perspective, creating such a dramatic illusion of depth in the painting. It is said that Giorgio Vasari (1511–1574), who chronicled the lives of the Florentine masters and accordingly is recognized as one of the first art historians, was so overwhelmed by the effect of Masaccio's perspectival foreshortening that he was convinced there was a "hole in the wall containing the make-believe chapel" (quoted in Kemp op. cit.).

But unlike Brunelleschi with his studious perspectival demonstrations, Masaccio was not primarily involved in a formal research exercise. He was executing a commission for an important client. And, whatever the role of Brunelleschi in the production of the *Trinity* fresco and in facilitating the perspectival effects achieved in this work, Masaccio's painting has harnessed this visual logic to achieve expressive and narrative outcomes in what is a sacred work rather than a technical exercise. The vast majority of artworks produced in Florence at this time were religious in theme. Indeed, the intellectual and creative investment in perspectival rendering was very much bound up with the view that this attention to geometrical method and optical detail in the realistic handling of pictorial space could "make literal the spiritual sense" (Edgerton 1975:19). Alberti's "window" by and large opens onto a biblical scene. For the Renaissance painters of Florence, "the theoretical, infinite space of the mathematician and physical space one sees before the eyes are one and the same thing" (ibid.:20). Each is a reflection of a divine order. That said, the production of such work seems to rest on a research process which facilitates a series of dramatic innovations in artistic practice. As such, alongside Brunelleschi's perspectival demonstration, it can be regarded as a prime historical exemplar of artistic research in action.

Brunelleschi and Masaccio, then, provide us with important exemplars of Renaissance artistic research. Important knowledge about both optical procedures and artistic strategy is gained through an art investigation. These exemplars, I would argue, continue to serve as models for creative arts investigation. In the case of Brunelleschi, we have a project where the objective seems to have been one of achieving a knowledge outcome in the first instance, rather than producing an art object. In Masaccio's fresco, on the other hand, we have a model of artistic research inextricable from the act of production of what we would now call "cutting edge" artwork.

In his own lifetime, Brunelleschi was not particularly known for his visual demonstrations, nor for being "the inventor of perspective". His reputation largely rested on his work on the Duomo and, in particular, its magnificent dome. It is only later, in the accounts of Alberti (1991) and Manetti, that his staging of this new artistic knowledge about perspective assumes a foundational, indeed mythic, role within western art. His demonstrations reveal perspective as the key "constructive principle" within an emerging aesthetic system – one which has survived to the present day largely due to lens-based forms of imaging in photography, film and multimedia.

However, perspective as a code of knowledge had other functions. Demonstration of a capacity to master perspectival form was to play its part in establishing the changing social status of the Renaissance artist and that of the artisan–engineer. Before the artistic revolution of Quattrocento Florence, painters had been regarded as one craft among many artisanal trades found in the late medieval world of guilds, masters and apprentices. The "Rise of the Artist" in the Renaissance period rested in part on the growth of humanistic learning, in which the revisiting of the classical past led to re-evaluation of the status of the artist – particularly one who could incorporate classical themes and designs in their work. As Andrew Martindale convincingly argues (1972:50), social prejudice against the artist as a craft-based artisan was only overcome when the re-reading of classical texts, containing high praise of artistic pursuits, became a common pursuit among the educated elite.

> The rehabilitation depended almost exclusively on admiration for sentiments and actions recorded in antique texts, and it was largely by dwelling on these that courtly society was able to convince itself that there might be honour among artists.

Painters themselves began to embrace the new humanistic learning and the sciences of geometry and optics that went with it. Thus, Brunelleschi's demonstration can be seen as a celebration of painterly intellectual achievement – the rendering of "space according to reason". Correspondingly, it represents an assertion of the scientific capacities and professional status of the artisan–engineer–artist. His public gesture of drawing art and science into a common sphere by staging his investigations in a civic place in front of the Duomo sought to transform art practice from its status an artisanal activity regulated by a guild system into a professional practice underpinned by scientific learning and theory. Brunelleschi is demonstrating the intellectual and professional standing of painting and architecture as branches of Renaissance science and asserting the social status of the artist as a gentleman.

5.3 Leonardo, science and invention

Leonardo provides us with a third model of artistic research practice. This is based on the use of modelling practices (an approach with which the other two artists would also have been familiar). Leonardo's models draw from his understanding of the design processes and machines employed by the Renaissance artisan–engineers (though his anatomical sketches seem to be based on direct observation and on occasion dissections of the human body and of animals). Leonardo was a pupil of

Verrocchio, the Florentine master painter and sculptor who had been entrusted with the manufacture of the globe designed to adorn the lantern of the dome of Santa Marie del Fiori. As a result, he had access to the cathedral's extensive workshops and appears to have assisted his master in the commission. In these workshops were stored the various machines developed by Brunelleschi for use in the construction and erection of the cupola, and Leonardo had the opportunity to study, sketch and use these lifting machines in the erection of the globe. In the modelling and machine work of Brunelleschi, Leonardo found an important exemplar to guide his research practice. Here was an artist responding to a perceived design need and engineering challenge via the construction of models and machines capable of being tested in practice. Such models provided the basis for the realization of complex architectural tasks. Such mega projects like the construction of the cathedral also became the site for the formation of new knowledge about materials, mechanical forces and the physical causes and effects which shaped the world.

Leonardo's later visual portrayals of man and machine, seen in his anatomical sketches (Figure 5.6) and engineering designs – like that for a proposed flying machine (Figure 7.13) – were in effect a way of doing science prior to the mathematical formalization of physical theory. In that sense, they are not just illustrating existing scientific knowledge but generating it. As Kemp argues, Leonardo's models can be conceived of as "theory machines" in so far as they guide an investigative practice that has shaken itself free from scholasticism. In effect, as with Brunelleschi's demonstrations of perspective, they perform the sort of proto-explanatory role later reserved for the algebraically expressed laws of physics. As Kemp notes (2006a:72), Leonardo "often calls his drawings 'dimostrazioni' (demonstrations) simultaneously demonstrating what could be seen and why it looked as it did. They were essentially models of nature, employing theory integrally with observation". So, what we find in Leonardo's work as a scientist is an art of visual modelling. This seeks to capture the polymath's acute observations of nature, whether of human and animal anatomy, mechanical and dynamic forces or astronomy. The models assist him in his quest to trace the causes and structure underlying all natural and human phenomena. This visual modelling was predominantly of a 3D character involving the building of prototypes, such as his mechanical devices, of which his flying machine designs are the most spectacular example (Figure 7.13).

Another way to characterize the investigative activities of Leonardo and the other Renaissance artist–designers is with reference to Gilles Deleuze and Felix Guattari's distinction between "royal" and "nomadic science", which we encountered in Chapters 3 and 4. The former refers to the institutionalized forms of science which enjoy state patronage and an intellectual legitimacy based on their theoretical and methodological systematicity, explanatory powers and technological import. The latter refers to forms of minor science (or "nomadic", "ambulant" and "itinerant" science) which explore matter, materials and real-world problems, often through various professional practices – such as design and architecture – rather than by applying theorems or generalized methodologies to abstract questions. Interestingly, Deleuze and Guattari identify military engineering and the science of metallurgy as

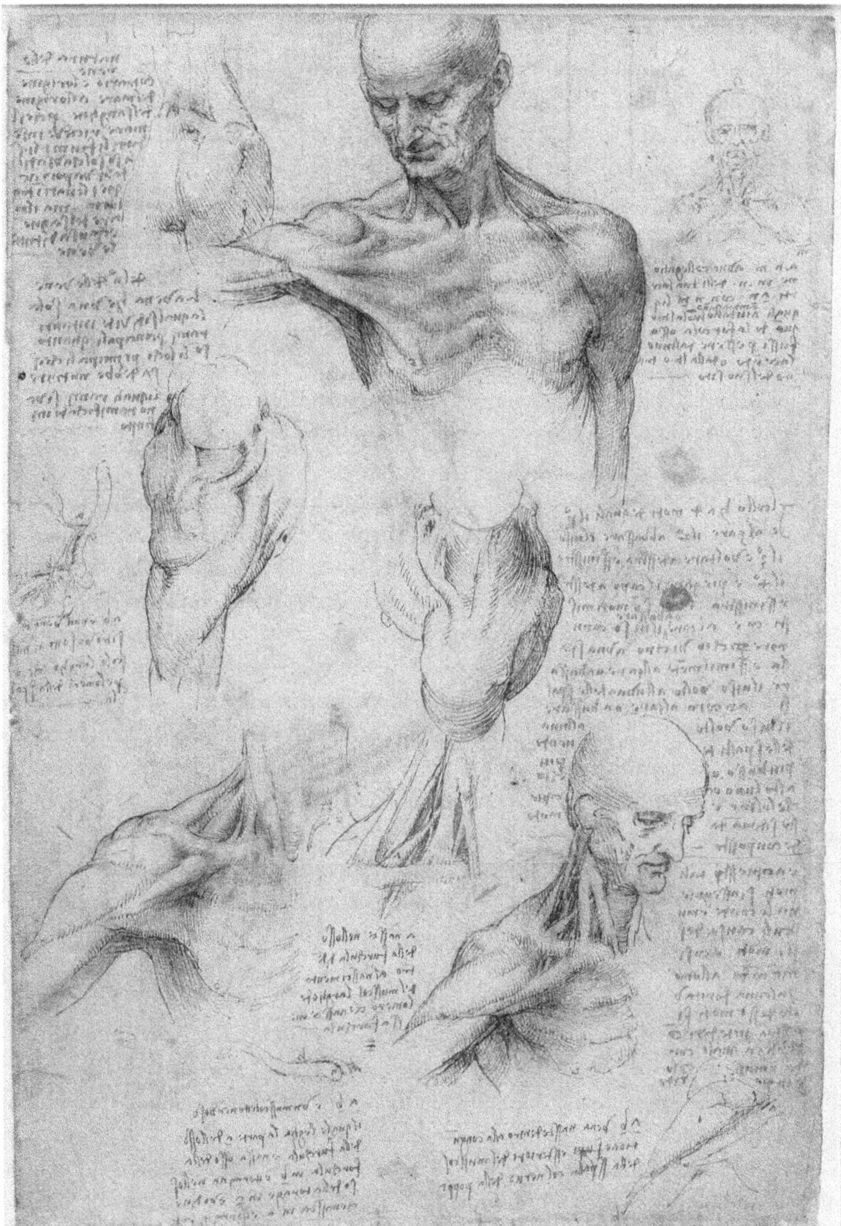

FIGURE 5.6 Leonardo da Vinci, *Superficial Anatomy of the Shoulder and Neck* (1452), Windsor Codex, Royal Collection

Source: public domain

fields where the forms of "approximate knowledge" and grounded research thrived. This science was harnessed to a range of real-world, problem-solving activities. Such problem solving is at the core of Leonardo's work as a scientist, engineer and designer, as it is within contemporary art and design research practice.

That said, Leonardo's "minor science" certainly advanced the art of scientific illustration and almost certainly prepared the way for the scientific advances of the seventeenth century. As Kemp observes (2006b:51), Leonardo developed a highly sophisticated practice of visually mapping the natural world, employing an extensive range of illustrative approaches in his work:

> fully rendered pictures of forms, sometimes in sequential views; solid and flat sections; see-through images; "exploded" views, in which joined forms are moved apart to show how their junctions are articulated; plans and elevations; and schematic images, ranging from simplified depictions to analytical diagrams in which lines represent such invisible things as lines of force.

As Kemp notes, these different modes of illustration are sometimes found on a single sheet, as in the large folded page, the "theme sheet" in the Windsor collection of the artist's notes and sketches produced by the artist around 1490. This graphic work suggests an enquiring mind at work that draws upon considerable observational and descriptive powers and which strives to arrive at a systematic analytical understanding of natural and mechanical phenomena, whether of a functionalist or causal character. In other words, Leonardo is engaged in a form of empirical science involving a rupture from classical natural philosophy. And indeed, the master artisan was always insistent that he was "a man without learning", guided by perception and experience rather than by scholastic doctrine (Kemp and Walker 2001:14):

> if you say that sciences that begin and end in the mind exhibit truth, this cannot be allowed and must be denied for many reasons, above all because such mental discourses do not involve experience without which nothing can be achieved with certainty.

Leonardo inherited the intellectually invigorated culture of the Florentine Renaissance. He makes his own contribution to this legacy by seeking to bring this art practice into alignment with a set of scientific aspirations. His surviving voluminous notes, annotations and diagrams deal with matters anatomical, medical, geographical and cartographical, mechanical and engineering, as well as with cosmological and philosophical issues, and of course he was concerned with reflecting on and understanding artistic process. Certainly, he employed a searching eye schooled in the attentive naturalism and perspectival accuracy of Florentine art. To this Tuscan intellectual and artisanal heritage, he was to add a proto-scientific concern with studied observation experimentation, design modelling, theorization and explanation, all of which flowed from his practice as an artist. In a sense, he is one of the first empiricists, believing that "all our knowledge has its foundation in our sensations" (Kemp 2006a:111). His work anticipates the scientific revolution of the seventeenth century, particularly in so far as he privileges methods of investigation based on observation and the testimony of the senses over scholastic learning and speculation. For him, vision is "the noblest of the senses". Accordingly, he seeks to link the exploration of natural laws to experimental and observational experience, including imaginative, probing design activity.

Leonardo may not have had the mathematical tools and astronomical devices and data available to Galileo to assist him in formalizing his experience as an experimentalist. Instead, his theorizing depends on visually expressed forms of modelling. In particular, he lacked a grasp of algebra and its application in advanced mechanics. In other words, he did not have a capacity to convert an understanding of mechanical matters present in his models into computations, similar to those of arithmetic or geometry, but with non-numerical and non-figurative mathematical objects, as employed, for instance, in an abstract theory of equations. His theorizations tended to take a visual form. Nor apparently did he have a facility in Latin. This would have given him access to classical scientific and philosophy texts, all of which were written in classical languages (or Arab) at this time. Latin was also, of course, the medium for the limited scientific communication taking place among intellectuals in Europe at this time.

Yet, while Leonardo's engineering and design projects rest largely on practical considerations and involve visual modes of explication based on artistic skill and insight rather than theoretical principles expressed in mathematical form, his sketches and notes suggest a researcher with an interest in the laws of nature and in causal forms of explanation. As the mature Leonardo asserts, "necessity is the mistress and teacher of nature, necessity is the theme and inventor of nature, the curb, the rule and theme" (Kemp 2006a:136). In his work, accumulated experience and documented experiment enable effects to be linked to general laws via causal chains. Discovery of such laws of nature could facilitate the prediction of events such as would greatly facilitate the engineer–designers striving to harness the forces of nature. For, as Leonardo had concluded, "in nature there is no effect without a cause; understand the cause and you will not need experience" (ibid.).

On the other hand, it is through the process of building particular models and constructing specific prototypes in conformity with existing knowledge of nature and artisan experience that we can demonstrate the mechanical dynamics of the natural world. The artist seeks, as Kemp intriguingly puts it, "to remake nature in his own works as the ultimate confirmation of natural truth" (ibid.). In other words, Leonardo's epistemic method is primarily that of the pragmatist who deems that if they can design a successfully operating machine then they are well on the way to providing a causally adequate account of the natural forces at play in a particular field.

Of course, it is only with Galileo and after him Newton that we can talk about a distinctive "royal" tradition of natural science gradually emerging with a fully resolved explanatory scaffolding and institutional formation. That is to say, a notion of science based on a unified theory of matter, the constituents and dynamics of which can be expressed in mathematical equations providing the basis for formulating the physical laws that uniformly govern the whole of the cosmos. That is, a science organized through learned societies and peer publications, with state support and technological application. Leonardo the polymath comes before the differentiation of science, philosophy, art, design and engineering. He also comes before the intellectual division of labour which will increasingly separate out theory and practice, science and art, intellectual and artisan. In the Aristotelian terms that Leonardo would have recognized, his work falls into the category of *techne* or useful knowledge, as opposed to *episteme* or systematic investigation into the abstract nature and causes of things.

But, in a sense, Leonardo's work as vernacular science defies these distinctions. Thus, we can see in his drawings and paintings a visually encoded account of the natural world which seeks to mirror and model its forms and dynamics. Indeed, he is critical of those who would elevate practice above theory (Kemp and Walker 2001:52):

> those who are in love with practice without science are like the sailor who boards a ship without rudder and compass, who is never certain where he is going. Practice must always be built on sound theory, of which perspective is the signpost and gateway, and without perspective nothing can be done well in the matter of painting.

Moreover, as he admits, "few painters make a profession of writing since their life is too short for its cultivation" (ibid.:13). That said, Leonardo himself produced an extensive corpus of writing throughout his life, and his notes often combine written material, sketches, analytical diagrams and computations. As such, his work provides a set of explanations of the natural world compatible with the incomplete scientific knowledge of his time.

However, he is painfully aware that painting and design, because of their visual rather than literary character, are judged by some to fall short of science, which he understands in a scholastic manner as "that mental discourse that originates in first principles". He acknowledges (52) that

> since painting does not achieve its ends through words, it is placed below . . . the sciences through ignorance, but it does not on this account lose its divinity . . . it possesses nobility in itself without the help of the tongues of others – no less than do the excellent works of nature.

5.4 Assessing the Renaissance contribution to the advancement of artistic research

The epistemic claims of the visual arts rest, then, not on abstract concepts and principles but on observation, experience, experiment and design and the capacity to naturalistically portray and capture nature. Leonardo subscribed to a scientific cosmology that held that all "created things relied upon the same set of natural causes (rules or laws)" (quoted in Kemp and Walker op. cit.: 57). Moreover, for him, these sets of laws can be divined in a painting or visual design as much as in an abstract scientific theory. As such, a painting or sculpture, or design, whether functional machine, imagined prototype, envisaged building project or urban plan, is a form of modelling (a "second nature") that postulates an objective reality governed by causal relations and universal laws which those models can portray.

The mathematization of scientific understanding after the Galilean revolution would in time drive a wedge between science and art, a division given further impetus in the Newtonian mechanical system and with the gradual emergence of positivism as a philosophy of science. And, of course, we continue to live with this chasm between scientific method and artistic practice to this day.

Certainly, in this early modern period a dynamic relation seems to have existed between science and art. The Renaissance's "geometricization of the world picture", to use art historian Sam Edgerton's intriguing term for the perspective paradigm, may well have influenced Galileo's own experiments on acceleration and projectile motion and his drive to provide a mathematical rendering of physical space and the matter within it. In 1609, employing a telescope, Galileo made a series of sketches of the moon. He used his grasp of both perspectival drawing and of the play of secondary light to produce a number of watercolour images of the cratered planet. These were later to serve as the basis for the engravings in his classic text *Sidereus Nuncius* (Starry Messenger) and set into motion his rejection of geocentrism and collision course with the Vatican.

To summarize, we have in our review of Renaissance art and design identified a number of models of artistic research practice. Three distinct methodological approaches can be divined: the *demonstrative*, concerned with revealing new forms of visual understanding; that based on *artistic innovation and experiment* and focused on making and sharing cutting edge work; and that involving the use of *modelling* to advance visual design and material construction and an understanding of causal and functional relationships. All three of these contribute to the sort of vernacular or nomadic sciences that emerge in the Renaissance within a distinctive art–science continuum. These approaches remain at the core of a tradition of artistic research which eschews the methodological paths of "royal science" for a more modest science that is ambulant and experiential, grounded and engaged.

1 In Brunelleschi's demonstrations of perspective, we seem to have an example of a practice-based research project where artistic and artisan means are used in an investigative project to arrive at knowledge outcomes expressed via a reported demonstration
2 In Masaccio's *Trinity*, we have an example of a "cutting edge" art project made possible by the artist's access to an emerging knowledge about perspective and their capacity to experiment with pictorial form, which facilitates the production of a unique work both embodying and advancing this emerging painterly knowledge
3 In Leonardo's work, the activities of drawing and model making enable designs to be elaborated, built and tested as prototypes, an iterative process that seeks

both solutions to material problems and to provide scientific understanding of the natural world

In reality, all three artists were familiar with each of these models of art-led investigation and alert to the different sort of strategies that might bring art practice and scientific enquiry into productive alignment. Their working practices drew on elements of each model. In the work of Renaissance artists, these methods interacted in significant ways within a vernacular science of art. These methods could be combined and given systematic expression within the common intellectual culture within which art and science developed in this period.

Notes

1 There was a period in the late 1960s and 1970s with the development of cybernetics in which artists such as Roy Ascot began to employ mainframe computing technologies to generate visual art. This, in turn, encouraged artists to engage with computer science and information theory (Shanken 2003). Ironically, with the domestication of information technology and development of the personal computer, artists are much less likely to have any analytical grasp of the scientific underpinning of information processing and the telematic technologies now employed so widely in their work.

2 One of the first and still most accomplished practice-based PhDs in the UK in the field of photography is by Caroline Rye (Napier University, Edinburgh, 2000). She experiments with the use of the camera obscura within performance art, reflecting on the transformation that take place when performance is recorded employing photographic or videographic means; see: http://citeseerx.ist.psu.edu/viewdoc/download?doi=10.1.1.462.121&rep=rep1&type=pdf.

3 Gunning (1990) alludes to the 1978 Brighton Conference organized by the International Federation of Film Archives as marking a renewed interest in the historical origins of cinema and in the earliest film productions in the period 1900–1906. These works are "primitive" in the sense that they do not adhere to the narrative, framing and dramaturgical cinematic conventions that we associate with the emergent post–World War I film industry in the US. Of course, it is precisely this unconventional character of the "cinema of attractions" that excites avant-garde film-makers pursuing creative moving image strategies beyond the confines of the classic Hollywood film text.

4 Martin Kemp quite explicitly throughout his work alludes to a "science of art", which he very much identifies with optical concerns within art practice and art's interaction with natural science.

5 Leonardo was by no means alone within the Italian artists of his time in terms of his multi-disciplinary abilities. For example, the Siennese artist and architect Francesco di Giorgo Martini (1439–1501) combined his work as a painter and a sculptor with scientific and engineering activities and with architectural theory and practice, achieving distinction in each field.

6 Damisch (1995) devotes over a hundred pages of detailed description and analysis in his book (for a critical assessment of his approach, see Wood 1995) to Brunelleschi's demonstrations, providing us with one of the most thorough discussions of an artistic project involving a model dialogue between theory and practice. In this sense, Brunelleschi, Ghiberti, Donatello, Masaccio, Leonardo and all the other journeymen of the Florentine Renaissance can be seen as involved in a massive, practice-inspired research project to reshape the world of art and to bring it into alignment with the new learning.

6

A SCIENCE OF ART?

Theory and practice in the modernist avant-garde

6.1 Introduction

We have to wait almost 500 years before an alignment of art and science occurs again within western history in any way parallel to that found in the Renaissance.[1] Up to the turn of the twentieth century, the notion of artistic research had scant academic recognition. Indeed, discussions of the creative arts rarely figured in the fierce methodological debates demarcating the natural and social sciences.[2] What little research there was in art academies was largely art historical, and it was generally assumed that artists made work largely on the basis of individual inspiration and by mastering a classical tradition rather than by employing any systematic research practice or set of empirically grounded theories. However, in the political, social and cultural upheavals of the first decades of the twentieth century, as art struggles to find its place in the tumult of industrial modernity and political revolution, artists again look to science for inspiration.[3] By the first decades of the twentieth century, visual art seemed to be an increasingly experimental affair. It also had theoretical aspirations. In that sense, at least, it was becoming like science.

The twentieth century saw successive waves of artistic innovation – Analytical Cubism's journey towards abstraction, Expressionism's concern with colour, perception and emotion, Futurism's with the dynamic image, Suprematism's pared back visual language of feeling, Constructivism's search for a unity of technology and aesthetics. Pictorial conventions and artistic sensibilities which had been the bedrock of western art since the Renaissance were now questioned, found wanting and abandoned. Increasingly, artists looked to science to provide a model for the seismic, paradigmatic shifts occurring within modernist art practice after 1870.[4] Art came to be conceived of as a process of invention and the studio as a laboratory in which theory and practice interacted and real-world problems could be addressed. The new visual languages and design methodologies could be codified and taught.

Once again, important exemplars of artistic research can be identified, and these may still have relevance in delineating the character of practice-based arts research today.

In this and the following chapter, I trace the emergence of a tradition of artistic research associated with the scientific, pedagogic and political aspirations of the modernist avant-garde in the first decades of the twentieth century. I examine the intellectual auspices of this tradition and explore the rapid development in the work of some of its leading figures after 1905, in particular Wassily Kandinsky and Kasimir Malevich. Of course, an argument could be made for repeating this exercise with regard to Vladimir Tatlin[5] or indeed for the artist–designer couple Varvova Stepanova and Aleksandr Rodchenko,[6] or for Olga Rozanova,[7] whose brilliant artistic career was tragically cut short by her death in 1918.

However, my concern is not to provide a survey of the art and theory of this period – there is already a not insignificant literature devoted to this task. Indeed, this work has recently been enthusiastically revisited in the context of the hundredth anniversary of the October Revolution in Russia. Rather, I choose to address the specific research and pedagogic aspects of the avant-garde tradition[8] and to tease out their significance for contemporary artistic research. In other words, I attempt to read this intellectual history against the critical challenges confronting current practice-based research, particularly in relation to critical theory.

The avant-garde research tradition I explore in these chapters uniquely combined theoretical concerns with an enthusiastic commitment to studio innovation. It also explored the social contribution of art. This was research concerned with the invention of new artistic methods and their widest possible application in societal transformation. Theory and practice chimed, as did creative innovation and political engagement. The historical crucible conditioning this ferment was undoubtedly the 1917 Russian revolution. The new art was an intellectual and cultural experiment being conducted within a broader social one. A revolution within a revolution.

Clearly, we live in a different era, which Alain Badiou has christened "the end of the age of revolutions" (2017:14). This is of course characterized by the collapse of communism but also by the subsequent prolonged series of economic and political crises that have occurred within a rampant capitalism. We make sense of art in a present shaped by a different technological infrastructure and set of socio-cultural conditions than those which obtained in 1917. Our thought is less philosophical and more sophist in character. It is less concerned with fundamentals and the pursuit of truth and more focused on the relativizing play of language and the problematization of truth claims. It tends to be dismissive of foundational schemes and vanguardist gestures. However, as we seek to advance a contemporary artistic research culture with its distinctive methodologies, interdisciplinary affiliations and patterning of theory, practice and social purpose, we may find it useful to revisit the modernist antecedents of artistic research, not least because these remain foundational for art and design pedagogy. They continue to influence introductory or basic courses in art and design and to inspire research-led practice.

In a period of artistic experiment and theorization unique in human history, lasting less than twenty short years (1910–1930) and mainly concentrated in two countries, Russia and Germany, a unique artistic culture emerged. In this, experiment and analysis, theory and pedagogy, ideology and political commitment would be paramount. From the outset, the avant-garde valorized the role of research and experimentation within their art. With the break from conventional modes of artistic depiction, they looked to science to provide a model for their practice. Artists identified the art process as a form of scientific enquiry and conceived of the studio as an experimental laboratory where innovative work could be made and basic hypotheses about the visual arts concretely explored. Some also sought to place the emerging abstract art within the broader development of western art, seeking to provide an analysis of the evolution of the distinctive pictorial language and creative strategies found in non-objective art. Each of these research tasks, whether conducted within studio practice or in a pedagogic context, was inextricably linked to the challenge of establishing the intellectual credibility of the new art, particularly in the face of public hostility to it. This historical moment saw art theory and practice in a dynamic relation. Each contributed to the forging of a putative science of art. Later, in the wake of revolution, this pattern of artistic "invention" would take on a more applied character and seek to advance a transformative design agenda for a new society, producing, in turn, a new conception of applied artistic research.

6.2 A science of art?

Members of the avant-garde, when describing their work and its critical context, talked repeatedly of a "science of art" and gave considerable value to the role of systematic research within art practice. As the painter, art theorist and pedagogue Kasimir Malevich was to express in his 1920s writings on the new abstract art, *The World as Non-Objectivity* (1976:270),

> art has become for me not only the means of solving aesthetic questions but also the means of cognizing through a whole series of experiments in colour, light, for and through their theoretical and philosophical forming. So, in this case painting is divided, like any scientific discipline, into theory, experience, and philosophy, the abstract division and the concrete one.

What could this idea of science mean? Why did the avant-garde come to conceive of the art process as akin to a scientific investigation? And what did this science of art look like in practice? And, perhaps most importantly, what purchase might these ideas of a research-led practice have on current attempts to fashion a tradition of practice-based arts research which is more than simply studio practice with notes and theoretical verbosity? To arrive at a theory of practice for the creative arts, we may, as I have suggested, have to look backward into history in order to move forward into the future.

The notion of science at work within the arts practice of the historic avant-garde was a complex one (as is that at play in contemporary art). It was shaped by a number of factors – the potency of late nineteenth-century science as a legitimating discourse, the refinement of art historical methods achieved in the late nineteenth century and, of course, after 1917, the alignment of art practice with an emerging Marxist doctrine with its own conceptions of science and of the political responsibilities of art.

The identification of artists with the objective stance and empirical methods of the natural sciences and their appetite for analytical and theoretical discourse can be read as a response to an age very much enthralled by the efficacy of the natural scientific method. So, it was hardly unusual that artists should seek to cast their claims for the new art in scientific terms. Indeed, this enthusiasm for employing a scientific frame of reference on the part of avant-garde artists grew after 1917 as leftist artists threw themselves into cultural reconstruction in Russia. The spiritualist concerns which had marked modernist art in Russia and Germany in the first decade of the century[9] gave way to a more sober and objective approach to art making and theorizing increasingly driven by design and production concerns. In turn, the idea of art practice was re-imagined as "laboratory work".

However, certain interpretations of how these avant-garde artists conceived of their "scientific calling" can be dispensed with. Despite the incessant talk on the part of avant-garde artists about laboratories, methods, research and analysis when describing artistic practice, this enthusiasm for things scientific did not involve the idea that the new art would literally follow the methods of positive science. Again, while various methodological developments and conceptual preoccupations within the history of art certainly left their mark on the thinking of the avant-garde, their conception of science did not involve a wholesale adoption of the rationale and methods of the emerging cultural and social sciences that the Germans christened the *Geisteswissenschaften*.[10]

Rather, the notion of a "science of art" seems to represent an attempt to develop a practice-based mode of theorizing the new art. Yet, this critical initiative, while it privileged the sort of insight that might arise from studio experiment and feverish creative innovation, sought to ground the understanding of modernist art in a new analytical visual language and in a notion of art as symbolic system. Such formalist concerns facilitated the drive to systematize the experiential knowledge of the practitioner in a manner compatible with the scientific logic of the day.

As we shall see, this intellectual impetus was most successfully pursued not in the esoteric artistic research laboratories and discussion circles, set up in Russia with state support after 1917 (such as INKhUK), but in the transformation of arts education that also took place during this period. The impetus towards developing a research culture for the creative arts was bound up in a broader mission to explain the complex and demanding work of the avant-garde to a new art public, to a new generation of students and, after 1917, to a new political class.

Like science, modernist art after 1900 seemed to progress in sudden great leaps and bounds of transformed practice and spurts of methodological innovation and theoretical insight. Like science, particularly the new science of atoms, electricity and relativity, art often challenged the taken-for-granted. To use Kuhnian terms, alongside the revolutionary science that had shattered the established Newtonian paradigm, a revolutionary art was emerging that sought to challenge the outmoded pictorial conventions and cultural assumptions of bourgeois society. Such was the narrative.

Indeed, the avant-garde art of the first decades of the century can be seen as the most vivid response within culture to the brave new world forged by post-Newtonian science. The classical conception of the world was that of an ordered space modelled in Euclidian terms. Within this world, matter moved in a consistent mechanical manner captured in Newton's laws. This picture was now challenged by the new physics of electromagnetism and subatomic particles and by non-Euclidian, curvilinear geometries.[11] Naturalism – the imitative rendering of the natural world within ordered perspectival space – seemed to be undermined by the collapse of the Newtonian world picture. Many of the avant-garde made a direct connection between the new forms of geometry and physics and the new art. They began to think of their discipline in scientific terms.

Painter Wassily Kandinsky grasped the potential import of the new physics of electricity and atoms for modernist art practice and its drive towards abstraction, arguing (1994:364) that

> a scientific event removed one of the most important obstacles from my path. This was the further division of the atom. The collapse of the atom was equated, in my soul, with the collapse of the whole world. Suddenly, the stoutest walls crumbed. Everything became uncertain, precarious and insubstantial.

In identifying the electron as a subatomic particle, science seemed to be decomposing the very world of solid bodies and physical matter which art had historically sought to realistically portray. Physical reality – now understood as a field of electrical pulses and magnetic fields – seemed to lose its objective quality. This new science seemed to sustain the drive towards a non-naturalistic art of abstraction. For, if there was not an objective world of tangible solids, then there could not be an art dedicated to replicating this world. In other words, non-objective art was consonant with the new science of atoms and relativity.

For Kandinsky, this scientific *weltanschauung* presented an opportunity to finally dispense with the baggage of a representative, mimetic art which had dominated western painting since the Renaissance. As he admits (Bowlt and Long 1980:103),

> modern painting is still limited almost exclusively to natural forms borrowed from nature. Its present task is to try out its own means of expression, its own forces, and to appraise them, to cognize them, as music has done from time immemorial (. . .), painting must use these means and forces in a purely painterly way in the creative process.

6.3 Founding fathers (and mothers) of twentieth-century artistic research

It is possible, I think, to trace the origins of contemporary practice-led artistic research in the programmes of basic visual arts research and pedagogy plotted by painters like Kandinsky and Malevich; in the Constructivist modelling of the design process developed by Lizzitzky, Tatlin, Rodchenko and Lazlo Moholy-Nagy; and in the experiments in film and film education pioneered by Sergei Eisenstein. These avant-garde artists can rightly be regarded as the "founding fathers" of practice-based arts research. Sadly, there were few mothers.[12] This cadre of artists largely began their careers as avant-garde painters (Eisenstein in theatre). Most came to embrace an interdisciplinary practice in synchrony with a new type of society. They played a pivotal role in developing the radical tradition of arts education associated with the restructured art and design institutions created in Russia after the revolution and with the Bauhaus Academy of Design established in Germany in 1919 (Kandinsky, of course, taught in both systems).

Herbert Read, writing in the late 1950s in the heyday of post-war modernism, and building on the pioneering work of Alfred Barr (1936),[13] produced the first comprehensive review of modernist art in English (Read 1968). In this, he acknowledges the methodological contribution of the modernist pioneers Kandinsky and Paul Klee. Of Klee, who played such a significant role in the development of the curriculum at the Bauhaus, alongside Kandinsky, Read says (1968:186),

> Klee always showed the greatest respect for the science of the art of painting. And most of his pedagogical work is concerned with practical details. Das bildnerische Denken (Pictorial thinking), his lectures at the Bauhaus and Weimar (a volume which, with its illustrations and diagrams, runs to more than five hundred pages), is the most complete presentation of the principles of design ever made by a modern artist – it constitutes the Principia Aesthetica of a new era of art, in which Klee occupies a position comparable to Newton's in the realm of physics.

Adding Kandinsky to this "pantheon", he continues (233):

> Klee and Kandinsky, in the course of their experimentation, had anticipated all the possibilities that lay ahead: every type of abstract Expressionism that was to be developed between 1914 and the present day, has somewhere its prototype in the immense oeuvre of these two masters. They went to the limits of plastic consciousness, in so far as limits have been given to that conscious in our age.

Hyperbola? Perhaps. In the context of a withering postmodern re-evaluation of the modernist project, Read's claim can appear an inflated one. And, certainly, the same argument can be made about the theoretical and pedagogic contribution of

Malevich, Tatlin, El Lizzitsky, Rodchenko and Moholy-Nagy, and the numerous others who made important theoretical, investigative and pedagogical contributions to the development of the creative arts in this period. The work of the Russian avant-garde was perhaps less well known to Read than that of the German Expressionists.[14]

Somewhat perversely, contemporary proponents of a practice-based visual arts research paradigm have by and large shied away from any serious engagement with the very historical tradition that might illuminate their quest to define the parameters of arts research and its problematic relation to science. Yet it can be argued that the early twentieth-century avant-garde research paradigm rested on firmer intellectual pillars than the instrumentalized versions of practice-based research prevalent in our universities and colleges today, with its mix of positivist rhetoric, bureaucratic modes of accountability and faddish poststructuralist sophism with little purchase on arts practice.

Arguably, Kandinsky is the key figure during this first period of avant-garde activity (1910–1921) with a foot in the artistic cultures of both Russia and Germany. Indeed, he is able to freely transplant ideas and practices from one country and culture to the other. He is the artist most at home with the tasks of theory, pedagogy and cultural policy. He is also the one with perhaps the most developed sense of a distinctive tradition of artistic research that can be attentive to the intellectual rigour and experimental practices of science, but responsive to its own lights, or what he calls art's "inner necessity".[15]

Born in Moscow, but brought up in Odessa, he was familiar with Russian traditions of folk art and icon painting. As a graduate researcher in law and political economy, he had conducted ethnographic research on the folk culture and legal customs of the peoples of the northern steppes.[16] Like a number of Russian artists, he drew parallels between primitive art and modernist experiment. After university, he opted to study art in Munich. There he established his artistic career and participated in a number of avant-garde movements, the most important of which was the *Blau Reiter* (Blue Rider) group. Kandinsky painted his first "non-objective", that is to say predominantly abstract painting, in 1910.[17] He also completed the first theoretical account of the new art, *On the Spiritual in Art*, in that year (though its publication was delayed until 1912). Repatriated to Russia as an alien with the outbreak of World War I, he vied to assume leadership of the avant-garde in Moscow. No armchair theorist, he played a key role in restructuring the art institutions of the new state after the 1917 Bolshevik revolution, working closely with Anatoly Lunacharsky, the People's Commissar of Education and Culture. In 1919, at Lunacharsky's request, he sketched a research programme for the proposed Institute for the Arts, INKhUK, attached to the Museum of Artistic Culture in Moscow. Surely this is the first example in the twentieth century of a coherent, institutionally based artistic research initiative. In parallel, he developed a new curriculum for the restructured art schools in Russia (though he was but one of a number of artists involved in this task). On a potential ideological collision course with both the new state and his fellow avant-garde artists whose work was increasingly moving in a productivist direction, he left Russia in late 1921 to take up a post at the Bauhaus Academy in Weimar. There he had an opportunity to apply his pedagogic theories in the new institution, working closely with Paul Klee, among others. With the Nazi takeover and closure of the Bauhaus, he moved to Paris in 1933, where he remained until his death in 1944.

The rapidity with which Kandinsky's work evolves between 1905 and 1913 is truly astonishing. It is, however, typical of the lightning development of the work of the Russian avant-garde during this period. In the case of Kandinsky, we can clearly trace a trajectory of artistic innovation which leads to a total rethinking of visual art. We can glimpse the evolution in his work between 1901 and 1913 in the sequence of paintings discussed below. Again, our concern here is not primarily to provide an art historical account of the evolution of Kandinsky's work as it moves towards abstraction, but rather to focus on the research-driven nature of this creative trajectory and its interaction with his theoretical reflections on the new art.

His 1901 landscape *Sketch for Achryka* (Figure 6.1 left) portrays the country house and gardens of a Moscow relative and is painted in something close to an Impressionist style, albeit with a different treatment of colour and pattern of brushstrokes that owes something to Fauvism. Kandinsky closely followed the development of modernist art in France and Germany while seeking to preserve a distinctive Russian sensibility in his work.

In the 1902 work *Old Town II* (Figure 6.1 right), portraying a Bavarian townscape, the Fauvist influence is more present. His 1906–1907 painting *Riding Couple* (Figure 6.2) employs a *Jugenstil* (the German version of art nouveau) motif and visual style. The treatment of the picture surface is pointillist and shows an awareness of post-impressionist strategies of visual representation and experiment.

By 1909, in the oil painting *Mountain*, he is experimenting with abstraction (Figure 6.3). Figurative traces are still to be seen, but these are increasingly dissolved within the abstract swirls of form and colour that characterize the painting; the artist has moved decisively beyond his earlier work seen at the time as that of a landscape painter and colourist (labels he always rejected).

We can trace a gradual, almost reluctant,[18] departure from figuration – apparent, for example, in the transition between the two drafts of the work *All Saints I* both completed in 1911. The first of these (Figure 6.4 left) is a reverse glass painting with a clear motif and recognizable figuration. However, the oil on canvas version (Figure 6.4 right) indicates a decisive move towards abstraction. The representational elements present in the first painting have been simplified and replaced with a rudimentary patterning of form and colour in which any figurative depiction is hard to divine – though we can still spot residual figures and symbols, like the angel sounding his trumpet, and detect outlines of some of the saints assembled below.

Kandinsky is not just exploring a range of post-impressionist and expressionist styles in order to acknowledge changes in artistic fashion, but is also pursuing a discovery path complete with a set of theoretical reflections that leads him inexorably

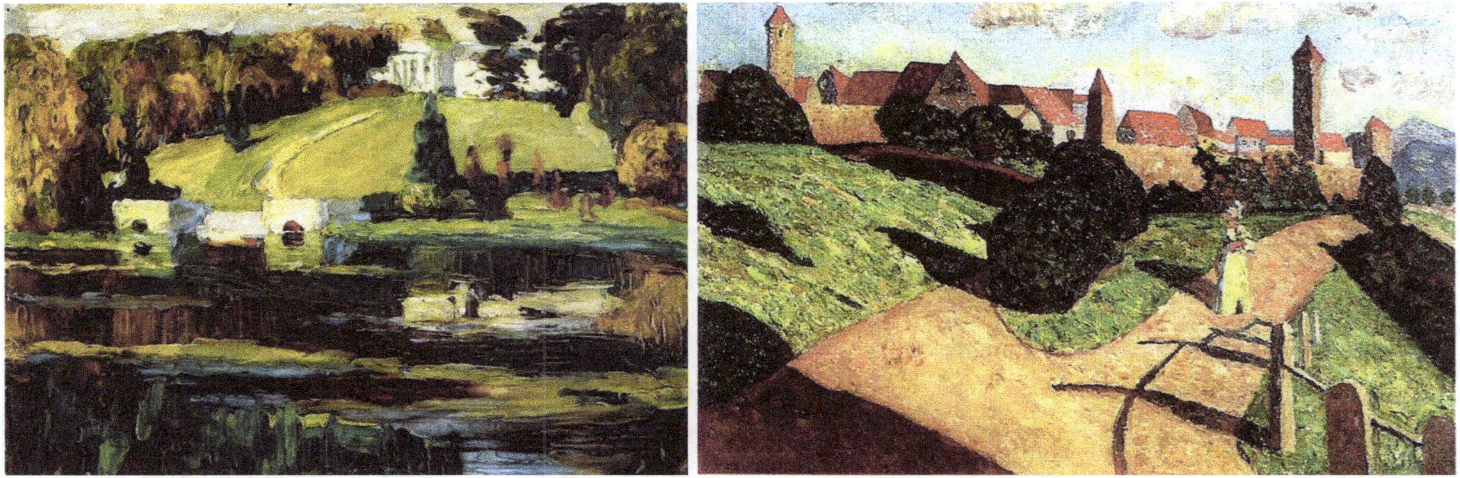

FIGURE 6.1 (left) Wassily Kandinsky, *Sketch for Achryka* (1901), oil and tempera on canvas board, 23.6 × 32.7 cm, Städtische Galerie, Lenbachhaus, Munich; (right) Wassily Kandinsky, *Old Town* (1902), oil and tempera on canvas board 23.6 × 32.7 cm, Georges Pompidou Centre, Paris

Source: public domain

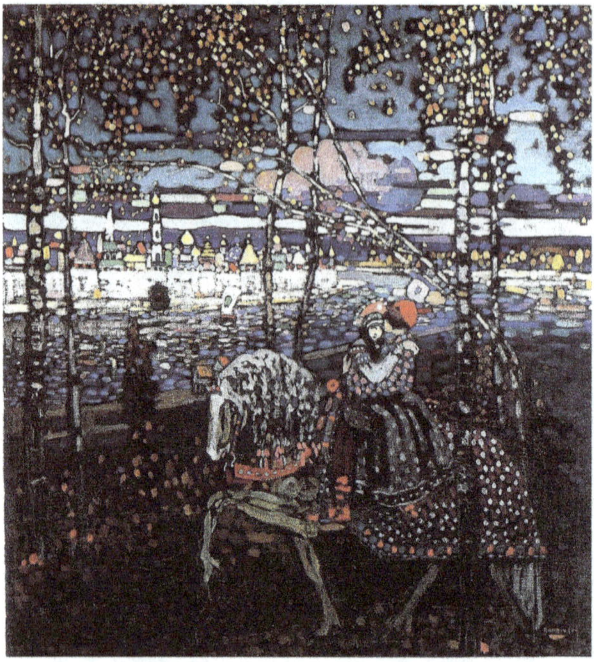

FIGURE 6.2 Wassily Kandinsky, *Riding Couple* (1906–1907), oil on canvas, 55 × 50.5 cm, Städtische Galerie, Lenbachhaus, Munich

Source: public domain

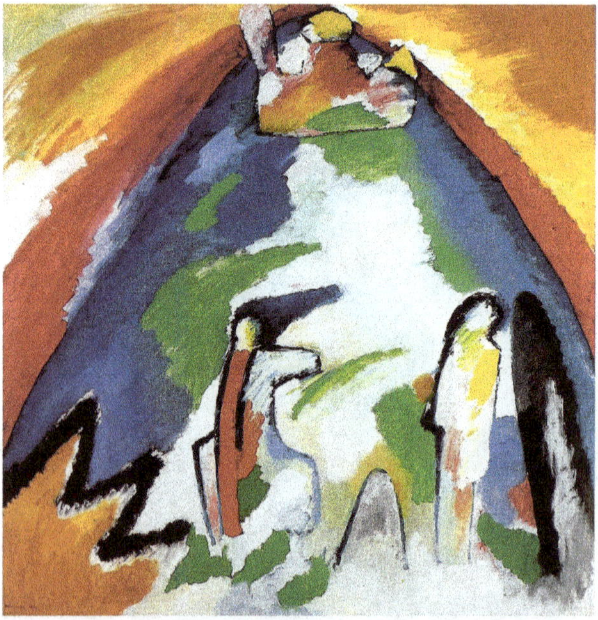

FIGURE 6.3 Wassily Kandinsky, *Mountain* (1909), oil on canvas, 109 × 109 cm, Städtische Galerie, Lenbachhaus, Munich

Source: public domain

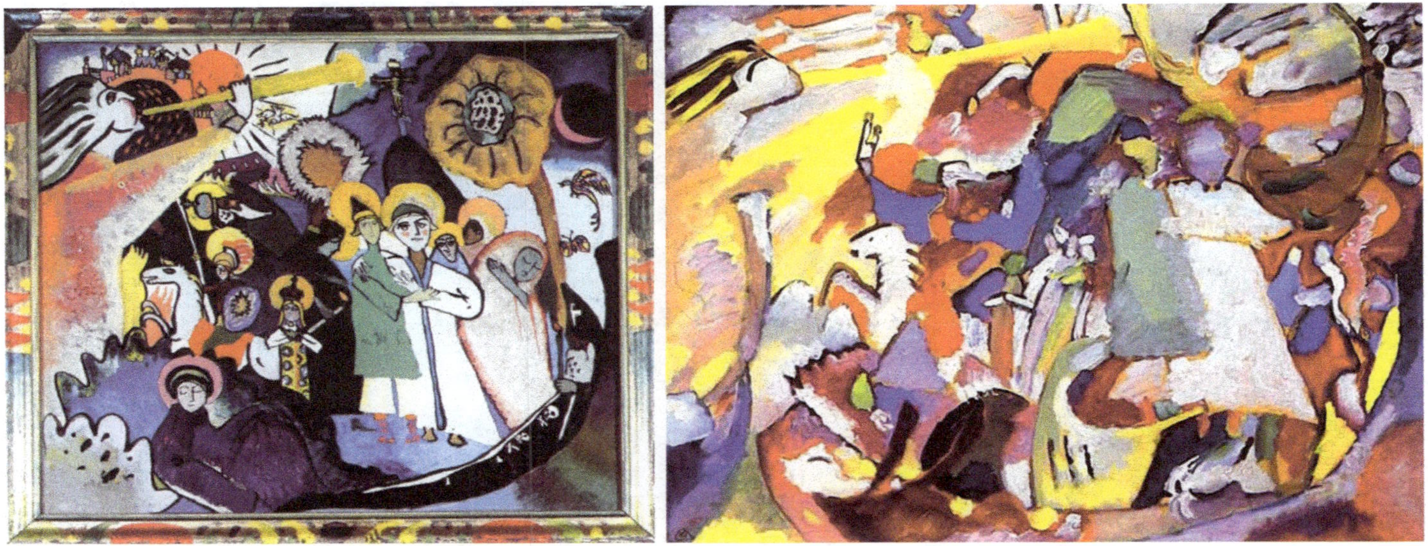

FIGURE 6.4 (left) Wassily Kandinsky, *All Saints I* (1911), reverse glass painting, Städtische Galerie, Lenbachhaus, Munich; (right) Wassily Kandinsky, *All Saints I* (1911), oil on canvas, Städtische Galerie, Lenbachhaus, Munich

Source: public domain

towards non-objective art. In that sense, his trajectory is a research-led one. As art historian and curator Jelena Hahl-Koch concludes (1993:168), for Kandinsky

> the road to abstraction was long and there were many incidental discoveries and backward steps *en route*; it cannot therefore be seen as a straightforward progression towards a single goal. It is more difficult to reduce Kandinsky's progress to a concise formula than that of his fellow artist.

But, by 1911, he and his fellow Russian artists were marching the road to abstraction, and at speed. Why were these dramatic transformations in artistic practice taking place with such speed and cultural force in Russia before the revolution? After all, in 1910 Russia was still a relatively backward country (feudal servitude, for instance, had only been abolished in 1861). However, perhaps because of its "uneven and combined development",[19] characterized by an experience of late but accelerated modernization of its economy, social structure and culture, the country was ripe for political revolution. The new art was eagerly embraced by those who saw it as a harbinger of a vast social transformation pulsating through Russian society after 1905. The immediate catalyst propelling Russia into artistic modernity was the importation of Post-impressionist and Cubist work into a Russian society which lagged behind western European society in cultural terms but had its own distinctive visual traditions not beholden to western perspectival and naturalist art. Artists saw the work of Cezanne, Picasso and Leger and were transfixed. They were, however, not overwhelmed by Cubism, nor did they simply ape its style. Rather, they revisited Russia traditions of icon painting and folk art and drew parallels between the modern and the archaic.[20] They collaborated with a new generation of poets and came into contact with new theories of language and strategies of cultural analysis. This set into motion an investigative dynamic at the heart of the Russian avant-garde project that was quite unique. A new and specifically Russia pictorial and visual discipline emerged in the work of Kandinsky, Tatlin, Goncharova and Malevich, and surely this can be regarded as the prime historical exemplar of a research-led, creative arts practice. As John Milner concludes (1983:86), "poets, painters and musicians were drawn together. From physics, linguistics, history, they sought an art with the broadest application, with no equivalent in the West. Their investigative approach to creative activity was committed to exploration and discovery". We can get some measure of the rapidity of the artistic developments that occurred within the Russian avant-garde during these years before the Revolution by looking at the parallel evolution of Malevich's work as a painter over the 11-year period from 1904 to 1915. Again, I offer this not as an art historical exercise but as an exemplar of a research-led creative practice involving a dialogue with, one the one hand, a range of cognate modernist work and, on the other, a thinking through of the challenge of the new art in a specifically Russian situation.

Malevich's work moves from impressionist-influenced early work, such as the pastoral study *Birkenhain* (Figure 6.5 left), a painting probably executed after he had the

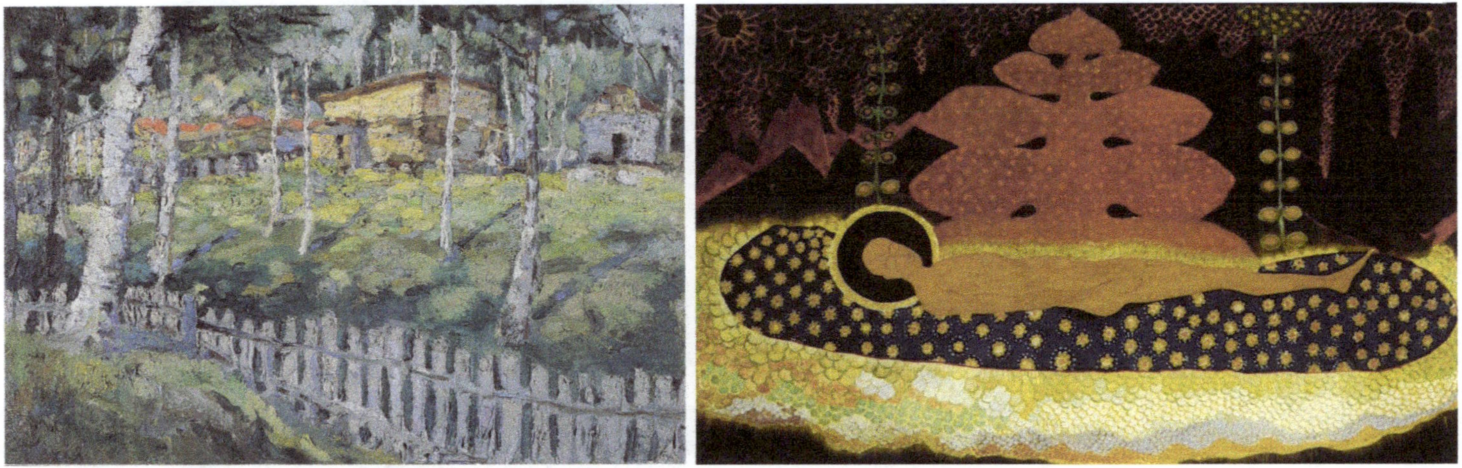

FIGURE 6.5 (left) Kasimir Malevich, *Birkenhain* (1905), oil on canvas, 34.5 × 49.5 cm, Albertine Museum, Vienna; (right) Kasimir Malevich, *Veil* (1908), gouache on cardboard, 61 × 26 cm, Tretyakov Gallery, Moscow

Source: public domain

opportunity in 1904 to view works by Monet and Cezanne in the Moscow collection of Sergei Schukin, to the Symbolist work of 1908 such as *The Shroud* (Figure 6.5 right). This painting is influenced by his interest in Russian iconography and folk art and by religious themes. After exposure to the work of Picasso in the Schukin Collection – which can be glimpsed in a 1914 photograph of his mansion (Figure 6.6) – he then breaks from Symbolism with experiments with the Cubist style of volumetric modelling, probably indebted to Leger's conical and tubular styling, while employing peasant motifs in *Reaper in Red* (Figure 6.7 right) and in a range of other figurative paintings produced after 1911, such as *Taking in the Rye* (Figure 6.7 left).

In these pieces, an evolution is perceptible in terms of his use of colour. The stripes of colour present in *Taking in the Rye* seem to assume a significance distinct from the figuration. Freed from any representational responsibility, they threaten to develop their own forms.

Then, in 1915, comes the dramatic move to pure abstraction, already apparent in his canvas *Painterly Realism* (Figure 6.8).

Suprematism burst upon the scene in December of that year with the exhibition of Malevich's *Black Square* painting in the *Last Futurist Exhibition of Paintings: 0,10.* at the Dobychina's Art Bureau located close to the Hermitage

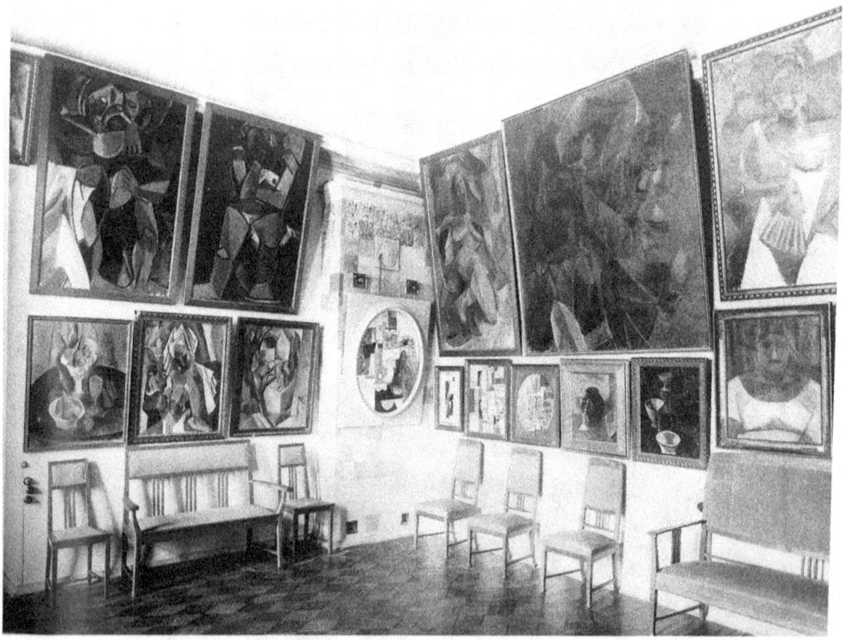

FIGURE 6.6 The Picasso Room in the Sergei Schukin Mansion, Moscow (1914)

Source: public domain

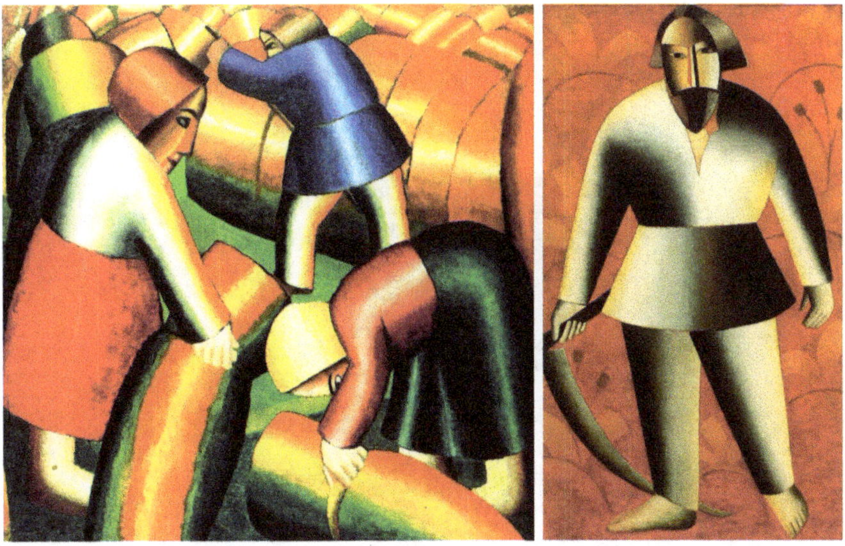

FIGURE 6.7 (left) Kasimir Malevich, *Taking in the Rye* (1912), oil on canvas, 72 × 29 cm, Stedelijk Museum Amsterdam; (right) Kasimir Malevich, *Reaper on Red* (1911), oil on canvas, Nizhny Novgorod State Museum of Fine Arts, Nizhny Novgorod, Russia

Source: public domain

Museum in St Petersburg. His painting (Figure 6.9 left), an unframed single black square on a white surface,[21] was hung in a corner of the gallery's main room (Figure 6.9 right) as the centre of focus of the exhibition.[22] This was undoubtedly a provocative gesture, as the corner location chosen for *Black Square* was the place (the "red corner") where a religious icon would normally be hung in a Russian household together with a votive light. It is hard to know if this move was intended as an act of blasphemy or one of recognition of the contribution of icon painting to the distinctiveness of a Russian art tradition which had never fully adopted western pictorial conventions. Whatever the allusions, the painting represented a rupture. In a sense, the *Black Square* was more than a painting.[23]

Art would begin again and from a zero point. Malevich presented his work as a complete break from the Futurism that had held the Russian avant-garde in its sway. "I transformed myself in the zero of form and emerged from nothing to creation, that is, to Suprematism, to the new realism in painting – to non-objective creation" (Malevich 1915:24). Such manifesto rhetoric was a feature of avant-garde activity before the revolution.[24] However, as Malevich anxiously noted in his 1915 manifesto (ibid.), we "are no longer Futurists, but we still don't know how to define ourselves".

FIGURE 6.8 Kasimir Malevich, *Painterly Realism of a Boy with a Knapsack: Colour Masses in the Fourth Dimension* (1915), oil on canvas, 71 × 44 cm, Museum of Modern Art, New York

Source: public domain

Malevich, like the other Russian artists who matured alongside him in the crucible of political revolution, was not simply responding to changing artistic fashions. Rather, he was striving to critically understand the evolution of modernist painting and to identify the distinctive place of the Russian avant-garde within this.[25] This meant, of course, demonstrating that his own work represented the most advanced position to date in this development. He believed that in order to establish his role as the founder of a radically new direction for art, he also needed to elaborate the philosophical and aesthetical underpinnings of the new style. The new art would be research-led. But the research impetus would be coterminous with a primary programme of innovation within art practice.

Art, he argued, would have to break from its "old habits of mind". In his essay "From Cubism and Futurism to Suprematism: The New Realism in Painting", written in 1916 in the aftermath of the Last Futurist exhibition, Malevich embraced the core anti-naturalist creed that Kandinsky had already endorsed (Bowlt 1988:122):

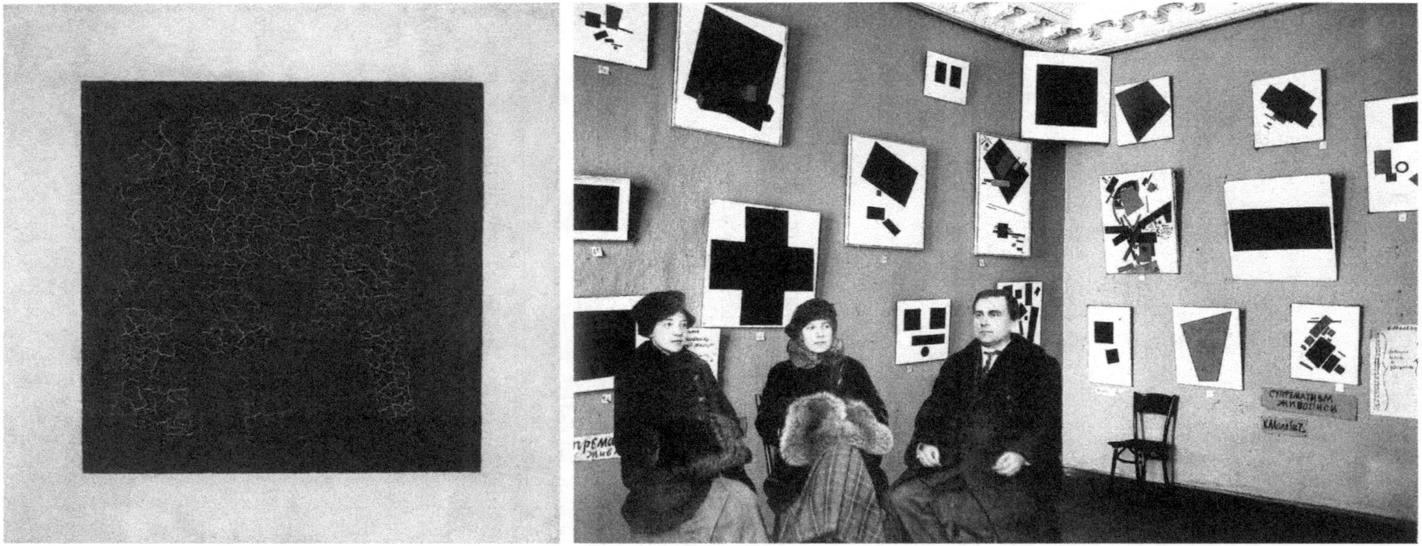

FIGURE 6.9 (left) Kasimir Malevich, *Black Square* (1915), oil on linen, 79.5 × 79.5 cm, State Tretyakov Gallery, Moscow; (right) Last Futurist Exhibition 0.10, St Petersburg (1915), installation of the Black Square, St Petersburg, with Kasimir Malevich photographed seated beside fellow exhibiting artists Olga Rosanova and Kseniya Boguslavskava

Source: public domain

"creation is present in pictures only where there is form which borrows nothing already created in nature, but arises out of the painted masses without repeating and without altering the primary forms of the objects of nature". The goal of the new art was to "make paintings" rather than "painting objects", an approach which restricted the plastic arts to merely repeating nature. As his fellow Suprematist painter, Mikhail Menkov, succinctly put it (Bowlt 1988:114),

> every art that is valued by its ability to repeat the visible is a defective art.
> Colour must live and speak for itself. Hitherto there was no such thing as pure painting; there were just copies of nature and of ideas.

The Cubo-futurist painter Olga Rozanova (seen seated in Figure 6.9 right) had also made the transition from figurative to abstract work. She stoutly defended the avant-garde programme of the Russian artists but recognized the scale of the communicational task involved in reassuring an anxious public about the new art. In her 1913 article "The Bases of the New Creativity and Why It Is Not Understood", written for the journal of the Union of Youth in Moscow (ibid.:106), Rozanova argues that

> the artist only fully consciously creates the painting when he does not copy nature, when he subordinates initial impressions of it to those informed by the whole psychology of modern creative thinking: what the artist sees + what he knows + plus what he understands and so on. The result of this awareness in the execution of a canvas is that through working constructively the artist brings into play the most important single contribution of all – the internal demands of the painting.

On the other hand, she realized that

> for the majority of the public nurtured by pseudo artists on copies of nature, the conception of beauty rests on the terms "Familiar" and "Intelligible". So when an art created on new principles forces the public to awaken from its stagnant, sleepy attitudes crystallized once and for all, the transition to a different state incites protest and hostility since the public are unprepared for it.

Clearly, a mission to explain was needed. And this required a new way of looking at art. Such a programme, if it was to achieve a measure of public acceptability, would, like its illustrious Renaissance predecessor, need an intellectual justification, and this would entail forging a systematic research basis for the new "artistic culture" and communicating this to a new generation of students.

6.4 Theory and practice in the work of Wassily Kandinsky

Kandinsky was probably the first Russian artist to grasp that avant-garde art practice, with its stylistic twists and turns and multi-various movements, manifestos and cultural interventions, had raced ahead of art theory and art historical study. He seemed to sense that it fell to him as the senior figure in the Russian avant-garde, and the one with the broadest intellectual formation and international experience, to address this gap, and he was to do so in his research, writing and teaching over the next decade.

Certainly, he understood that if the new Russian non-figurative art was to gain some measure of public acceptance, its intellectual auspices would also have to be established (and its international profile raised). And that meant not only abstractly invoking the authority of science but also providing an analytical framework for both artist and spectator within which the art could be comprehended. As he was to later argue in *Point and Line to Plane* (1947:17), "the general viewpoint of our day, that it would be dangerous to 'dissect' art since such dissection would inevitably lead to art's abolition, originated in an ignorant under-evaluation of these elements, thus laid bare in their primary strength". Kandinsky accepts that as art moves towards abstraction, it becomes more investigative in its practices – in other words, research-led: "consciously or unconsciously, artists across the arts are gradually concentrating on their own materials, they are investigating and researching them; they are appraising the inner value of those very elements which bid the artist create art". However, one has to dig deep into Kandinsky's *On the Spiritual in Art* and probe beyond its Theosophical mysticism to get a sense of the painter's analytical and scientific aspirations. It is really only in the second part of his book, in the section "About Painting", where he discusses colour and its relation to form and draws upon the scientific knowledge of the day about the colour spectrum and perception, that we get any concrete sense of his proclaimed search for a "grammar of the visual". Slowly, however, he gropes toward a language of visual form (Bowlt and Long 1980:69).

> The language of painting is created from two primary elements – colour and form, from their combination, inter-subordination, inter-gravitation and inter-repulsion. We are now mastering this language syllable by syllable. We are forced to create the individual words in this language.

In so far as he had a model for such a language for painting, it was undoubtedly music with its clear notational systems and evolving body of theory and practice (Kandinsky 1977:9):

> A painter, who finds no satisfaction in mere representation, however artistic, in his longing to express his inner life, cannot but envy the ease with which music, the most non-material of the arts today, achieves this end. He naturally seeks to apply the methods of music to his own art. And from this results that modern desire for rhythm in painting, for mathematical, abstract construction, for repeated notes of colour, for setting colour in motion.

Time and time again Kandinsky, an accomplished musician familiar with the parallel revolution taking place in musical composition around Arnold Schoenberg's experimentation with atonality, returns to the idea of composition as an abstract creative endeavour and to the theme of developing a vocabulary of the visual equivalent to musical notation with its refined capacity to codify pitch, dynamics, rhythm, harmony and counterpoint. The visual arts, if they are to bring a measure of order to the modernist experiments in colour and form, need a set of "hard and fast rules, principles resembling those of the figured bass" (ibid.) (the figured bass or *Generalbass* was a system of musical notation popular in the Baroque that provides a written bass line with chordal information on a numbered basis, facilitating improvised musical activity within the fugue).

Kandinsky expresses his envy of music's notational systems and long-established tradition of theory and aspires to develop a notation for "scoring" basic visual forms and colours and their interplay. His interest in exploring the parallels between modernist musical developments and those in the non-objective visual arts was shared by his contemporary at the Bauhaus, Paul Klee, and many avant-garde painters.

In 1910, such an intellectual basis for the visual arts seemed a long way off. He expresses his frustration (1994:92), saying "it is impossible to establish a very definite theory or to create a mechanical figured-bass for painting". Moreover, he accepted that to seek to do so in a purely formal manner could lead to a situation where theory had no real relationship to actual practice. Accordingly, in *On the Spiritual in Art*, he concludes with a definitive statement on the relation between theory and practice (Bowlt and Long 1980:81), "it is premature to contemplate the intricate details of the theory. Theory never advances before art and will never drag practice behind it". The matter, he feels, can only be resolved via art practice rather than by theoretical speculation. It is through studio practice, improvisation and experiment that artists will gain the confidence to handle the new modalities of colour, line and plane and in time build their experience into a theoretical understanding with applicability right across the spectrum of the creative arts. Indeed, that was the key lesson that all the avant-garde artists had learnt. As arts practitioners, *they* were the ones responsible for the development of their theoretical practice. And while as artists they were open to a range of theoretical discourses and putative methodologies, the problems of theory would be solved in the studio, in the classroom and on the streets. We are still learning this lesson today.

Kandinsky's life had always been one in which theory and practice, research and innovation, writing and teaching were locked in a reverberating conversation. In the period between 1910 and 1930, his studio practice, pedagogic concerns and theoretical work evolved in tandem. This dynamic was at the core of an emergent paradigm of artistic research and is apparent in the work of disparate visual artists: that of the painters Kandinsky and Malevich; in the design and photography of Rodchenko and Moholy-Nagy; and in the film practice of Eisenstein. Like Eisenstein, Kandinsky seems to have possessed a refined capacity to reflect upon and analyse his own artistic practice and then to draw upon such understanding in his work as a teacher of art.

This capacity for analytical reflection is apparent, for example, in his published discussion of certain of his paintings in what he calls his "Supplementary definitions", written in 1913 and 1914 when he was still in Munich. These were statements about individual works published in Germany in his 1913 text *Rückblicke* (Reminiscences). His short, self-critical pieces go beyond what passes for artist statements in much current art, where such writing often seem to have a self-justificatory character and a public relations function with an eye more to market position rather than critical elucidation. Kandinsky, on the other hand, provides for his reader a sense of the artistic processes involved in his work as it moves towards abstraction. In these statements, he gradually feels his way towards a number of general conclusions about the interaction of colour, form and expressivity within abstract art and shares these insights with his readers. Of course, these are questions which he had already been exploring in his book *On the Spiritual in Art*.[26]

For example, he provides us with an insightful description of his 1913 *Painting with White Border (Moscow)*, a work which he tells us expresses his vivid memory of a return to a snowy Moscow in November 1912 (Figure 6.10). The picture "was the outcome of those recent, as always, extremely powerful impressions, I had experienced in Moscow – or more correctly of *Moscow* itself" (Hahl Koch 1993:205).

His "Supplementary definition" for this painting provides a detailed account of the artistic process involved in its composition. He traces the numerous designs, sketches and drawings which informed the various iterations of the painting,

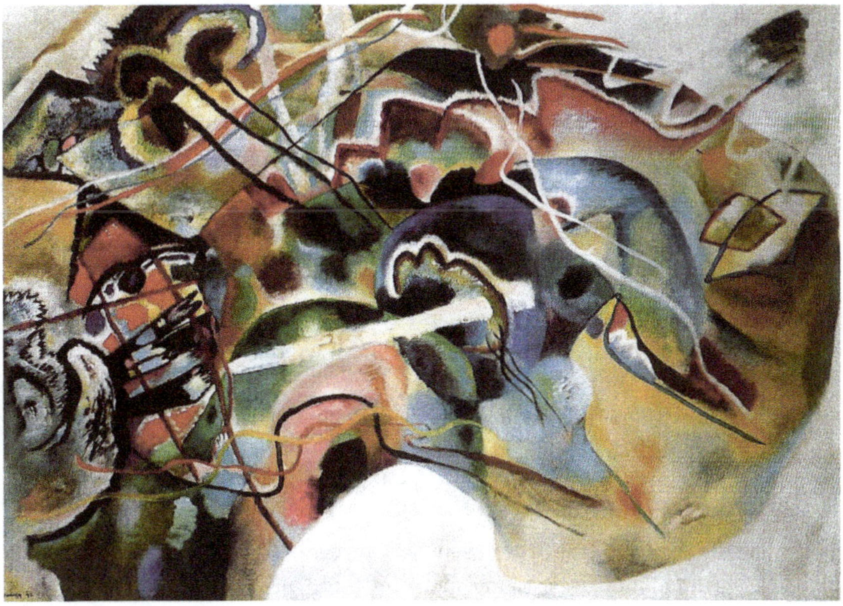

FIGURE 6.10 Wassily Kandinsky, *Painting with White Border (Moscow)* (1913), oil on canvas, 140 × 200 cm, Solomon R Guggenheim Collection, NYC

Source: public domain

analysing the pictorial form and use of line and colour. Kandinsky identifies the first design for the work as being "very concise and restricted". This draft seems to have been based around a number of figurative elements, including a troika or horse-pulled sledge, pictured cutting through the Moscow snow. But as he notes, the picture was soon transformed in a more abstract direction, with figuration giving way to hieroglyph, then to pure abstraction (1977:389):

> In the second design, I succeeded in "dissolving" the colours and forms of the actions taking place in the lower, right hand corner. In the upper left remained the troika motif, the three-horsed sled. This is what I call the three lines curved at the top, which with different variations, run parallel to one another. The lines of the backs of the three horses in a Russian troika led me to adapt this form which I had long harboured within me and which I had already employed in various drawings.

From his account, we get a real sense of an artist struggling with a new visual language. He identifies the basic elements of the composition: the play of movement across the canvas; the masses and forms assembled; the relationship between blurred and contoured lines; the running over of colour beyond the boundaries of form as colour planes function independently of graphic outline. He describes the slow progress in the evolution of the work as he seeks to resolve these pictorial qualities within the piece as it progresses towards more abstract outcomes. This resolution only becomes possible when he focuses on the framing of the piece (389): "after nearly five months that I was sitting looking into the twilight at the second large-scale study, when it suddenly dawned on me what was missing – the white edge". He throws himself into the final version of the work on a fresh canvas, having resolved the size of the work, and begins to treat the white edge as "the solution to the picture", subsequently naming the picture after it (ibid.:391).

> I treated this white edge in the same capricious way it had treated me: in the lower left a chasm, out of which a white wave that suddenly subsides, only to flow around the right-hand side of the picture in lazy coils, forming in the upper right a lake (where the black bubbling comes about) disappearing towards the upper left-hand corner, where it makes its last definitive appearance in the picture in the form of a white zigzag.

His supplementary definitions are exemplary pieces of richly descriptive, practice-based art writing. They indicate the source of some of the forms employed in the painting but report his striving to challenge painting's dependence on descriptive graphics and a naturalist use of colour. This is not just an exercise in casual reminiscence by an artist looking back at his life. Rather, in communicating to his audience the pictorial approach employed in a particular work, his discourse seeks to mobilize an analytical language and explore a range of expressive possibilities already sketched in his theoretical writing.[27]

However, he also warns of the dangers of over-interpretation of the abstract imagery he is now working with – particularly with regard to the temptation to provide either thematic, symbolist or literary "readings" of the paintings. It is the expressive potency of form and colour as painterly means that interests him, not their role as mystical hieroglyphs. Any student of painting concerned with a reflective analysis of their art process and product would, I suspect, still find his supplementary definitions a useful guide to their own research endeavours, as they would his strictures about over-theorization of work.

One of the problems that research students working on creative visual arts projects seem to face today is having an agreed descriptive language of process and of visual form available to them to discuss their own work. In the past, this was largely provided within art colleges by the discipline of art history. For film students, it was provided in traditions of frame and shot analysis, which facilitated detailed description of the visual character of a particular film, as compositional analysis did for photography students discussing their images. The last decade has seen the increasing marginalization of art history within the contextual studies element of the art school curriculum (and in photography and film courses). The discipline had traditionally provided much of the visual language and interpretative methodology for artistic research. With its diminished role within arts pedagogy, something has been lost in terms of students' capacity to discuss their work in a shared language of the visual.

Moreover, history of art has now morphed into a broader study of visual culture and representation – as has film studies to some extent. Both fields employ complex theories of signification, deconstruction and spectatorship.[28] They are more concerned with the strategies of representation at work in a "text" than with the creative construction, expressivity and authorial voice present in the artwork. Unfortunately, this powerful set of analytical methods for "reading" the artwork has often driven a wedge between the cultural theorist and artist–film-maker, who find their authority to talk about their own work and its expressive intent diminished by this discourse. The "linguistic turn" in cultural studies has downplayed the role of artistic creativity and individual style and authorship and, indeed, as Barbara Stafford has cogently argued, marginalized a consideration of the material, visual and sensate dimensions of art. We have failed to attend to the haptic and optic specificity of artistic languages.[29] The image is not simply reducible to the semiotic structures of textual discourse. Nor are linguistic methods of analysis adequate to understanding the contemporary artwork in all its diversity, as event, performance and proposition.

Interestingly, the danger of literary reductionism within art criticism was a theme pursued within the avant-garde after 1917 as it sought to develop new bodies of theory to engage with non-objective art. For example, the art historian and founding member of INKhUK, Nikolai Punin, in a lecture given as part of a crash course for student teachers of art, warns his audience (Bowlt 1988:173) that

> critics are essentially literary people. Generally speaking, the critic throughout art history has been an artist or a writer *manqué*. But unfortunately, in the fields of painting, sculpture and often architecture, critics are normally not artist *manqués*, but *literati manqués*. They introduce into their appreciation of works of art the sum total of their literary convictions that one way or another they did not manage to realize in works of literature. Hence, we can understand that works of art that contain some literary elements or pseudo literature in general are glorified most by critics, since these are the works that are, above all, intelligible to these literary *manqués*.

Kandinsky, searching for a new form of art criticism, does regard art as a language of sorts. However, for him, as we have seen, it is more of a musical than a natural language such as a linguist might study. He emphasizes the expressive and communicative dimensions of art as a language rather than focusing on the formal linguistic features of language as a system of signification. Like Malevich, Kandinsky after 1913 was becoming familiar with the formalist theories of language being advanced in Russia at this time. However, his interest is more in the experiments in poetic language being conducted by the Zaum poets[30] than by any grand theory of signification which might accommodate art within a symbolic system (Kandinsky himself wrote and published poetry, as did Malevich). Despite his familiarity with the formalist analysis of language and symbolic form deployed in the study of literature in Russia at this time, Kandinsky's analysis has no explicit semiotic dimension. Saussure's work was as yet unknown in Russia, as was that of the American pragmatist and semiotician Charles Saunders Peirce. The semiotic work of Roman Jakobsen and dialogical framework ofVolisinov and Bakhtin lay in the future.

6.5 Art as expression

As Kandinsky reminded his readers, abstract art was not painting by numbers. Abstract geometrical art was not new and had existed for centuries, for example, in Persian and Islamic art. Like his contemporaries Malevich and Tatlin, he was also mindful of the various traditions of Russian icon painting, which had never operated within Renaissance codes of perspective, nor adhered to a naturalist use of colour and a realist rendering of the image. Moreover, if non-objective art was just an analytical visual code of signs capable of endless mathematical combinations, it could degenerate into "mere geometric decoration, something like a necktie or a carpet" (Bowlt and Long 1980:93). At this time Kandinsky found it hard to envisage the sort of design applications of the new visual language later proposed by the Suprematists and exploited by the Constructivists and Productivists. So, for Kandinsky, the analysis of non-objective art facilitated self-clarification on the part of an artist of the visual means at their disposal. However, his fundamental concern is with the synthesis by the artist of these visual elements in order to achieve expressive force and communicational impact for their painting. As he argues with regard to the activity of composition (1994:81),

> feeling is everything, especially at the beginning. Only through feeling can artistic truth be attained. A loose construction is possible in purely theoretical terms, but that extra ingredient (the real soul of creativity and, therefore, relatively speaking, its essence) can never be created by theory. (...) True results can never be attained through cerebral activity or through deductive calculation – even with the most exact dimensions or the most accurate balances.

Kandinsky's concern as a pedagogue is with developing both the analytical and communicative awareness of the artist. His theorizing seeks to foreground the sensate, affective and communicative dimension of art, a position which stands in stark contrast to the cognitive, deconstructionist and culturally relativist ethos of contemporary art criticism. Kandinsky in his writing invokes notions like the soul and the spirit, terms that in a more materialistic age we are now uneasy with. He seems to regard artistic language as universal in form and not culturally bound (although interestingly he acknowledges that musical theory is). Today, most of us would dispute this universalism.

After he moved to the teach at the Bauhaus, Kandinsky increasingly drew upon Gestalt psychology to provide an account of perception and cognition to undergird his theory of artistic communication. This move encouraged a line of research enquiry that was to eventually to see important interactions between art theory and psychological research (Vygotsky 1925/1971; Gombrich 1960/2002; Arnheim 1966).

Yet long before his exposure to Gestalt psychology, Kandinsky was groping his way towards a distinctive theory of artistic communication. In a catalogue essay written as early as 1910 for an exhibition in Odessa involving the work of a number of Munich artists who would later form the *Blau Reiter* group, Kandinsky provides us with a sketch of a model which acknowledges the centrality of expression to art practice (1994:87):

> A work of art consists of two elements, the inner and the outer. The inner is the emotion of the soul of the artist; this emotion has the capacity to evoke a similar emotion in the observer.
>
> Being connected to the body, the soul is affected through the medium of the senses – the felt. Emotions are aroused and stirred by what is sensed. Thus, the sense is the bridge, i.e. the physical relation between the immaterial (which is the artist's emotion) and the material, which results in a work of art. And again, what is sensed is the bridge from the material (the artist and his work) to the immaterial (the emotion in the soul of the observer).

Kandinsky traces the communication flow at work here: "the sequence is: emotion (in the artist) > the sensed > the artwork > the sensed > emotion (in the observer)" (ibid.). This model may have been elaborated to understand the German Expressionist work of art and its reception. With its concern with physical resonance and tonal expression, it may also be derived from similar models employed in understanding musical communication (as we have seen, Kandinsky often looked to music as a template for the visual arts). Indeed, it may be tempting to see his model as anticipating later attempts to model communication as a message flow between a sender and a receiver. Certainly, in a later piece of writing produced in Paris in 1935, Kandinsky (1994:773) uses terms which seem to anticipate such a Shannon–Weaver type communicational model, talking as he does about the artist as a "transmitter" of meaning and affect and the spectator as "the receiver" of messages.

However, his concern is not primarily with the cognitive dimension of this communication, as is the case with classic information flow models, but rather with the role of feeling and indeed spiritual communion in facilitating the nexus between artist and viewer. As he argues, neither slavish adherence to a calculus of visual forms, nor a purely rational appropriation of the work by the spectator, can hope to facilitate an adequate artistic experience (ibid.:773):

> just as all the recipes in the world can never succeed in creating a work of art by themselves, so too they can never supplant the kind of feeling that is virtually necessary in order to "understand art" – one's head is not a bad device but an "unfeeling" head is worse than a "headless" feeling in art.

This sort of modelling of artistic communication has been largely marginalized within contemporary visual arts theory with its predominantly semiotic concerns. Contemporary critical theory, as we have seen, is primarily concerned with issues of representation and discourse rather than with modes of artistic expression and communication. In his stubborn cleaving to an account of art rooted in the experience of the artist, Kandinsky anticipates our current concerns with the limitations of such semiotic and culturalist approaches to visual art.

Kandinsky envisages the modality of communication between artist and viewer as sensate and immediate and not bound by the conventions of a natural language. It is, in other words, envisaged in expressive and sonic terms (communication via vibratory resonance rather than via a process of signification linking speaker and listener in a chain of mediated meanings). While the encounter between the artwork and its audience is considered as dialogical in form, the artist is afforded a primacy in the generation of meaning and affect. So, clearly, there is no acknowledgement of the "death of the author" here.

For Kandinsky, the artist mobilizes a visual language which is possible to analytically describe and thus to learn and master in some detail. However, in the communicative nexus with their audience, the artist is primarily setting in motion a chain of affect and response rather than navigating a discursive terrain. Similarly, the spectator in their reaction to the artwork viewed responds to this expression of affect through the activation of their own feelings rather than coolly deconstructing the work as text. Much of Kandinsky's research, both in INKhUK and at the Bauhaus, was concerned with the psychology of artistic communication, particularly the effect of the employment of colour by the artist.[31] The embodied artwork is the sensate medium of communication rather than any linguistic utterance that might accompany it. Nor is the work itself held to primarily constitute a semiotic structure. Rather, it is an event.

It is perhaps more appropriate to see Kandinsky's model of expressive communication as more akin to recent attempts within an anthropology and phenomenology of the senses (Howes 1991; Classen 1993; Le Breton 2017) to acknowledge the role of the haptic and expressive in social interaction and cultural exchange and to foreground the sensory interface as the foundation of all human communication, including art making. This, of course, is a methodological move which has brought parts of the social sciences closer to arts practice.[32]

Malevich, who as we have seen seemed to compete with Kandinsky for intellectual authority and leadership of the Russian avant-garde, shared Kandinsky's research interest in the relationship between colour and form and how each of these combine to impact viewer experience. However, he took what was probably a more realistic position about the contribution of science to these investigations, concluding (1976:127) that

> the problems and aims of optical science are quite different from those of art and painting. It is one matter to examine the form of an object and its deformation on the optical level, and another to perceive phenomena on the artistic level; the artist may and, indeed should take an interest in the achievements of science in the field of colour and form, but he must be careful. In any case the exploitation of the results of scientific analysis concerning the interrelations between colour and form will depend on its quality and nature, as well as the artist's ability to make use of the latter in his work without the work reminding us of a column from a scientific paper.

This is a problematic that practice-based research is still wrestling with. Today we seek to distinguish between the knowledge outcome produced through a programme of methodologically guided artistic research and the actual artwork produced. We are inclined to believe that the latter has to possess an integrity and intrinsic value within the field of art. Practice-based arts research involves engagement with both the knowledge and the art object. The live art project is the crucible of the research investigation but has its own creative and professional dynamic. On the other hand, its development, production, reception and evaluation involve processes of critical interlocution and scientific documentation which can facilitate a knowledge outcome. All of these methodological problems of combining theory and practice, analysis and synthesis, art and science, are present in the work of the avant-garde artist–theorist–pedagogues we explore in this text. So, attention to their work might well assist us in clarifying the issues that arise within artistic research.

We can then talk about an *expressive* model at work in Kandinsky's theorization of the art process. It might be useful to compare and contrast this with the *representational* one which currently dominates theories of art and visual culture and with which art students are expected to wrestle in order to provide a theoretical contextualization of their practice. I suspect that Kandinsky's model still underlies the conceptions that many creative artists – painters, photographers, film-makers, composers, writers, choreographer and dancers – have of their work. The expressive possibility is what motivates them.

The expressive model sketched by Kandinsky differs in significant ways from the sort of contemporary "theory" dominating arts research, which now focuses primarily on questions of representation and discourse. We need to grasp these differences if artistic research is to develop a body of theory today appropriate to its field of practice. The evidence is that at present we have a body of theory that often seems hostile to practice.

The contrast between the two models can be summarized in tabular form (Table 6.1).

It is hard to know what Kandinsky, or indeed any his fellow avant-garde, would make of contemporary "theory". In general, Kandinsky, like Malevich, was fairly

TABLE 6.1 The distinctive features and foci of an expressive theory of art practice in comparison with a contemporary theory of artistic discourse

Expressive theory of art practice	*Contemporary theory of artistic discourse*
Construction as the core art activity of the practitioner	Deconstruction as core activity of the critic–theorist
Integrity, "inner necessity" of the artwork as the primary concern of the artist	Artwork seen as a porous and polysemous text, in part constituted by the "reader"
Formal language of artistic means	Art as a semiotic system
Exercise of plastic imagination as core value	Exercise of effective textual strategy as core value
Primacy of free association of images	Primacy of play of signifiers
Artwork as autonomous structure	Artwork as set of open-ended discursive relations
Artwork as intuition and expression	Artwork as signification and coded product
Communication of affect	Mobilization and interrogation of meaning
Search for truth	Production of truth
Attending to a work	Reading a text
Primacy of the sensate	Prison house of language
Creative sovereignty of the artist	Death of author/artist as subject of the work
Voice and intention of the artist central to the meaning of the work	Interpretative work of the reader central to the meaning of the work
Producerly composition	Readerly text
Sacred/transcendental	Profane/immanent
Theory of practice	Practice of theory

contemptuous of critics who did not produce work themselves and live with the existential risks this entails. As he argued, in artistic research, practice must lead theory. Any attempt today to ground creative arts research either in the sureties of scientific method or indeed in critical sophistry could be a fruitless one. For, as Kandinsky would have argued, it is the theory of practice developed among the creative uncertainties and risks of the studio and field that is likely to make the telling contribution to artistic research and creative achievement.

6.6 Founding texts and axioms of modernist artistic research

Kandinsky constantly redrafts his treatise *On the Spiritual in Art* over the next 15 years before publishing *From Point and Line to Plane* in 1925 in the series of texts published by the Bauhaus aimed at establishing the research and pedagogic basis of the new art (the list of publications can be seen in Figure 6.11 left). In this redrafting, he is able to draw upon his accumulated pedagogic and indeed research experience. His theoretical work becomes more grounded as it seeks to provide a formal analysis of the new art language and the new creative strategies of abstraction which might be of some use to teachers and students grappling with the new visual sensibilities and design challenges. These two treatises effectively book-end this unique period of avant-garde experimentation. Sixteen years of unparalleled innovation and social and political engagement – an unfinished project in Kandinsky's case – eventually overwhelmed by the rise of fascism.

We can, then, list the key works that found a modernist tradition of artistic research: Kandinsky's theoretical prospectus for abstract art, *On the Spiritual in Art*, and his "mature" Bauhaus text, *From Point and Line to Plane*; Paul Klee's *Pedagogical Sketchbooks*, drafted like Kandinsky's treatise to encapsulate his teaching on the Basic Course at the Bauhaus in Weimar and Dessau,[33] but also to report on his accumulated insight into the new visual language of modernist design combining "exactitude and intuition"; the Bauhaus-published texts of Piet Mondrian and Theo van Doesburg on visual design; Malevich's 1916 Suprematist manifesto, *From Cubism and Futurism to Suprematism. A New Painterly Realism*, plus his curriculum prospectus for the UNOVIS project in the Vitebsk Technical School of Art, later adapted in Petrograd, and his 1927 publication for the Bauhaus, *Non-Objective World*, summarizing the recent history of modern art read from the perspective of a painter–practitioner; Moholy-Nagy's 1925 Bauhaus text, *Painting Photography Film*, exploring a range of lens-based media beyond easel painting and thereby extending fine art practice in significant ways (1969); Sergei Eisenstein's writing on film aesthetics written while teaching at the Soviet Film School GIK and later gathered together in collections like *Film Sense* and *Film Form*;[34] and Johanes Itten's *The Art of Color: The Subjective Experience and Objective Rationale of Color*, first drafted during his time at the Bauhaus and eventually published in English in 1961 in the US.

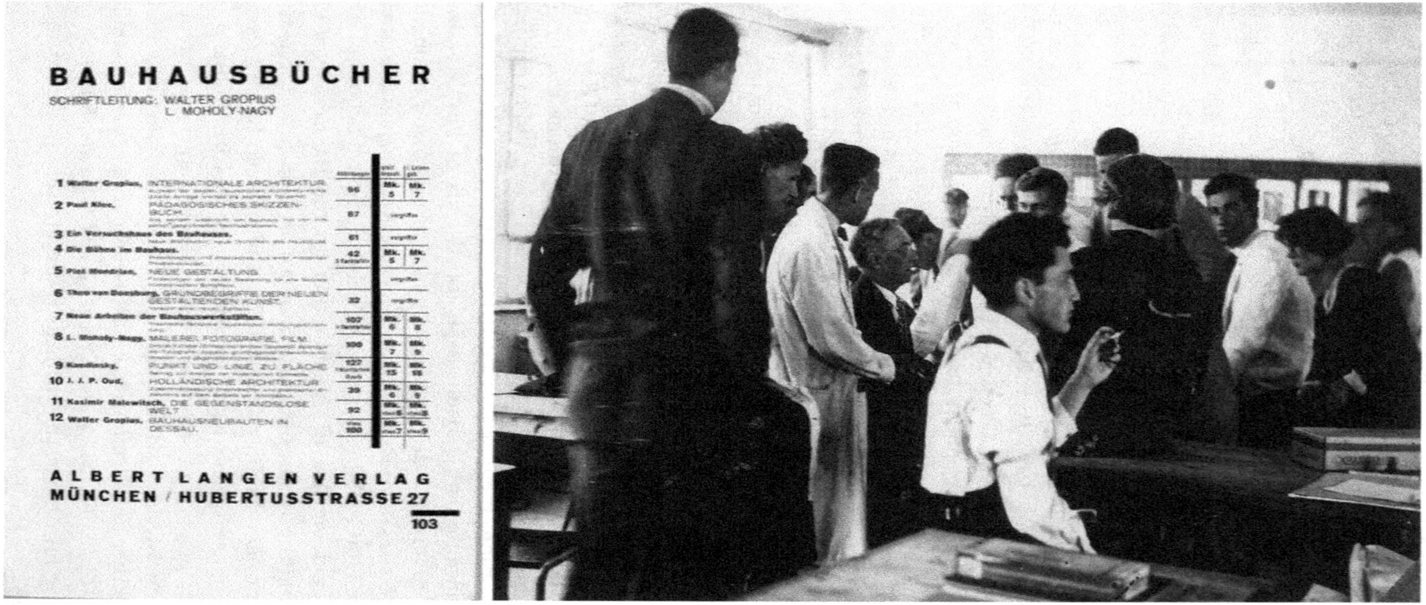

FIGURE 6.11 (left) List of books published in the Bauhaus Books series from Malevich's *Non-Objective World*, Bibliotheque Nationale, Paris; (right) Wassily Kandinsky surrounded by students in the Bauhaus (1928), Bibliotheque Nationale, Paris

Source: public domain

Art pedagogy was core to this critical research trajectory. The modernist pedagogic initiative found its institutional form in the restructuring of art and design education in Russia after the 1917 revolution, in particular in the State Higher Technical Workshops (SVOMAS), subsequently renamed VKhUTEMAS, and in the parallel foundation of the Bauhaus Art Academy in Germany in 1919. In this new type of art and design academy, teaching was to be done on a cross-disciplinary basis resting on a core foundational programme of teaching and research-led project work. This Basic Course introduced students to a new language of abstraction and to an experimental approach to the use of materials and media. This acquired facility with abstract visual form was then to have the widest possible application in advancing a design practice appropriate for an industrial society across a range of arts disciplines – painting and sculpture, but also theatre, performance and film, in what we now call product design, and, of course, architecture.

In the restructured art institutions, programmes of studio research and pedagogic innovation were developed hand in hand, bringing theory and practice into a productive alignment. All of this was done in a situation of material constraint and of political volatility. A massive programme of experiment in education and culture was conducted within a revolutionary social experiment. Indeed, we can say that the research and theoretical activities of the avant-garde emerged as much out their experience of the classroom and the studio (and from the self-organized research and development activities of the students of those institutions) as from any set of philosophical or methodological precepts.

Of course, it is largely the pedagogic programmes of the Bauhaus and the Russian VKhUTEMAS that we remember today. Indeed, the theory and research programmes developed by modernist pioneers like Kandinsky, Klee, Moholy-Nagy, Malevich, Tatlin, Lizzitsky and Rodchenko went on to influence art college pedagogy throughout the world in the period after World War II, and traces of this thinking are still to be found in many of the foundational and basic courses offered to arts students today. However, the theoretical writing and research programmes instigated by these avant-garde artists have largely been ignored.[35] Postmodern critics have been resistant to the theoretical overtures of any avant-garde, not least one so closely associated with a revolutionary political situation.

There has been a tendency within art historical study to read *From Point and Line to Plane*, and indeed the other theoretical works produced by the Bauhaus masters and by their Russian equivalents, as purely pedagogic works and to evaluate their intellectual worth purely on this basis. My argument is that the pedagogic milieu was certainly the context within which these artist–theorists honed and systematized their theoretical insights. But these had been gained through artistic practice, experiment and grounded analysis. In works like Kandinsky's *Point and Line to Plane* or in Paul Klee's *Pedagogical Sketchbooks*, we find a theoretical summarizing and communication of this experience representing a high point of artistic research *per se*. These artists strove to formalize the emerging visual language of abstraction. In a sense, to abstract was to theorize. However, the concrete goal became to transmit the fruits of this research and experience to a new generation of artists who, in turn, would be encouraged to explore the design applications of this new sensibility.

We can see a practical application of this in Kandinsky's approach to teaching the introductory drawing course that he provided at the Bauhaus in Dessau. It was conceived not as an introduction to life or nature drawing but to encourage observational skills in students and to develop their analytical capacities and ability to use abstracted visual forms in their work. As Kandinsky noted in a 1928 paper, "Analytical Drawing" (1994:729),

> the teaching of drawing at the Bauhaus is an education in looking, precise observation, and the precise representation not of the external appearances of an object but of constructive elements, the laws that govern the forces (= tensions) that can be discovered in given objects and of their logical construction – an education in clearly observing and clearly reproducing relationships, where two-dimensional phenomena are an introductory step to the three-dimensional.

Kandinsky gives us an example of such drawing exercises in *From Point and Line to Plane* in his graphic rendering of a series of movements by the German modernist dancer Palucca. Working from a photographic image, the basic visual form/gesture is abstracted (see Figure 6.12). The task as Kandinsky described it was "to subordinate the whole complex to one simple overall form, thereby facilitating the representation of the whole construction by means of the most concise possible schema" (ibid.:728).

The analytical and visual skills and insights acquired by the student then can be utilized across a wide range of research and visual development tasks and applied design activities. However, in pedagogic and artistic terms, the student's lived experience remains primary. Reflecting on his experiences of teaching at the Bauhaus in a 1929 essay on the "Psychology of the Creative Personality" (1994:740), Kandinsky summarizes the situation.

> I demand of my students that they think very precisely, that they execute purely cerebral exercises with exactitude; we even discuss the works they submit in purely theoretical terms. But I always emphasise with particular force that this theoretical path . . . is only a way of arriving at the "content," which is why I place particular value upon the lived experience of "tensions". Theory is (especially today) indispensable and fruitful. But woe betide him who would create a "work" in this way alone.

Certainly nothing comparable has emerged in art education and research since then, either in terms of empirical reach or seriousness of intellectual purpose. Nothing comes remotely close to this sustained attempt to integrate theoretical insight with the dynamics of practice. And it is for this reason that I would suggest that we might all profit from the closer study of this corpus of thinking and practice.

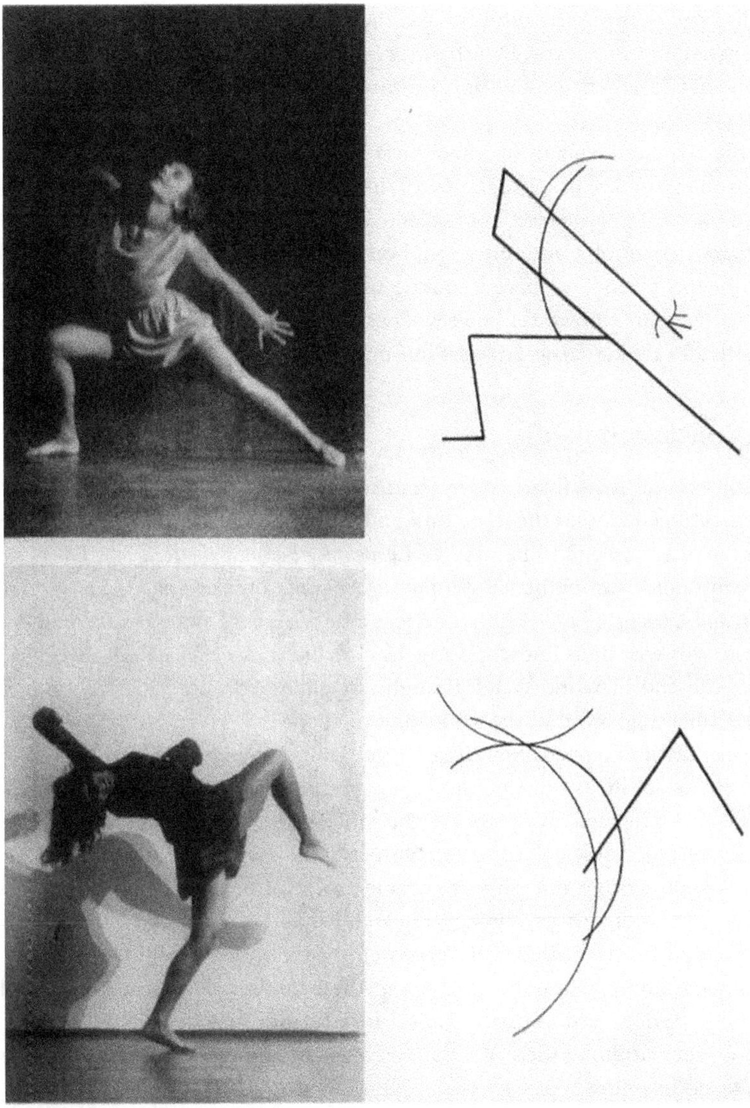

FIGURE 6.12 Photograph of the movement of the dancer Pelucca by Charlotte Rudoph (1926) and Kandinsky's abstracted drawing of this from *Point and Line to Plane*, Bibliotheque National, Paris

Source: public domain

We shouldn't over-rationalize these works. Nor romanticize them. They are of their time. They combine analytical concerns with an acknowledgement of the expressive and intuitive dimensions of art making. That is their strength. Sadly, this balance is largely absent in contemporary artistic research, which rather keeps away from any theory of emotions or of expressive effects. However, Kandinsky, like

Klee, strives to bring the intuitive and the analytical aspects of art into alignment, acknowledging this involves "a balance of creative forces, which may be divided up under two schematic headings – intuition and calculation: 'practical' science" (1994:535).

This is the constant in his theoretical work; art is a practical science in which analytical clarity has to be aligned with life experience and a will to communicate with an audience. However, non-objective art, he believed, was uniquely placed to facilitate a range of communication, both cognitive and expressive with its audience. In this period, we have a unique situation where art theory and innovative practice were in a powerful synergy. Theory was tested via programmes of artistic research and given practical import in a major pedagogical and cultural experiment.

6.7 Conclusion

The theoretical texts listed above seem to be equivalent to those canonical texts which emerged around the same time in disciplines like sociology and psychology, works such as Emile Durkheim's *The Elementary Forms of the Religious Life* (1912) or Wilhem Wundt's monumental *An Investigation into Developmental Laws of Language, Myth, and Conduct* (1900–1920). The text that is perhaps closest in form and sensibility to this writing is Ludwig Wittgenstein's *Tractatus Philosophicus*, a work drafted before the end of World War I, though not published until 1922, and which has more than a suggestion of the art manifesto about it.[36]

Every discipline needs founding figures and to identify its key texts and works, for these establish its core methods and perspectives and advance its methodological distinctiveness. Why would this not also be the case in artistic science? If practice-based research is to establish itself as a viable academic field, then it needs, like any other discipline, to attend to its historical origins and theoretical auspices. Sociology or psychology research students are expected to know about the foundation and historical development of their discipline. Indeed, this is part of their training as researchers. They cut their methodological teeth by studying the methods deployed in the canonical investigations in their discipline and then adopting or critiquing these. Why are creative art and media students not encouraged to do likewise?

In the work of the revolutionary avant-garde, we seem to have an originating moment in the creation of a research culture for the creative arts. Many of the key ideas of contemporary art – the idea of an expanded and public art practice, that of the studio and gallery as a laboratory, of art process as investigation which can be documented, of art's open engagement with science, the primacy of concept over medium and interdisciplinarity over craft – were in effect pioneered by the avant-garde during this period. This was a project riven with contradictions, which attempted to reconcile the competing claims of the rational and the intuitive; the analytic and synthetic; the critical and the creative; the autonomous and the socially engaged; the material and the spiritual. Most of these antinomies are not far from the surface of contemporary arts education.

Notes

1 Of course, the modernist engagement with science and the research process is of a different character to that found in the Renaissance (although some Russians greeted the early twentieth-century flowering of avant-garde art as a new renaissance and its leading figures were well aware of the parallels with the Florentine Quattrocento renaissance).
2 The philosopher Wilhem Dilthey (1833–1911) argued that while the study of man could be given a scientific form, this would have to be attentive to human creativity and meaning-making and attend to the role of culture in shaping man. He sought to distinguish between the positive methods of the natural sciences and the cultural approach of the human sciences, seeking to elaborate the philosophical distinctiveness of the latter.
3 That is not to say that artists were not interested in matters scientific prior to the twentieth century. Certainly, in the last decades of the nineteenth century, the Pointilists, Seurat and Signac and French post-impressionist painting, more generally, were greatly interested in the physics of colour and the psychology of perception.
4 As Noemi Smolik (2006) reminds us, both the notions of a "renaissance" and indeed a "resurrection" in Russian culture were invoked by intellectuals and artists in the country after 1905 as a broadly based movement of cultural renewal that sought to engage with modernity through specifically Russian cultural traditions and sensibilities.
5 Vladimir Tatlin (1885–1953), like Kandinsky and Malevich, enthusiastically embraced French and German modernism, but also sought to also address the distinctiveness of Russian visual traditions in his work. Like Kandinsky and Malevich, he embraced the revolution and found himself playing a leading role in the administration of the arts sector after 1917. While not a prolific writer and theorist like the other two figures, his work has a clear investigative basis in so far as it foregrounds work on materials and constructive principles. As such, he plays a key role in the development of a Constructivist model of art and design research and development.
6 Rodchenko and Stepanova also made the creative journey from figurative to abstract painting and onwards to Constructivist design activity, playing a significant role as arts administrators, curators, researchers and pedagogues while advancing a laboratory model of design research and development.
7 Rozanova's work developed in parallel to that of Malevich, passing quickly through Realist, Impressionist, Cubist and Futurist stages before arriving at Suprematism. She worked in collage, linocut and other print forms, as well as with easel painting, and developed an interest in graphic and book design and later industrial design. She played an active part in the art institutions restructured after the October Revolution, drafting a plan for establishing museums of industrial art and design across Russia. She died of diphtheria, having exhausted herself working on the decoration of Moscow for the first anniversary of the revolution.
8 There is not a significant literature on this topic, though Rainer Wicks and Gabriele Graw's *Teaching at the Bauhaus* (2001) provides a substantial account of Kandinsky's pedagogic activities in the Bauhaus. Christine Lodder, in her *Russian Constructivism* (1983), gives us a useful account of the research and pedagogic developments in Russia. The publication of the collected writings of Kasimir Malevich (edited by Troels Andersen) and of Kandinsky (edited by Lindsay and Vergo) provides access to the thinking of these two major figures on matters scientific and pedagogical.
9 Kandinsky, like Paul Klee, Piet Mondrian, Malevich, Tatlin and a number of other avant-garde artists, was significantly influenced by Theosophy, and in particular, the thinking of Rudolf Steiner. At the core of Steiner's thought was an attempt to plot a synthesis between science and spirituality. That was its attraction to avant-garde artists. For Kandinsky, the argument that science and spirituality could be in fruitful unison had considerable appeal, not least because it was envisaged that art would facilitate such a synergy.
10 Kandinsky was one of the few avant-garde artists who had any practical experience of the social sciences, having studied law and political economy and conducted ethnographic research into peasant folk culture in northern Russia (see note 16).

11 El Lizzitsky, for example, had a background in engineering and was well versed in mathematical methods, as was his fellow Constructivist Nuam Gabo, the sculptor. In a pamphlet in 1924, he sought to trace the parallel transitions between the modelling of space in mathematics and in painting.
12 While women did play an important role in the avant-garde movements in Russia and Germany, historians of art have been slow to acknowledge the contribution of figures like Natalya Goncharova (1881–1962), Olga Rozanova (1886–1918), Varvara Stepanova (1894–1958), Lyubov Popova (1889–1924), Alexandra Exter (1882–1949) or Nina Kogan (1887–1941). See Udaltsova, Nadezhda and Larissa Zhadova (1980). The 2017 Tate Modern exhibition *Red Star Over Russia: A Revolution in Visual Culture 1905–55*, perhaps conscious of this occlusion, included a room focusing on some of the artist couples who worked together: Aleksandr Rodchenko and Varvara Stepanova; Gustav Klutsis and Valentina Kulagina; El Lizzitsky and Sophie Lizzitsky-Küppers. Still not altogether a satisfactory strategy for addressing the problem.
13 Alfred Barr (1902–1981), the art historian and curator, produced an initial survey of modernism after a 1927 research trip to Germany. This appears to have been prepared to assist his teaching at Wellesley College. Throughout the 1930s, after his appointment as Director of the Museum of Modern Art in New York, he produced a range of monographs and catalogues dealing with a number of modernist artists, including Picasso, Matisse and the Cubism, Dada and Surrealism movements.
14 Writing as he was at the height of the Cold War, Herbert Read did not yet have available to him the ground-breaking surveys of the work of the Russia avant-garde later provided by Camilla Grey (1962) and John Bowlt (1976). That said, Read was involved in a 1957 publication on Nuam Gabo, writing an introductory essay to the collection. This piece, in addition to surveying Gabo's work, included key Constructivist documents, such as the 1920 *Realistic Manifesto* and the 1921 *Program of the Productivist Group*. Read also corresponded with, and supported, Camilla Grey in her research on the Russian avant-garde, eventually published in her *Great Experiment*.
15 Kandinsky first used this term in his *On the Spiritual in Art* (1912/1977). It functions as sort of categorial imperative for Kandinsky, requiring the artist to develop an original voice, expressed in a distinct style of its time and place, but true to art as a calling – what he calls "the element of the pure and eternally artistic" (1977:35). His idea of "internal necessity" bears the marks of his involvement in German Expressionist art in so far as it privileges both the expressive powers of the artist and the autonomy of art process over political and societal engagements. This was a position that was always going to put Kandinsky into conflict with the more politically minded members of the avant-garde in Russia and Germany.
16 Kandinsky was sent to northern Russia to conduct fieldwork by the Imperial Society for Science, Anthropology and Ethnography, basing himself in Vologda, located over 400 kilometres north of Moscow. He later published an ethnographic paper on the culture of the Zyrian people of the area. During this trip, he was exposed for the first time at close quarters to the folk art of the region with its use of vivid colour and simplified forms. He acknowledged this ethnographic experience shaped his art and outlook.
17 Art historians continue to argue over which of his paintings should be afforded this status, though the consensus is that the 1910 untitled watercolour currently in the Musée National d'Art Moderne, Centre Georges Pompidou, Paris, France, is the strongest contender.
18 Kandinsky strove to take his audience with him as he moved towards abstraction, warning his fellow artists about the danger of abandoning image and object: "the removal of an object in art makes very great demands on the inner experience the purely painterly form, that therefore an evolution of the observer in this direction is absolutely necessary" (Kandinsky 1994:370).
19 This concept in Marxist political economy is usually associated with Leon Trotsky, who in his *History of the Russian Revolution* (first published in 1932) explores why a fully-fledged

socialist revolution was possible in a relatively "backward" country like Russia. As he writes, "the privilege of historic backwardness – and such a privilege exists – permits, or rather compels, the adoption of whatever is ready in advance of any specified date, skipping a whole series of intermediate stages" (Trotsky 2008:4). Less developed nation–states often in a precocious way adopt the most modern available technologies and indeed cultural forms as they anticipate their accelerated development. The concept seems to have applicability in understanding the way that the Russian avant-garde not only quickly assimilated the modernist artistic language of western European art after 1905, but leapfrogged this tradition, racing ahead of it in its innovations and engagements to produce a cultural formation which was, in Trotsky's words, "an amalgam of archaic with more contemporary forms" (ibid:5).

20 John Milner (1983:41) identifies this dynamic in the work of Vladimir Tatlin:

> Tatlin's exploration of the icon brought him to conclusions curiously compatible with aspects of Parisian cubism. A flattened picture surface, an elaborate pictorial structure to which imagery was subsidiary, and the separation of colours, all concerned the cubists as much as Tatlin.

21 An X-ray examination of the painting, the surface of which has deteriorated significantly over the last hundred years, has revealed that the *Black Square* was painted over an existing abstract colour composition. Malevich was inclined to view his iconic painting as a work in progress or indeed as a hypothesis rather than a unique artwork and was happy to reconstruct the painting on a number of occasions after 1915.

22 Its mode of hanging in the Last Futurist show was in the manner of a counter-relief with three-dimensional aspirations and may have been influenced by Tatlin's Corner counter-reliefs, the first of which had appeared as early as 1914 and which had already laid down a challenge to easel painting.

23 Some commentators (Tupitsyn 2002) have seen in the *Black Square* Malevich's acknowledgement of the potential impact of film on the visual language of painting; certainly its structure – a white canvas which frames the central black square – gives it a semblance of a cinema screen onto which the square had been projected. Malevich also seems to nod in the direction of film montage in the spatial relationship of the *Black Square* to a number of other, smaller black and white Suprematist paintings installed below and to the side of the main work.

24 For example, in December 1912 David Burliuk, painter, polemicist and curator, published his statement "Kubizm" in the aptly named anthology *A Slap in the Face of Public Taste*, an anthology of experimental poetry, critical articles and programmatic statements on the new art and poetry. These publications, which often accompanied the avant-garde exhibitions and events, could be at times be extravagantly expressed (Bowlt 1988:70).

25 As John Gage argues (1999:218), "the idea of painting as a progressive series of visual discoveries is, of course, as old as art historiography itself" and "has become part of the mythology of specifically modernist art". Certainly, the rapidity in the evolution of the art of the Russian avant-garde after 1905 must have encouraged this progressivism among Kandinsky and Malevich.

26 In *On the Spiritual in Art*, he has already mapped out his key argument (Bowlt and Long:77): "art cannot contain exclusively material forms. It is impossible to transmit material form with 100 percent accuracy (. . .) the conscious artist, dissatisfied with the conventionalization of the material object, aspires to impart expression to the object he is depicting".

27 Thus, what we are being offered is not simply a Rückblick (a reflection or reminiscence), but rather a rational reconstruction of a work in which intuitive understanding is replaced by discursive reasoning.

28 Indeed, the theoretical machinery of the new art history was initially rehearsed within film theory (particularly accounts of spectatorship) before being imported into art historical analysis.

29 This critical position has been taken up by Barbara Maria Stafford (1996) in her critique of the semiotization of art historical study and neglect of the materiality of the visual, and by Vivian Sobchack (2004) and Laura Marks (2000) with regard to the haptic properties of cinema.

30 The Zaum poetry movement, represented by figures like Velimir Khlebnikov and Aleksei Kruchenykh, championed an approach to poetry and cultural production based on linguistic experiment with sound symbolism pushing language beyond normal semantics. Zaum as a form of inflected language can be said to exist at the limits of linguistic meaning and to probe these. The movement significantly influenced a range of avant-garde visual artists like Tatlin and Malevich who were seeking to understand art as a language, as it did those theorists developing the formal study of language as a signifying system, such as Roman Jakobson. As a teenager, Jakobson had assisted Malevich in his work. For an account of the movement and its relation to the visual arts, see Janacek (1996).

31 In so far as Kandinsky seeks to provide a research basis for his model, it is with reference to a psychology and indeed physiology of affect.

32 A number of researchers identify the "strong affinities between the work of anthropologists and contemporary artists" (Schneider and Wright 2013) and, in particular, between conceptualism in art and an anthropology revitalized by its engagement with the sensate basis of all human interaction.

33 Beeke Sell Tower (1981) draws our attention to the publication of Klee and Kandinsky's original Bauhaus teaching notes, which indicate the breadth of their pedagogy well beyond what is reported in their publications. As she notes (163),

> a detailed comparison of their course notes would show that Klee and Kandinsky frequently discussed the same problems without quite arriving at the same conclusions, and, conversely, they occasionally arrive at similar results by pursuing different lines of argument.

34 While Eisenstein was undoubtedly influenced by Constructivist art and some of its key concepts, like faktura, collage and fabrication, and while he borrowed the stylized and intense language of gesture developed in in the Constructivist theatre of Meyerhold, his is a different route. And while his major contributions to the theory of film, namely the theory of montage and of the sound–image relation, bear the hallmark of Constructivist thinking, his theoretical and research work is developed within a set of unique professional considerations.

35 Many contemporary art historians write off the theorizing activities of the Russian avant-garde as "pseudo-scientific". Indeed, T.J. Clark (2001) goes as far as to brand Malevich the "divine idiot". This harsh judgement fails to appreciate the humanist model of science most of the other avant-garde artists were working with and, in particular, their practice-based formulation of a "science of painting".

36 Indeed, the philosophical structure of the *Tractatus* has more than a passing resemblance to the fundamental premise of Suprematism, namely the reduction of the subject and the object to a zero point. As Wittgenstein argues, "the subject does not belong to the world: rather is a limit of the world" (5.632). The objective world is famously described (1 and 1.1) as "all that is the case" and as "the totality of facts, not of things". Indeed, Wittgenstein uses a visual analogy to indicate the non-substantial character of the self, arguing (5.633) that just as the eye is the limit of the visual so that "nothing in the visual field allows you to infer that it is seen by the eye", so the transcendental subject is the limit or zero point of the world.

7

UTOPIAN VISIONS AND MATERIAL REALITIES

Artistic research and the revolution

7.1 Introduction

As Russian society is transformed after 1917, artistic research increasingly takes an objective, scientific and applied form.[1] A generation of sublimely talented and dedicated artists initiate a range of innovative artistic practices that pioneer new forms of creative research and novel strategies of social and political engagement. Having enthusiastically endorsed the revolution, the avant-garde threw itself into the task of redesigning a new society. The praxis they developed utterly transformed our visual language and understanding of the role of art in society. Under the political circumstances of the time, the theory and practice of the avant-garde was propelled towards questions of design, pedagogy and public culture. Investigative process and analysis became paramount within the plastic arts.

This was true both in Russia and Germany. However, in Russia this dynamic took a more ideologically charged form, and for a while, artists had access to the – albeit – depleted resources of the new state. Avant-garde artists, whatever their reservations about "applied art", and indeed about the ideological demands of the Party, committed themselves wholeheartedly to reform of the country's art institutions. They came to play a key role in the restructuring of Russia's museums and galleries and its art and design schools. While initially based on the experiments and theories of the abstract painters, their aesthetic programme sought wider cross-media application and social utility. This cadre sought to formalize the visual language and practices of non-objective art, and the great educational experiment they initiated provided the means to do so.

In this chapter, I identify two distinctive models of artistic research present in the varied work of the Russian avant-garde. Firstly, a "science of painterly culture" concerned largely with fine art[2] considerations and typified by the work of Wassily Kandinsky and Kasimir Malevich. Secondly, a Constructivist model associated with figures like Rodchenko, Lizzitsky and Tatlin, which became more concerned with a range of practical interests. The two models were often in tension – sometimes within the same individuals. Avant-garde artists wrestled with the challenge of aligning their fundamental beliefs about the expressive and indeed spiritual character of art practice with their commitment to developing a science of art for an industrial society and a revolutionary polity.

However, the main historical narrative is clear. A model of artistic research, based on a model of non-objective art and a core theory of expressivity, gives way to a social utilitarian approach to practice, now grounded in the methodological practices of "laboratory work" and concerned with advancing a design practice for an industrial society. The idea of the art and design process as laboratory work involving a research team working together is, of course, one of the significant legacies of the Russian avant-garde.

Today, practising artists are largely marginalized from the making and implementation of arts and cultural policy – as they often are from leadership roles within art education, galleries and museums. After 1917, members of the Russian avant-garde found themselves drawing up new curricula and research plans for art, redesigning museum spaces and exhibition practices, acquiring and commissioning work for galleries and representing their country in international cultural relations activities. This practical experience in running artistic affairs gave the avant-garde a certain institutional ballast and capacity to turn theory into practice. This, together with their ability to develop a unique form of art educational experience, created a unique situation in which a research culture for the creative arts could flourish.

7.2 Artistic research for a revolutionary moment

The leftist avant-garde had enthusiastically backed the Revolution. Now they were rewarded for their political loyalty by being given a leading role in the restructuring of the country's art institutions. After 1917, the museums in Russia had been reorganized by the People's Commissariat for Enlightenment (*Narkompros*), the renamed Ministry of Culture and Education. In Anatoly Lunacharsky, the newly appointed "Commissar for Enlightenment", the avant-garde had a champion. And despite the generally rather more conservative cultural tastes of the Bolshevik leadership and party, the avant-garde under Lunacharsky's tutelage found itself able to play a key role in the country's artistic life. Lunacharsky immediately saw the potential of harnessing the new art for the revolution (Zhadova 1982:81):

> Despite our impoverishment, despite the impoverishment of Russia, we are on the way to a flowering of the arts whether we want it or not. . . . A need has arisen to change the aspects of towns as quickly as possible, to express the new experiences in works of art, to get rid of a mass of sentiment obnoxious to the people, to create new forms of public buildings and monuments. And this need is enormous.

A generation of avant-garde artists – Kandinsky, Malevich Tatlin, Rodchenko, Stepanova, Rozanova, Lizzitsky – assumed leading positions in both the new art educational institutions and a radically restructured museum and gallery sector being refashioned to share a range of contemporary art with the new Russian publics. They were soon playing a leading role in designing a new curatorial policy and practice and were given the responsibility for purchasing and commissioning contemporary work for display in the new museums being developed across Russia.[3]

The Cubo-futurist artist Vera Pestel found herself in 1918 organizing art classes for the children of Moscow through IZO (the Department of Visual Arts within the Commissariat of Education). She comments excitedly on the new public role of the avant-garde (Zhadova 1982:84), noting that

> apart from commissions, not one of these artists had ever worked until then in any organization. . . . Now artists could decide their own fate completely independently. . . . We were full of hopes, and of the most fantastic projects.

At Lunacharsky's instigation, Kandinsky drew up an outline research plan and constitution for INKhUK, a planned institute for the visual arts in Moscow. The research programme adopted sought to explore "the objective criteria of artistic value in so far as this is defined as a professional value" (Lodder 1983:79). The overall cultural policy objective seems to have been to build on the impressive body of innovative artwork that had been produced in Russia in the previous decade and to theorize this with a view to "settling questions concerned with the science of art in all its aspects" – quite a lofty ambition.

However, Kandinsky was determined that the new institute would be more than a talk shop for artists. The primary aim of INKhUK was identified as "the investigation of the analytical and synthetical basic elements of the separate arts and of art as a whole" (ibid.), a formulation that has Kandinsky's theoretical stamp all over it, as has the programme of research adopted within the INKhUK laboratories set up under his plan.[4] These labs were predominantly concerned with investigating "the effect of the artistic means employed on the experiences of the man apprehending it", basically an early type of reception study. They sought to operationalize an empirical research strategy to explore the process of artistic communication and the efficacy of various means of artistic expression. Kandinsky sets out an agenda for an ambitious cross-disciplinary creative arts research programme (1994:458). He identifies the key research questions to be explored:

> What is the effect on the psyche of:
>
> a) Painting in its colour–volumetric form?
> b) Sculpture in its spatio–volumetric form?
> c) Architecture in its volumetric–spatial form?
> d) Music in its phonic and temporal form?
> e) Dance in its temporal and spatial form?
> f) Poetry in its vocal–phonic and temporal form?

This programme would involve laboratory trials, psychological tests and the administration of questionnaires designed to assess – rather unsuccessfully, as far I can see – the effects of the deployment of colour and form by the visual artist on the sensibilities of the spectator. In so far as Kandinsky seeks to provide a scientific basis for his model, it is with reference to a psychology and indeed physiology of affect.[5] In any case, this research programme was never fully implemented in INKhUK, in part due to the increasing tensions within the institute between Kandinsky and a rival group of Constructivists led by Rodchenko who wished to see artistic research move much closer to the processes of product design for industrial production. But Kandinsky had laid down a marker within the broader scientific community in Russia, insisting that art be recognized as a valid field of research activity alongside other branches of "pure" science.

That said, it can be argued that Kandinsky's greatest contribution to the advancement of a tradition of artistic research lies not in the elaborate interdisciplinary research programmes drawn up for INKhUK, but in his own innovative studio work and writing and, above all, in his pedagogic activity and theorizing in Russia and later at the Bauhaus. For him, the training of the creative arts student is at the same time the training of a researcher. At the core of his reformed arts pedagogy is the idea that arts practice is research-led. The development of methodologies of studio innovation and the adoption of an analytical language of the visual are the vehicles for such research. They are also the tools for the formation of a student as an independent learner and creative agent.

Malevich, like Kandinsky, combined his pedagogical and research activities. Like Kandinsky, he sought to develop an institutional base and intellectual validation for his research plans. In 1923, after a spell spent reorganizing the art and design school in Vitebsk in Belarus (see below), he was appointed as Director of the Museum of Artistic Culture in Petrograd. This position enabled him to set up GINKhUK, an arts research institute with a core theoretical department and studio and laboratory provision. Like INKhUK in Moscow, it aimed to "develop ideas found in painterly systems and to make a practical study of these systems" (Malevich 1976:317).

In Petrograd, Malevich had the opportunity to develop his vision of creating a research institute for the visual arts that would be based around the study of the practices of painting. The Institute of Artistic Culture, which he now headed, announced an ambitious brief. It would carry out "research work in the sphere of scientific registration of various aspects of Art cultures with regard to their organic, physical, psychic, formal, theoretical, ideologically philosophical, sociological formation" (Malevich 1976:230).

Malevich had from the outset identified Suprematism with a scientific outlook, seeing it as a new field of research and scientifically informed cultural practice rather than just a style of painting. In a pamphlet published in 1920 in Vitebsk, he writes (Zhadova 1982:286), "one of the foundations of Suprematism, in terms of experience and practice, is Natural Science, which makes it possible to overcome the world of books and replace it by experience and action, whereby all will share in universal creativity". For Malevich and the Suprematists, art and science are parallel investigative endeavours, save that art deals predominantly with the sensate, the affective and the experiential. As Boris Arvatov, artist, critic and founder of the radical journal LEF, was to summarize the relation (ibid.:68), "just as science proceeds by means of hypotheses in the realm of abstraction, so painting will maintain its function of constructing hypotheses in the realm of the concrete". Accordingly, "painterly culture" was to be regarded as a field of invention and studied empirically. Malevich's concern was to plot the key developments in the evolution of modernist art and to conduct a range of investigations – aesthetic, psychological and cultural – into aspects of the new art, exploring the potential application of this knowledge to architecture and design and other practical matters, such as education and cultural provision. The collection of the Museum for Artistic Culture, together with the output of Malevich's experimental studio, would provide the "data set" to prosecute a form of analysis that would directly engage with, and seek to theorize, painterly practice. While details of the work that took place in GINKhUK remain obscure, we get a sense of an ambitious programme of practice-based visual arts research that evolves parallel to a range of pedagogic activities and curatorial initiatives.

Certainly after 1923, Malevich was pioneering a new model for the art museum in tune with the spirit of the times. As he proclaimed in one of his polemics (ibid.),

> instead of collecting all kinds of old stuff, we must organize laboratories, for a worldwide, creative, building apparatus, and from its axes will emerge artists of living forms, not of dead representations of objects. Let the conservatives go to the provinces with their dead baggage, with the lecherous cupids of the bygone debauched houses of Rubens and the Greeks. We will bring I-beams, electricity and the fires of colour.

Interestingly, his thinking about the status of the museum seems to anticipate current curatorial debates and emerging gallery practices. As we have seen, curators have been concerned with facilitating audiences' understanding of contemporary art processes and how various research practices facilitate these. The contemporary gallery wishes to reconceptualize itself as an arts laboratory, and this approach is often presented by curators like Hans Obrist as a novel one. In fact, the Russian avant-garde pioneered all these curatorial practices and notions of artistic research and public engagement a century ago. So, it may be useful to focus on this less known aspect of the artistic culture of revolutionary Russia.

After 1917, the new art museums would be concerned not only with the exhibition and preservation of artworks but also with artistic research and education and with what we would now call "outreach" activities. They aspired to become what we would recognize today as "art centres". No longer would the museum function as a traditional mausoleum of beaux-arts relics. Instead, it would be a creative institute, a "living organism", with an arts laboratory at its centre[6] conducting creative arts research and facilitating educational programmes and publications.

Even before the revolution, avant-garde artists had grasped that if abstract art was to establish itself, it would require a different approach to exhibition practice that went beyond the bounds of easel painting and traditional modes of hanging work.[7] After 1917, the search was on to find innovative ways of displaying art. The traditional narrative order of the museum was soon challenged by artists. Hundreds of artworks had been purchased through the acquisitions committee of *Narkompros* (chaired by Rodchenko) for exhibition in the new museums.[8] How should this work be shown? How could exhibition be harnessed to social pedagogy?

Writing in her diary in 1919, Stepanova captured the revolutionary mood (Rodchenko and Lavrentiev 1989:42).

> Our art has to be taken out into the streets, onto the fences, onto the rooftops. . . . That is why icons, signboards and cheap popular prints were developed in our culture. I think we ought to show our own Russian pictorial culture which unfortunately we cannot value much when we try at all costs to follow the West.

Stepanova sought to bring the skills of the new graphic art practitioners to bear on the design of the Museum of Artistic Culture in Moscow and to have this innovative visual culture work recognized by including it in the Museum alongside the fine art exhibited. She envisaged the use of "icons signboards, popular prints (lubok) – to show the connection with graphic works; Impressionism, Futurism Cubism; Orphism; Suprematism; Non-objective creativity" (ibid.).

In the journal *Artistic Life*, published by the Visual Arts Section of Narkompos, Kandinsky suggested that the chronological narrative which guided traditional museum exhibition practice might have to be deconstructed (1994:437): "the usual type of museum is reminiscent of some chronicle that draws its thread from one event to the next, without seeking to penetrate their inner meaning". Instead, in the new museum,

> our desire is to show the development of art not in chronological, chaotic fashion, but on the contrary by giving strict prominence to two aspects of artistic development:
>
> 1 New contributions of a purely artistic nature, i.e., the invention as it were of new artistic methods
> 2 The development of purely artistic forms, independent of their content i.e., the element, as it were, of craft in art.

So, within the presentational logic of the new museum, chronological order was out, as was organization of work on the basis of its national origin. The focus was to be on the history of artistic invention and particularly on the forms of innovation found within modernist art and design, now seen as equivalent to scientific advances. The installation of the new works of art was to be accompanied by a range of explanatory displays and examples of contemporary visual culture, revealing a continuum between pure art experiment and its application within popular graphic and decorative art. The displays sought to share details of artists' working methods and explore their use of media. Audiences were presented with work in progress. Artists were invited to report from a field of scientific investigation. Malevich, in particular, employed a series of explanatory tables[9] to plot the development of modern art, regarding his graphics as part and parcel of the work presented (an example of which is seen in Figure 7.1 left). As he put it, these were "museum collectable".[10]

In other words, the avant-garde had a keen interest in what we now call exhibition design, and a number of creative and political experiences fed into this: the massive displays of street art produced after 1917 to mark public commemorations, not least by Malevich's students in Vitebsk, who decorated the whole of the town in Suprematist designs for the third anniversary of the October Revolution (see Figure 7.2 right); the designs by Gustav Klutsis (Figure 7.3 left) and others, such as Lizzitsky's Lenin Tribune (Figure 7.10), for publicity kiosks and mass communication platforms for ideological work which envisage using graphic art, photography, film and public address systems;[11] the innovative forms of graphic design and militant poster art found in Suprematist and Constructivist publications designed by Rodchenko, Mayakovsky and Lizzitsky (Figure 7.3 right).

Alongside this explosion of visual communication activity inspired by the revolution, Malevich's curatorial and research activities in the Museum of Artistic Culture in Petrograd after 1922 seem quite restrained. His primary focus seems to have been the laboratory work conducted in GINKhUK and the taxonomy of non-objective art which emerges from this. This may reflect his attempts to protect himself from the political and ideological maelstrom that swarmed around Suprematism in the early 1920s and his growing sense that the revolutionary tide was already ebbing in Russia. Indeed, his enthusiasm for employing scientific rhetoric to describe the work of his research team after 1923 can be seen in this light. Scientific discourse came to function not only as an "enabling" rhetoric but also as a defensive one within Soviet society.

This interdisciplinary research programme developed in GINKhUK bore all the hallmarks of Malevich's long-term intellectual interests in a formal analysis of artistic systems. Spectral and colour studies were proposed, together with the use of X-ray techniques to examine the canvas of historic paintings. The study of pedagogic methods was given a central place. The ethnographic study of art was also identified as an approach which could address the impact of climatic, cultural and sociological factors upon the structure of artistic systems and artists' behaviours.

As with Kandinsky's research at INKhUK, various methods of social and psychological investigation concerned with the reception of artworks were employed, including questionnaire surveys and interviews. Moreover, collaborative projects were developed with a range of scientists – physicists, physiologists, psychologists – in order to complement the study of painterly practices and underwrite the scientific claims that Malevich was anxious to make for his institute's work. All in all, an impressive elaboration of what artistic research in the modernist era might look like.

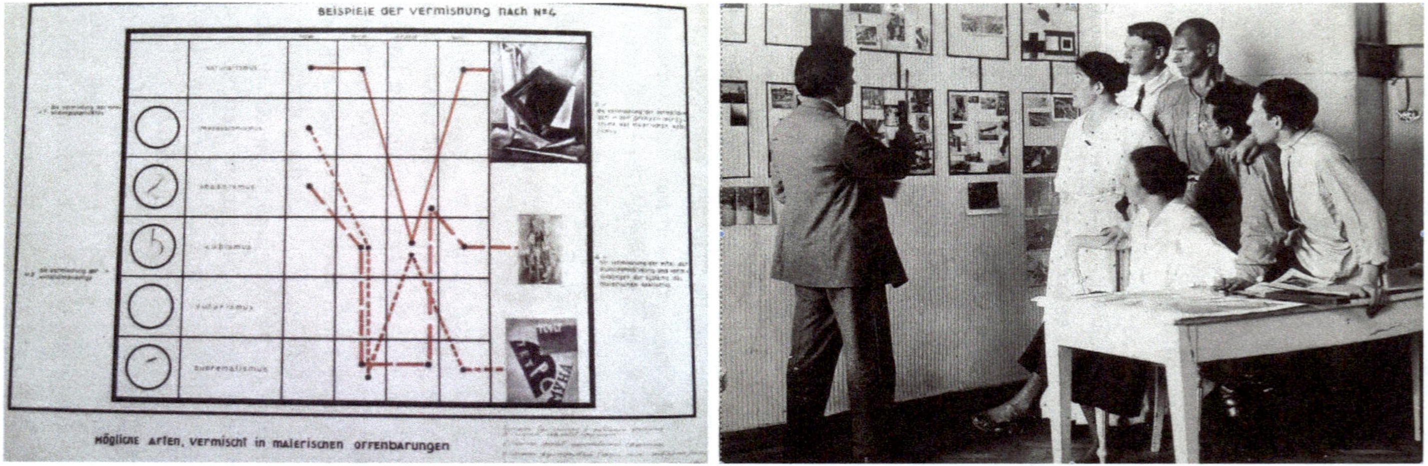

FIGURE 7.1 (left) Kasimir Malevich, pedagogical chart (1926/27); (right) Malevich demonstrating his research to his students in GINKhUK, St Petersburg (c.1926)

Source: public domain

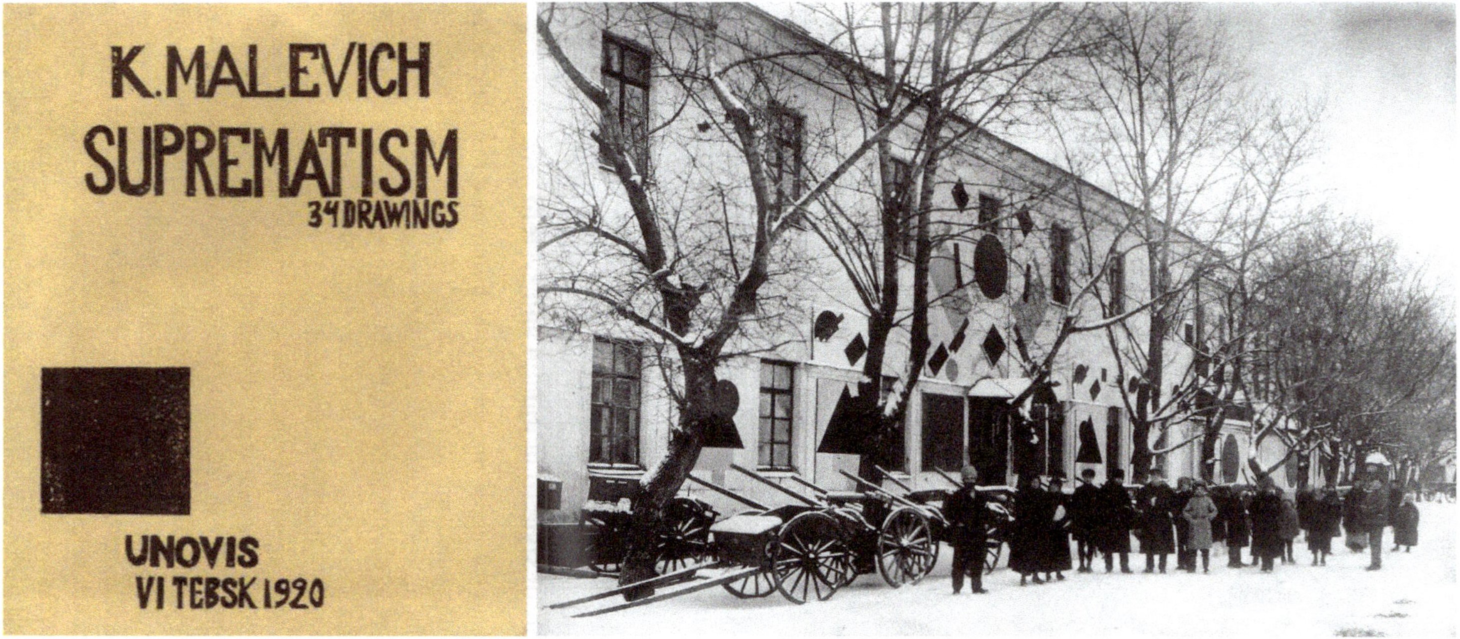

FIGURE 7.2 (left) Kasimir Malevich, design for UNOVIS publication (1922); (right) UNOVIS wall paintings, Vitebsk (1920)

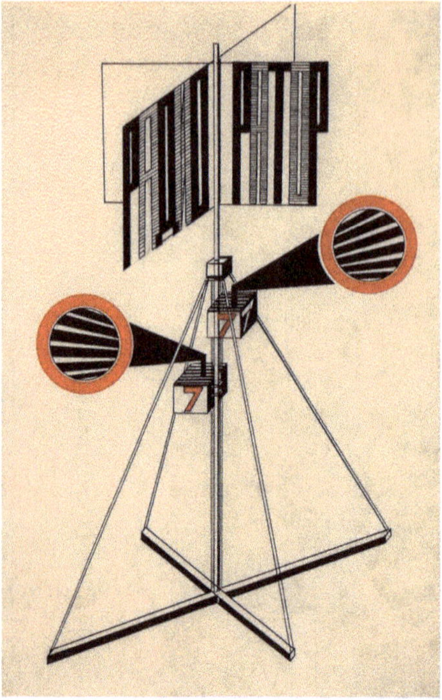

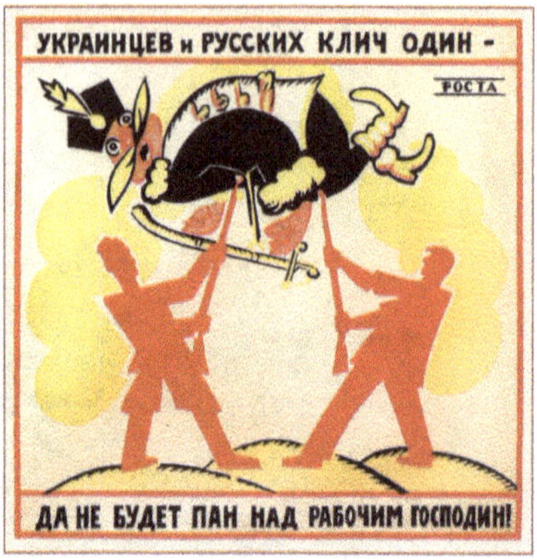

FIGURE 7.3 (left) Gustav Klutsis, design for loudspeaker system (1922), George Costakis Collection, Athens; (right) Vladimir Mayakosky, Civil War poster (1924)

Source: public domain

7.3 The auspices of non-objective art: from the analysis of style to language and system

What, then, was the underlying theory which informed Malevich's research programme for the visual arts? In his key text, *The Non-Objective World*,[12] Malevich identifies painting as the methodological gateway to a "science of artistic culture". His research promises us (Malevich 1959:11) "an investigation of individual visual effects in art and the ways in which they are achieved, in order to discover the causes of observable changes in art and artists".

Arguably, painting was the most developed branch of the plastic arts at the turn of the twentieth century in terms of its commitment to formal innovation. For painters like Kandinsky and Malevich, and even for Constructivists like Rodchenko and Tatlin (who after all had begun their careers as painters), an understanding of the development of the discipline provided the key methodological access point to a more general understanding of the processes of invention across a range of arts disciplines like architecture, design and, subsequently, lens and screen-based arts. In that sense, the deliberations of the painters were foundational for all artistic research practice.

The priority for Malevich, in particular, was to plot the rupture in pictorial conventions wrought by abstract art, particularly with the advent of Cubist modelling. This would require developing an account of the evolution of pictorial systems and the visual elements at play in these. Malevich's analysis cuts against any idea that painting as a field of knowledge is subject to gradualist development within a grand European tradition. Indeed, for many of the avant-garde, such as the painter and sculptor Vladimir Tatlin, Russian art was seen as outside of that tradition with a direct relation to traditions of folk art and icon painting. Moreover, for Malevich, transitions from one paradigm of artistic practice to a succeeding one were seen as occurring suddenly. Artistic innovation lead to "additional elements" being introduced into a painterly system which destabilize it, leading to the emergence of a new, more advanced pictorial paradigm. As he argues (1978:299),

> in each new art, each painterly system has its own characteristic forming element, with the help of which, the artist embodies and forms various sensations and this or that content.
>
> We also call these forming elements additional or deforming, when they turn one system into another, for example, Cubism into Suprematism.
>
> So, having made an analysis of form we looked at the fact that all the systems of new art have their own distinct structure, textural relations between form and colour, various principles, methods and *Weltanschauung* in general.

Malevich sought to identify the "additional element" which represented the defining specificity of each modernist movement and which served as the motor of transformation from one movement to the next. Malevich's notion of the additional element is a complex one. Indeed, it is difficult for us today to fully make sense of the various explanatory charts he drew up to illustrate his analysis of the evolution of modernist art based on his theory of the additional element (such as that represented in Figure 7.4 left). That said, his historical intent is clear. The charts, with their vertical axis of modernist artistic styles and horizontal chronological axis, seek to schematize his research into the historical development of modernist art from Cezanne. They trace the interconnection between the different forms of modernist art practice up to and including Suprematism, the point from which his analysis is leveraged. The three-panel photomontage (Figure 7.5 left) which Malevich produced in 1925 to identify the characteristic visual styles of Cubism, Futurism and Suprematism is informative (a version of this can be spotted being demonstrated in a tutorial room in Figure 7.1 right). The Suprematist panel reveals, in particular, how the aerial perspective available from an aircraft may have conditioned his own use of form and colour to model contrasting planes and to give an illusion of depth in his abstract paintings.

Moreover, this pedagogically motivated theorization of art practice has left its mark on the academic study of modernist art. It seems likely that Alfred Barr, who was to pioneer the study of modernist art in the US, was familiar with Malevich's work and his pedagogic charts. Certainly, the front cover of Barr's pioneering study

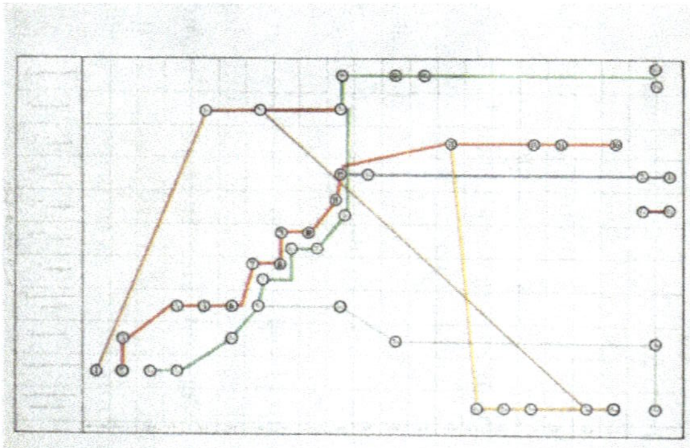

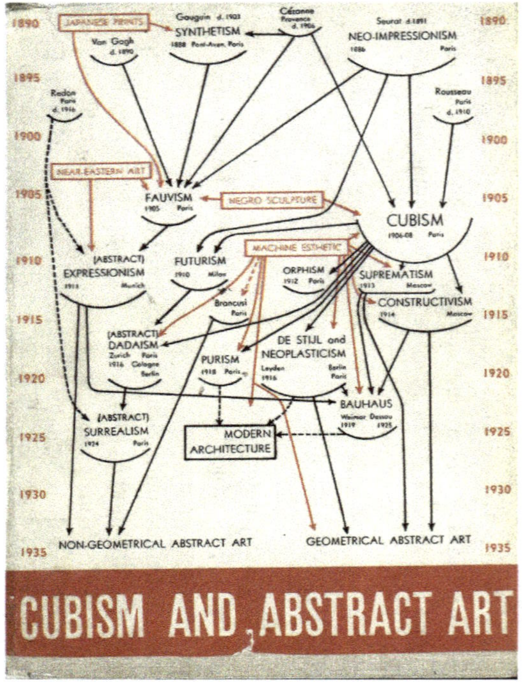

FIGURE 7.4 (left) Kasimir Malevich, pedagogic chart indicating the evolution of modernist painting (1926); (right) Alfred Barr, front cover from his book, *Cubism and Abstract Art* (1936)

Source: public domain

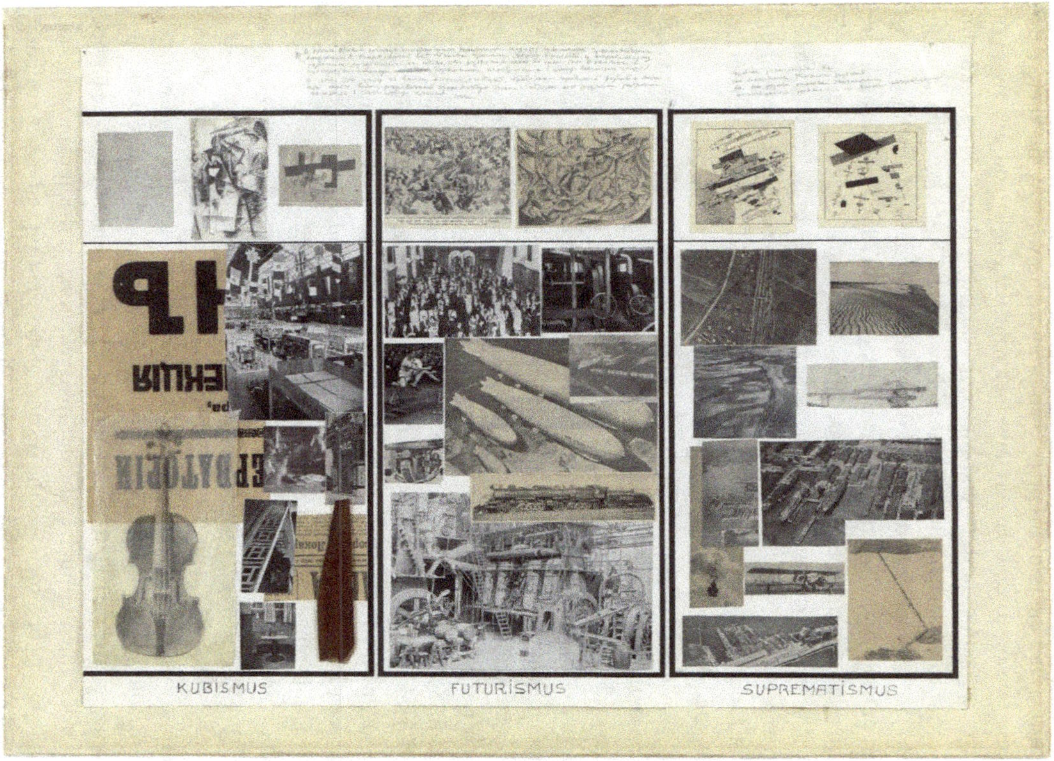

FIGURE 7.5 Kasimir Malevich, explanatory photomontage indicating the stylistic features of Cubism, Futurism and Suprematism (1925–1926), Museum of Modern Art, New York

Source: public domain

of modernist art Cubism and modern art, published in 1936 with its diagram of the evolution of modernist art (Figure 7.4 right) bears a certain resemblance to Malevich's schema (Figure 7.4 left). Barr, while registered for a PhD in art history at Harvard, had travelled to Germany in 1927 to familiarize himself with the work of the Bauhaus and more generally to learn about the avant-garde culture and new arts pedagogy that had emerged around the German–Russian cultural axis. Malevich also visited the Bauhaus in 1927 to promote his work. While there is no record of the two meeting, Barr certainly met with Kandinsky and was aware of the research programmes both he and Malevich had led in Russia. He would also have had access to Malevich's just published book, *The Non-Objective World* (Figure 7.6 left). So, to some extent at least, the auspices of the academic analysis of modernism pursued in the US from the early 1930s lies in Alfred Barr's familiarity with an emerging tradition of practice-based visual arts research closely tied to the pedagogic aims of the Bauhaus. The subsequent appointment of Walter Gropius to the School of Architecture at Harvard and of course Barr's to the directorship of the Museum of Modern Art in New York cemented this bond between German arts practitioners and American historians and theorists of modern art.

Malevich's schema, let us recall, is the theorizing of an artist rather than an art historian or aesthetician. It is an attempt to systemize the collective experience of an avant-garde living through an artistic revolution and trying to make sense of its own practice. Malevich's "science" involves a working through of ideas in a medium rather than the adoption of a theoretical or philosophical template extrinsic to art practice, however sophisticated.

Malevich traces the additional or distinctive elements characteristic of Post-impressionism, Symbolism, Expressionism, Cubism, Futurism and, of course, Suprematism. He seeks, in other words, to establish a creative trajectory which leads inexorably to his own non-objective art.[13] Earlier modernist movements – and as we have seen by 1915, he and many of his colleagues had experimented with painting in a wide range of Post-impressionist, Symbolist and Cubist styles – were identified primarily as precursors of Suprematism. It is on the basis of this experimentation that the schematization becomes possible.

His theory seeks to explain the abrupt transitions that occur between successive modernist stylistic systems of painting. Cubism, as manifested in the work of Picasso or Braque, represents for Malevich the most dramatic rupture from the established naturalistic paradigm. With Cubism, a new pictorial norm is established with additional elements – such as the distorted rendering of the human body, face and physical objects, or a focus on the texture of the painting and its construction from multiple materials other than paint. Such elements directly challenge the normative structure of naturalism, both in its picturing of the human form and with regard to the transparency of the picture plane. Initially the challenge is resisted (Malevich 1959:14).

> The new norm destroys the norm of the generally accepted order in painting, of the established rest between the straight line and the curved line.

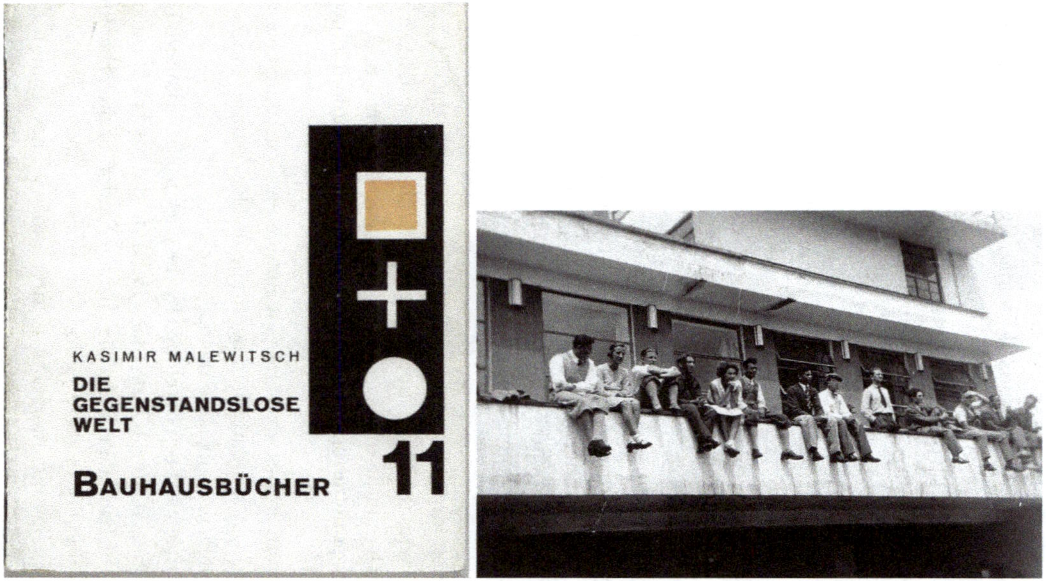

FIGURE 7.6 (left) Front cover of Malevich's *Non-Objective World*, designed by Lazlo Moholy-Nagy (1927); (right) Students relax at the Bauhaus c.1931, Bauhaus Stiftung/Pompidou Centre, Paris

Source: public domain

> And, consequently society raises the question of retaining the norms of the established order and it thinks about the isolation of the painters captivated by the new norm.

However, in time, the new paradigm, with its additional elements – such as the decisive breaking from a naturalist modelling of the body through various representational distortions – gains adherents. The additional element, like a troubling new scientific hypothesis or experimental finding, challenges the traditional paradigm, and a new, "revolutionary" paradigm emerges. In the first instance, this gains acceptance within a specialist, radicalized professional subculture – an avant-garde. Only later does it command wider public support. Eventually, the new paradigm or artistic system gains societal recognition and institutional acceptance, becoming in time the new normative structure of art.

Malevich's approach has more than a passing similarity to Thomas Kuhn's idea of scientific progress, discussed in Chapter 2. For Kuhn, scientific development is a discontinuous process characterized by shocks as new experimental evidence undermines established positions. "Revolutionary science" attacks "normal science" and new paradigms emerge.

Both Kuhn's schema and Malevich's account of the development of modernist art are probably indebted to Marxist historiography with its theories of crises, contradiction and historical transformation. As Lenin had argued in one of his last articles, "Better Fewer. But Better" (1923/1964:490) in a formulation that seems to perfectly encapsulate the dynamic of Suprematism,

> great revolutions grow out of the contradictions between the old, that which is directed towards the destruction of the old and the very abstract striving for the new, which must be so new as not to contain even a grain of the old. And the more abrupt the revolution, the longer such contradictions last.

Malevich's work utterly breaks from art's past and presents a startling new vision of art's future. Suprematism as an artistic programme involves seismic, indeed revolutionary, paradigm shifts in form, content and "style".

The concept of style had been developed by the German art historians Heinrich Wöfflin (1864–1945) and Alois Riegl (1858–1905) on the basis of detailed studies of the Renaissance or Baroque era and the characterization of the prevailing artistic language and sensibility of these periods. Significantly, these German scholars brought a series of scientific aspirations to art history.[14] They fashioned, in the words of Patricia Railing (1989:56), a "theoretical and practical art history based on the creative language of the artist – the artistic means of colour, line, form, space, movement". Needless to say, academic art history in the early twentieth century was still almost exclusively dealing with the art of the distant past and not with the bewildering world of contemporaneous art practices. However, in the hands of a generation of avant-garde artists striving to understand the dramatic transformations in art practice wrought by modernism, these art historical tools could be used to fashion new theoretical insights into modernist practice. Malevich's theory of the additional element seems to be one such attempt.[15]

In 1907, the art historian Wilhem Worringer presented his doctoral dissertation at the University of Bonn: *Abstraction and Empathy: Essays in the Psychology of Style*. For the first time, an academic art historian addressed the advent of modernism in art – albeit elliptically.[16] Worringer argued that mimesis, or realistic depiction, is not inherent to visual art. Moreover, stylized, non-naturalistic art, such as found in French Cubism and German Expressionism, is as it is not because of a culture's inability to master the conventions of naturalistic representation, but because it reveals social and psychological factors and contradictions present in a culture or historical period. In other words, artists seek to develop an expanded visual language in order to represent objects in a more nuanced and inflected manner. Indeed, in *fin de siècle* Europe, an unease with modernity had led to growing existential doubts on the part of many artists and to an upsurge of interest in spirituality. This sensibility motivates a search for new forms of cultural expression. Thus, artistic representation – and most significantly, the expressionistic and increasingly non-naturalistic art of the avant-garde – could be seen as symptomatic of a troubling new world view.

As Hal Foster cogently argues, drawing on the work of Mikhail Bakhtin and Pavel Medvedev,[17] "the becoming-academic of art history at the end of the nineteenth-century" may well be intimately connected to the "becoming-autonomous of modernist art during the same time" (1996a:97). Indeed, it was the new reflexivity of modernist art and the abiding concern of artists with the "constructive aspect" of the artwork which provided the methodological basis of the new "scientific" art history. The tail of art practice was now wagging the dog of art historical analysis. Modernism's attention to "alien" or "primitive art" with its non-objective and expressive character provided the stimulus for art history to focus on the "will" and expressive intent of the artist as the motivational force that breathed affective life into artistic form and style (we explore the relationship of Primitivism to colonialism in Chapter 8). Art historians sat up and took notice. Practice had run ahead of theory. The intellectual responsibility of providing an analysis of the new art now lay largely in the hands of avant-garde artists themselves. The conjuncture was one in which the working, experimenting, reflecting artist was in a privileged position to advance a cutting edge programme of contemporary artistic research. Malevich took up the challenge.

The historical schema he developed to make sense of development modernist art was first rehearsed in the 1916 *Futurism to Suprematism* publication. It was later refined during his period as an art educationalist in Vitebsk from 1919 to 1922, where his theories formed the basis of the restructuring of the curriculum in the Technical Art Institute in Vitebsk after 1919 by the UNOVIS group of Suprematists and Constructivists. It was later worked on with his colleagues in the Formal–Theoretical Department of GINKhUK, and the research tables he drew up, with the assistance of Anna Leporskaya (seen among the student group in Figure 7.1 right),[18] were displayed in various exhibitions, including a 1927 exhibition of his work held in Berlin in the *Grosse Berliner Kunstaustellung*. A visit to the Bauhaus that year secured publication of his text *Non-Objective World* as part of the series of monographs within which Kandinsky's *From Point and Line to Plane* had already been published.

However, while Malevich was undoubtedly familiar with the empirical methods that art historians were employing, his methodological focus lies elsewhere. The origins of his idea of the additional element and his concern to treat modern art as a system probably lies in the theories of the literary formalists, with whom Malevich seems to have been in constant communication after 1915.[19] In terms of critical interlocutors, Malevich seems to have been particularly influenced by the literary theories ofViktor Shklovsky (2016). Shklovsky had analysed the development of literature as a discontinuous process marked by the sudden appearance of new literary genres involving radical breaks in style. Clearly, within the visual arts Suprematism represented the most radical rupture from naturalistic art. Here was a body of bewildering but vital work characterized by dramatic pictorial innovations that turned the world of representational, naturalistic art upside down.

Shklovsky was no stranger to the idea that the challenge of art was to "make strange" and contest the taken-for-granted.[20] He grasped that Suprematism represented the most virulent attack on the body of the old art. Responding to Malevich's theories as well as his paintings, he invoked a medical metaphor to describe Malevich's iconoclasm. "The Suprematists", he commented in a 1920 response to Malevich's publication *From Cézanne to Suprematism*, "did for art what chemistry did for medicine: they isolated the active factor in the remedies" (quoted in Zhadova 1982:326). One could only hope that contemporary practice-based research could in its investigations get at the "active factor" in contemporary art.

Malevich, like many of the Russian avant-garde, conceived of the new art in terms of a model of artistic innovation akin to pure scientific experimentation and theorization. His enthusiasm for what we would now call "laboratory work", "hypothesis-painting" and "project art", as pursued in the Petrograd Research Institute (Zhandova 1982:59), anticipates both the conceptualism of the 1960s and more recent attempts to foreground the processual and research aspect of art practice. We can see in the work of Malevich, like that of Kandinsky and later Rodchenko, the emergence of a tradition of practice-based arts research. This initiative, while involving interlocution with a range of critical sources, is primarily grounded in the studio practices, visual investigations and pedagogic concerns of the working artist.

7.4 Malevich as pedagogue: the UNOVIS experiment

Getting at the "active factor" in contemporary art was certainly at the core of the curriculum Malevich developed within the Vitebsk Architectural and Technical Institute after 1919, together with the UNOVIS collective. The UNOVIS group of artists and students, acting in the heat of political revolution, had effectively taken over and restructured the Institute along rigorously modernist lines while enthusiastically embracing the collectivist spirit of communism.[21] Staff and students had taken their art outside the walls of the college to decorate the streets of Vitebsk for events like the First of May labour celebrations, painting walls, the facades of buildings and trams and erecting banners, posters and street decorations. From the outset, UNOVIS envisaged art as an "expanded practice". A young Sergei Eisenstein, still a soldier and not yet embarked upon his career as a film-maker, passed through the town of Vitebsk on his way to the front in 1920 during the civil war and commented on the street decorations (1971:432).

> This town is particularly odd. The red brick is particularly odd. The red brick of the main street is covered with white paint. Green circles, orange squares, and blue rectangles swarm over the white background. This is Vitebsk 1920. K.S. Malevich's brush has travelled over its brick walls. "The squares are our palette" these walls shout out.

Socially and politically engaged and manifest in public spaces and communal events, Suprematist art was at home with the transformative spirit of the revolution. When Malevich arrived in Vitebsk, the college was already in ferment. As one student, Ivan Gavris, later recalled (Zhadova 1982:306), "there was complete freedom of artistic activity in every workshop". But as he more soberly concluded, echoing his mentor's view,

> freedom alone is not nearly sufficient to ensure fruitful creative activity. Creative activity must develop in accordance with the method of a given system. Everything that is unsystematic and aimless is doomed to decay because it has no ground to grow in.

Malevich set about reorganizing the school around a pedagogic structure that reflected both his analytical understanding of the development of modernist art and his own painterly exploration of form and colour (developments which quickly led to the departure of its previous director, symbolist painter Marc Chagall). The UNOVIS curriculum implemented in Vitebsk was envisaged as applicable across the Soviet art and design sector, and Malevich travelled widely across Russia promoting his ideas for educational reform.

From the outset, UNOVIS was not simply an art and design academy. It was also a research institute and laboratory for the study of the visual arts. In other words, pedagogy and research practice were in alignment. As Malevich grasped, teaching matters were now central to establishing the new art (Zhadova 1982:87): "trends like creative inventions can be taken further but method must follow them and teach the masses to appreciate and develop the things that are contemporary".

The new teaching programme undertook to give its students a "modern consciousness based on the innovations of Cubism, Futurism and Suprematism". It employed what Malevich called "a scientific formal theoretical approach". This rather catch-all term focused on providing students with an understanding of the visual language of abstraction and encouraged them to adopt a scientific mind-set in their work. These analytical concerns were initially rooted in painterly practice but were envisaged as applicable across a range of arts disciplines, including design and architecture (we see Malevich with his painting class in Vitebsk in Figure 7.7).

Students developed a working knowledge of the "new colour pictorial world formulation" and explored the application of this new visual language to decorative composition (in the fields of graphic, theatre design, textile design) and via constructed three-dimensional "architectonics", which might have subsequent application to architecture.

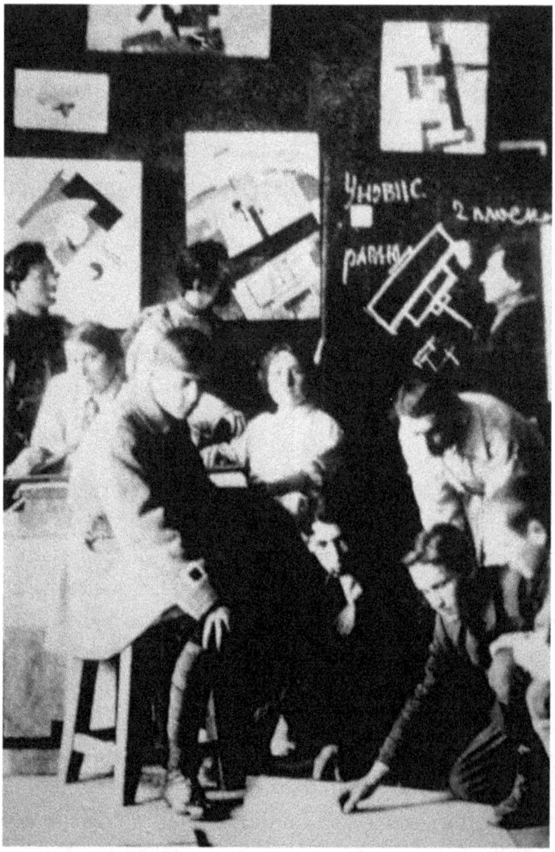

FIGURE 7.7 Malevich's painting class at UNOVIS (1920) (he can be seen by the blackboard)

Source: public domain

The basic curriculum introduced students to methodological questions of form and structure within art, initially via drawing procedures concerned with the graphic abstraction of objects (not dissimilar from Kandinsky's later Analytical Drawing classes at the Bauhaus).

Indeed, we get a concrete sense of the activities promoted within this curriculum in the Trial Drawing classes introduced at Vitebsk in March 1920. These sought to demonstrate, in the words of one student, that "cubism is not a fantasy or a senseless activity and that we are actually doing something and progressing" (Chasnik 1922). The workshop seems to have involved discussing various cubist forms such as a Picasso archetypical portrayal of a violin (an example Malevich was to use in his *The Non-Objective World*). Malevich reminded the students (1978:435) that for Picasso, "objective nature is merely *the starting point – the motivation – for the creation of new forms*, so that the objects themselves can scarcely, if at all, be recognized from the pictures". Tutors in the workshop provided accompanying elucidations about the pictorial part played by each of the component forms. Students were then expected to experiment with these abstracted forms, producing work aimed at exploring a basic analytical visual language. They were expected to be attentive to such matters as:

Centre of composition
Basic vertical division
Contrasting plane
Maximum sharpness of juxtaposition
Ordered formulation of the whole
Integrated pictorial tone
Entire composition elaborated as a whole

The most cogent rationale of the critical methods employed in UNOVIS seems to have been provided by one of Malevich's students, Nina Kogan.[22] She subsequently became a tutor on the introductory course in Vitebsk and seems to have played a leading role in street decoration activity in the town as well as working on theatrical set and costume designs and designing toys and other items constructed within a Suprematist visual tradition. Kogan taught in the college's "Cubist Studio", and her 1920 paper "The Principles of Painterly Abstraction" discusses Malevich's pedagogical method (Kogan 1920:85).

> The central idea upon which the method is based is that the art of painting is not an imitation of anything in the world around us . . . it should construct something totally new . . . the artist should copy neither nature nor anything else that already exists but should freely create new forms. Until the present, in order to study and develop, the painter has used the forms of objects, since objects were seen as something complete, and he forgot that the object and its form are the result, the completion, of a creative activity.

Kogan rehearses the analytical processes involved in the new non-objective art (ibid.).

> The elements of form are line, both straight and crooked, plane and volume. The elements of colour or texture are surface – shiny or dull, splintered or blistered. The texture of the surface is the structure of the surface. We do not copy the object, we break it down into elements and recombine them into a single painterly organism. We make a composition (. . .) we must make a construction out of them.

This approach seems to have been prosecuted in her Cubism workshop and across the UNOVIS curriculum. Within this painting was clearly the foundational discipline but sought application across a range of areas of design.

Indeed, in the hands of El Lizzitsky, who also worked alongside Malevich in Vitebsk, these formal tools provided the basis for a radical theory of art and design with the widest possible application to architecture, the built environment and the design challenges of mass communication and industrial production for a new society. Suprematist painters began to abandon the easel and the picture plane and began to explore work which while still in the first instance being largely in two dimensions, begins to portrays constructed geometrical forms (what Lizzitsky calls *Prouns*) assembled to produce visionary architectonic works (Figure 7.8 left) or graphic applications of the new visual language (Figure 7.10 right).

Lizzitsky and a number of other like-minded colleagues teaching in Vitebsk began to transform the curriculum of the college in a more utilitarian direction and to develop the workshop provision that could facilitate the laboratory work envisaged as the core of this new artistic approach. Indeed, Malevich as the leader of UNOVIS was perhaps more open to engaging with Constructivism than Kandinsky proved to be during his tenure as Director of INKhUK in Moscow. He played a central role in the agitational art activity conducted in Vitebsk in 1920 and 1921, as he had in Moscow in 1918, taking art out of the studio and into the streets. Indeed, Malevich had always been open to the idea that Suprematism was more than just a painting movement and was happy to envisage a role for three-dimensional work within his creative prospectus.[23] Malevich also recognized that formal experiments in painting could only take the new art movement so far. Perhaps bowing to the strong Constructivist current that was manifesting itself among his UNOVIS colleagues in Vitebsk and within INKhUK (of which he was a key member), he admits (1976:34) that

> an invention of art work becomes available or understandable to the general public very gradually through its practical employment or its mass production. Solutions of the most complex problems – the result of the invaluable creative activity of superior people – become general property and prepare the way for new creative activity.

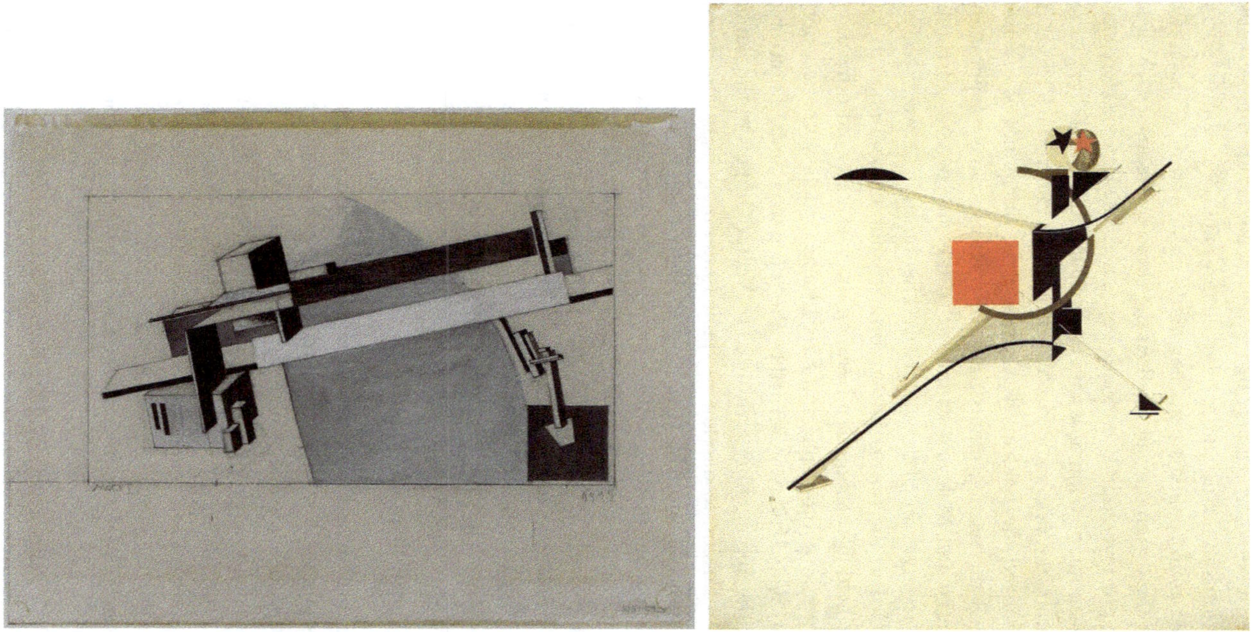

FIGURE 7.8 (left) El Lizzitsky, *Proun 1A (The Bridge)* (1919), pencil and gouache, 15 × 19.5 cm, Berado Collection Museum, Lisbon; (right) El Lizzitsky, *New Man* (1923), from the series Figurines: "Victory of the Sun"

Source: public domain

The historical momentum of the revolution was encouraging the application of modernist skills of visual analysis and experimental construction across a wide variety of fields from industrial and product design, to architecture, to design for stage and performance. UNOVIS members had experimented with the decorative use of Suprematist motifs in textile and graphic design since 1919. However, this was initially envisaged as an applied art strategy and left the historic division between fine art and craft and design unchallenged. As the above quote suggests, Malevich as a member of the pre-revolution avant-garde continued, like Kandinsky, to cling to a rather elitist conception of painting as a meta-discipline informing all the other arts.

Certainly, within the pedagogic context in Vitebsk and in the VKhUTEMAS in Moscow, or indeed the Bauhaus in Germany, the disciplines of drawing and painting remained the starting points for such interdisciplinary application. The "purists" Kandinsky and Malevich continued to use the language of painting as a shorthand for creativity in all the visual arts. The dynamic between analysis and synthesis, grounded in the formal innovations of painting, continued to be seen as the key to facilitating collaboration between the visual arts and other art forms such as theatre, music, photography and film. In effect, such synthetic work in its imaging elements would still be governed by the "deep structure" of non-objective painting. Kandinsky, appointed in early 1922 by Walter Gropius to the Bauhaus as a professor with a mission to develop an applied interdisciplinary impetus within the academy, continued to insist that "every one paints at the Bauhaus" (Kandinsky 1994:702). As he put it, "in the Bauhaus Painting is treated as a co-organizatory force, i.e., the student is led beyond the bounds of painting, albeit by means of its logical character, toward the synthetic Work". Painting was king. Its non-objective stance, research-based trajectory and expressive concerns were seen as underpinning the exploratory methods and interdisciplinary perspectives encouraged in the new art institutions.

7.5 The Constructivist challenge

Even before the revolution, the expanded notion of painting emerging among the avant-garde was pushing beyond the constraints of easel painting towards a more general conception of visual arts practice dealing with materials and constructed forms and with a range of media, including photography, film, print and graphics. Constructivism, with its emphasis on function, method, laboratory investigation and iterative design process, completely changed the terms of the debate. It was no longer a case of applying visual styles originating in the work of painters to achieve decorative outcomes. What was envisaged was the wholesale abandonment of a fine art aesthetic in favour of a design logic. This would involve a "back to basics" trajectory dedicated to fabrication, testing, product innovation and social purpose. The formal language and compositional methods of visual art might still be in play, but would now be allied with modes of scientific and organizational enquiry focused on constructing and testing prototypes for potential industrial application. What was envisaged, in other words, was a design process that would feed into industrial mass production, thereby transforming the lived environment of a revolutionary society.

What did these Constructivists stand for? What was their paradigm of artistic research? How has this influenced contemporary design practice?

Christine Lodder (1983:7) gives us a useful summary of the Constructivist approach.

> The Constructivists used the term "laboratory work" to describe formal investigation – usually in three dimensions but sometimes two – which was undertaken not as an end in itself, nor for any immediately utilitarian purpose, but with the idea that such experiments would eventually contribute to the solution of some utilitarian task. Theoretically, therefore, "laboratory work" consisted of artistic explorations into the component elements of form, material, colour, space, and construction, which were initiated not for their own sake, but with the longer-term purpose of establishing the objective bases of artistic criteria, and the general laws of their interrelationships, so that these could be exploited later in the design process.

Constructivism as a movement emerged from developments within non-objective art and carries the stamp of its origins in painting and relief fabrication. However, Constructivist laboratory work as a form of task-oriented research differed significantly from the sort of esoteric research programmes envisaged by Kandinsky and Malevich. In many ways, the intellectual lineage of such pure artistic research, with its focus on affect, lay in nineteenth-century romanticist concerns with the science–art nexus and with art as a privileged mode of cognition and communication. Constructivism entailed a completely different model of artistic research and education, one based, as we have seen, on laboratory work and on working closely with industry.

We can overemphasize the applied ethos of Constructivism. In many ways, it remained imbued with the non-utilitarian and utopian spirit of Suprematism. In time, it was to be succeeded by Productivism, which sought to harness all art practice to the design challenges confronting Soviet society. In that sense, it is an intermediate movement. So, for example, the models constructed by like Tatlin, Rodchenko and Lizzitsky – such as Tatlin's iconic *Monument to the Third International* (a partly completed workshop model is seen in Figure 7.9 left and in a street display in Figure 7.9 right), or Rodchenko's mobiles of the early 1920s (Figure 7.11) and Lizzitsky's *Prouns* and architectonic maquettes of the same period (Figure 7.8 left) – cannot be regarded as simply design prototypes. They are, after all, imaginative models and can be regarded as "theory machines", the term Martin Kemp uses to describe Leonardo's models and capacity to use design intelligence to understand physical phenomena. Interestingly enough, Vladimir Tatlin took inspiration from Leonardo's flying machine project (Figure 7.13) and designed a similar apparatus (Figure 7.12), envisaged as a recreational device which might make

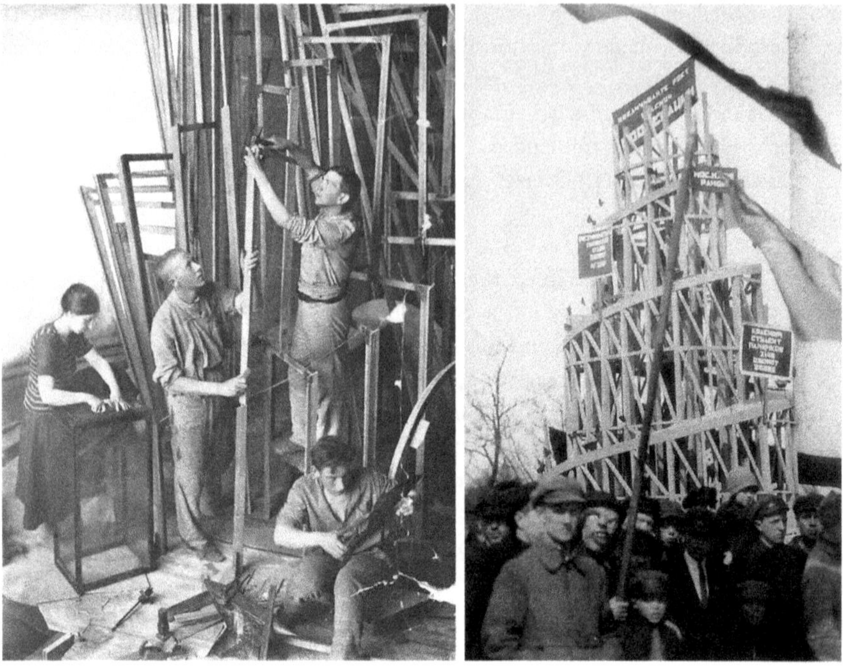

FIGURE 7.9 (left) Tatlin and students building a model of his *Monument to the Third International* (1920); (right) Tatlin's unfinished model in a street display (1920)

Source: public domain

flight accessible and understandable to Russian families. For the Constructivists, as with Leonardo, science and engineering, art and material application are melded together into an investigative practice in and through model making and design experiment. Moreover, for the Constructivists, this experiment is social and political as well as technical in so far as it envisages new forms of living and not just technological advance.

The Constructivist "laboratory" lay, then, somewhere between the traditional artist's studio and the design office of an industrial concern. Constructivists defended the creative autonomy of the artist while celebrating the intellectual curiosity and experimental methods of the scientist and embracing the purposiveness and professional discipline of the industrial designer. And, indeed, the "live in" studio that Rodchenko developed in the attic of a large Moscow house after 1922 with his life-long partner and working colleague Varvara Stepanova (seen at work in her studio in Figure 7.14 left) created a prototypical Constructivist space, facilitating creative innovation and functional design tasks in what remained an artisanal setting.

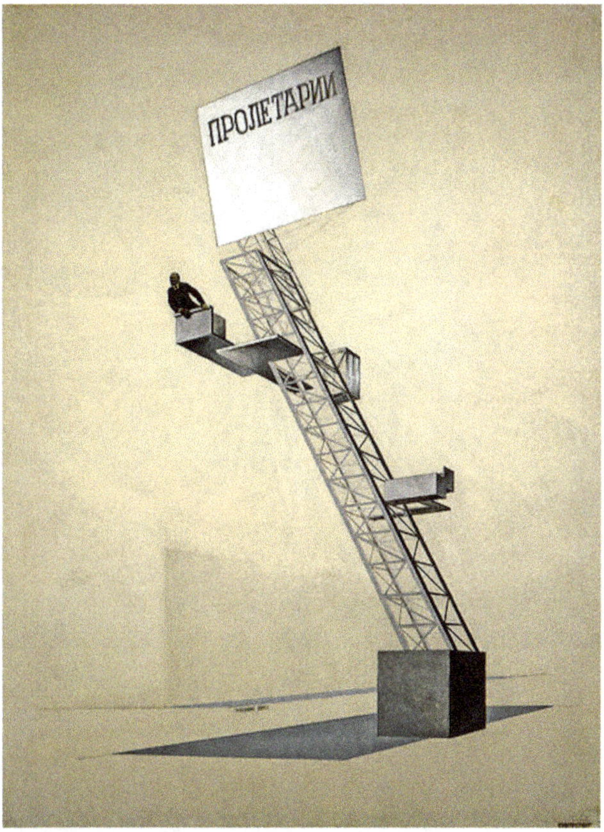

FIGURE 7.10 El Lizzitsky, *Lenin Tribune* (model for a publicity tower) (1920), State Tretyakov Gallery, Moscow

Source: public domain

As the writer Vasily Katanyan recalled (Rodchenko and Lavrentiev 1989:12),

> it was a huge studio. Tables, wardrobes and curtains served to divide it into a bedroom, dining room, and two work places for artists. And of course, a geometric photo-laboratory. Rodchenko–Stepanova, anti-easel painters, or whatever term you like: stage designers, book designers, advertising designers, designers of lighting, producers. The contemporary word "designer", though, was not known then. Instead of easels and palettes they turned their tables towards the light and acquired drawing boards, rulers, set squares, Indian ink and jars of gouache. There was also a press to dry the photomontages.

However, it was probably in the transformation of the workshops of the VKhUTEMAS undertaken by Rodchenko after 1921 (seen in Figure 7.14 right) that the Constructivist laboratory as a collective practice really took shape. In 1921, in

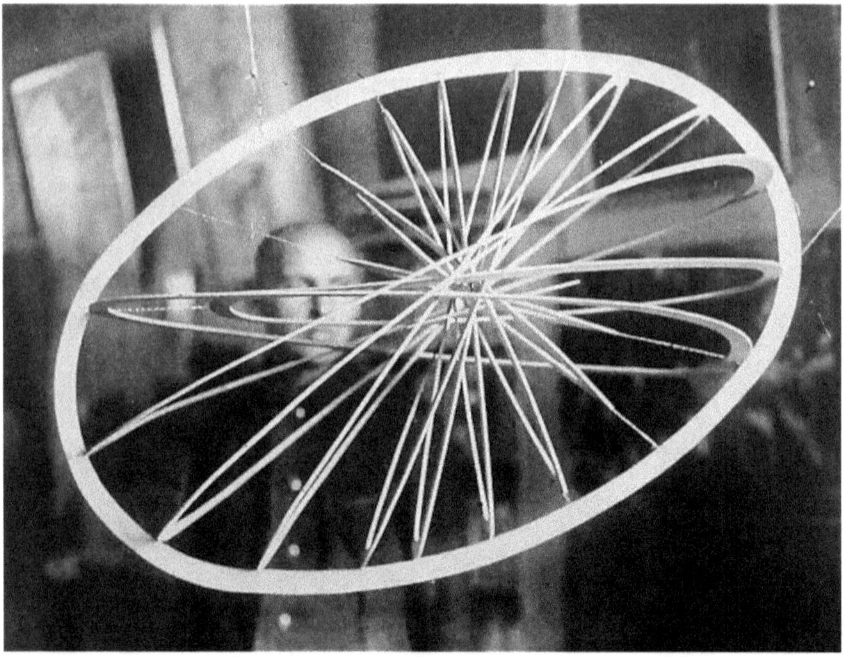

FIGURE 7.11 Aleksandr Rodchencko, *Spatial Construction*, with artist in background

Source: courtesy of the A. Rodchenko–V. Stepanova Archive Moscow

the context of the restructuring of the State Free Art Studios, Rodchenko was appointed as a professor to the metals workshop of VKhUTEMAS. He soon found himself leading the Basic Course, the area of the curriculum in most art and design colleges where pedagogic innovation was most likely to be focused. And just as the pedagogic encounter was the context for the articulation of a theoretical understanding of the new art, the instructional and experimental workshop became the prototype for the new design studio. Stepanova gives us an example of her partner's early work in VKhUTEMAS, indicating the playful and experimental approach he intended to employ in teaching basic awareness of materials and colour within design (ibid.:43):

> A month has passed since Rodchenko was appointed Professor in the Fine Art Department of the VKhUTEMAS. It seems he is gaining in popularity. His pupils are interested in him and some are persistently trying to be taken on by him. He set a still life there which was particularly interesting. All the ingredients of the course he has taken from home. It consisted of a porcelain pharmaceutical mortar, bits of glass, black paper on a board, and two sheets of pink and red glossy papers folded in pleats to be used as a background, which made the whole still life look very colourful. This still life surprised everyone!

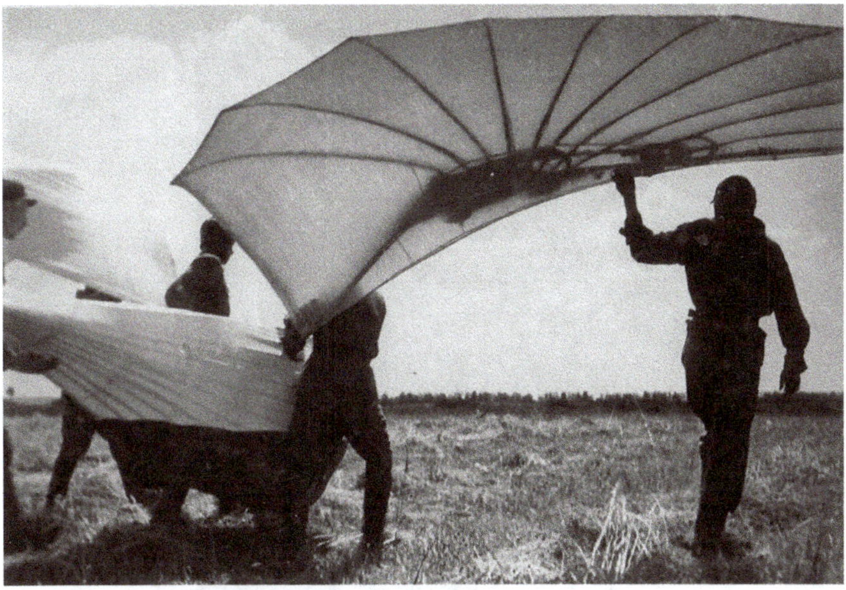

FIGURE 7.12 Tatlin's flying machine on display at a glider show in Moscow (1933), State Tretyakov Gallery, Moscow, unknown photographer

Source: public domain

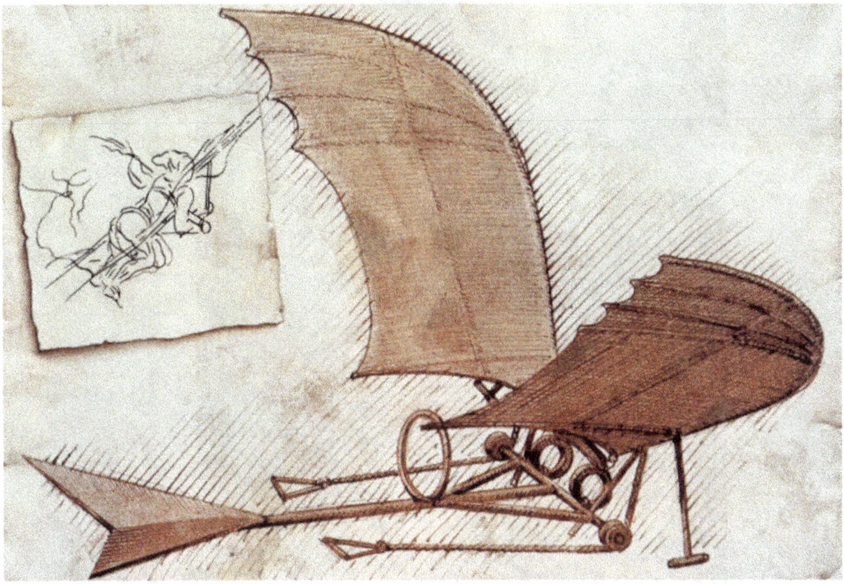

FIGURE 7.13 Leonardo da Vinci, sketch for a flying machine (c.1500), Biblioteca Ambrosiana, Milan

Source: public domain

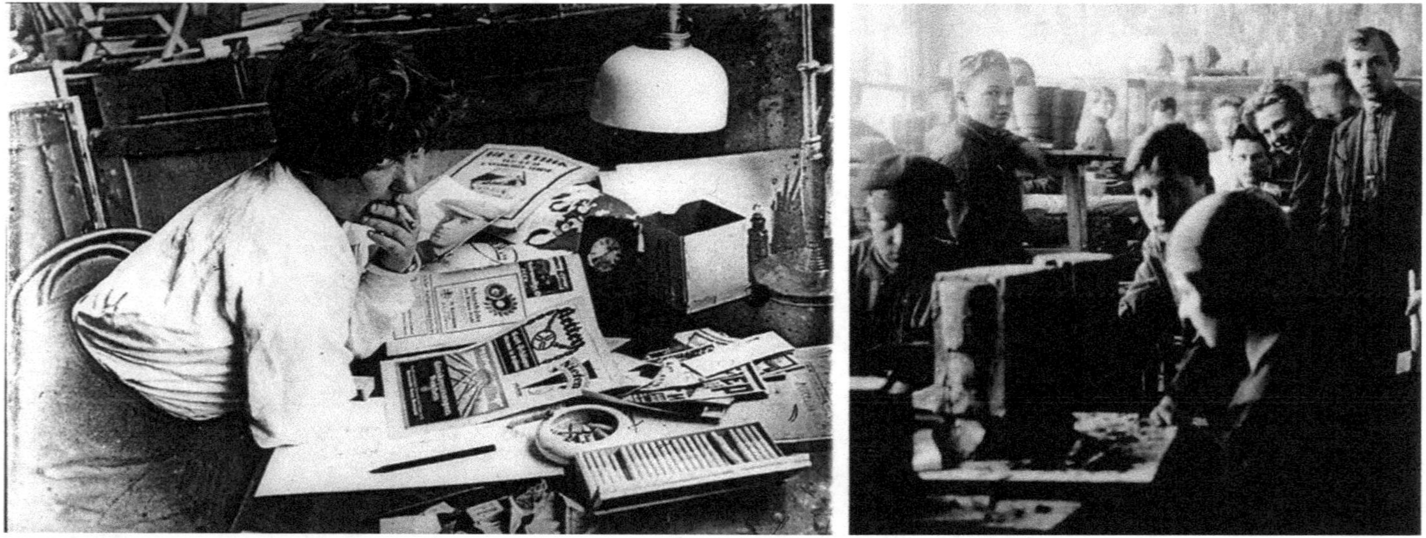

FIGURE 7.14 (left) Stepanova at work in her studio (1923); (right) design workshop at Vkhuktemas (1925)

Source: courtesy of the A. Rodchenko–V. Stepanova Archive Moscow

Constructivism managed the transformation of the design studio from the atelier model (and aesthetic whims of the individual artist) to a laboratory where groups of artist–researchers pursued lines of concrete design investigation informed by basic research into material, form, function and user need. But, as the above quote suggests, the playfulness of the fine art studio was preserved!

However, as Rodchenko sternly implored his fellow artists and students (12), in what was probably a swipe at Malevich:

> work among other people, for others and with others. Down with monasteries, institutes, workshops, studios and islands. Consciousness, EXPERIENCE, goals, CONSTRUCTION, technology and mathematics – these are the brothers of contemporary art.

In the VKhUTEMAS, a distinctive engineering and professional design ethos began to emerge. This focused on the workshop investigation of materials and on real-world design challenges involving the building and testing of prototypes. As Rodchenko recalled (Milner 1983:194),

> I was head of the metal workshops at the VKhUTEMAS, and in charge of projects. Out of the same department which once made mounts for icons, votive lamps and other church plate, there began to emerge constructors producing electrical devices, metal objects of daily use and metal furniture.

As Rodchenko declared, "all new approaches to art arise from technology and engineering and move towards organization and construction" (Lodder 1983:88). The emphasis was now on scientific objectivity, project experimentation and methodological exactitude, with the acknowledgement that "inventors are artists and in essence the artist is an inventor" (Rodchenko and Lavrentiev 1989:14). Indeed, the key notion in Constructivism became that art was invention and as such could be modelled on applied scientific method – that is, not on the pure sciences of physics and mathematics which inspired Kandinsky and Malevich, but on the applied sciences of engineering, design and manufacturing, driven by practical concerns.

Soon the Constructivist group within INKhUK found themselves on a collision course with Kandinsky, whose research plans for the Institute were based very much on his own distinctive artistic vision and methodological concerns, as sketched in his *On the Spiritual in Art*. Kandinsky chaired the Praesidium of INKhUK and had the ear of Lunacharsky, a cultured man who had lived in France prior to the revolution. As one familiar with both French and German modernist art, he acknowledged Kandinsky's international reputation and integrity. However, Kandinsky was coming under considerable pressure from the Constructivists bloc, led by Rodchenko, to adopt a more productivist prospectus for Moscow's new art institutions.

Within INKhUK – which despite Kandinsky's best efforts seemed to function more as a radical think tank for contemporary artists rather than as a formal research institute – this radical design approach was becoming the dominant current. Embracing as it did a social utilitarian ethos that chimed well with the revolutionary era, Constructivism was poised to replace Kandinsky's academicism as the most advanced thinking on art in Russia. History was on the side of the Constructivists (for a while, at least).

7.6 From Suprematist painting to Constructivist imaging

The younger artist–pedagogues like El Lizzitsky, Rodchenko and Moholy-Nagy (in Hungary and Germany) had cut their artistic teeth in non-objective painting. Now they operated with a model of art practice which sought practical application and in which research practice was harnessed to design iteration and the exploitation of new media. The investigative focus of Constructivism became the exploration of materials and of constructive elements assembled with a defined purpose, rather than a concern with abstract painterly form or with the psychology of artistic expression and communication. The emphasis was no longer on the artwork as a *composition* serving an aesthetic end, but as a *construction* increasingly conceived as a potential design application with social utility.

Indeed, Malevich partly concedes that painting can no longer claim a predominant role within the visual arts. He himself more or less ceased painting on canvas after 1920, preferring the forms of public art – posters, banners and installations – executed on the streets of Vitebsk. He was also increasingly drawn to three-dimensional experimental work that opened up a dialogue with architecture. In the early 1920s, he also became increasingly interested in both photography and cinema and wrote about these media throughout the second half of the decade, including discussions of the work of Eisenstein and Dziga Vertov.[24] He also planned an experimental film with the German artist Hans Richter.[25] However, his major interest remained the relation of these lens-based media and the forms of visual representation they facilitate, to the abstract language of non-objective painting. It is the latter, he believes, which continues to set the parameters of visual innovation.

Needless to say, with the introduction of reproductive media like photography and film, typography and graphics, into art practice – media with their own distinctive histories and professional identities – these debates took on a whole new form. New critical tools were needed to make sense of these developments. Photography and film were now at the core of a wider debate about creativity, mass communication and public culture.

For example, Rodchenko, like his contemporary in Germany, Lazlo Moholy-Nagy, had made the journey from painting to photography. Both insisted that photography brought something new to the new abstract aesthetic sensibility. As Rodchenko argued (Rodchenko and Lavrentiev 1989:56–57),

> The new fast and real reflection of the world is in photography. It seems to me that considering its potential, photography should begin to depict the

> world from all angles of vision and develop our ability to look at it from all sides.

In his 1934 programmatic statement *Photography Is an Art* (Rodchenko 2008:10), he makes the case even more boldly.

> Photography has broken free from being secondary and initiating the techniques of etching, painting or carpet making. Having its own way, it is blossoming and fresh breezes that brings a scent that is particular only to photography. New possibilities lie ahead.
>
> The multitude of its aspects are as complex as fine drawing, more interesting than photomontage. The contrasts of perspectives. The contrasts of light. The contrasts of form. Viewpoints that are not possible either in drawing or in painting. Viewpoints with exaggerated foreshortenings and pitiless texture of the material.

Similarly, Moholy-Nagy, who in 1925 published his ground-breaking *Painting Photography Film* in the Bauhaus book series, acknowledged (1969:7) that

> the camera has offered us amazing possibilities, which we are only just beginning to exploit. The visual image has been expanded and even the modern lens is no longer tied to the narrow limits of the eye; no manual means of representation (pencil, brush, etc.) is capable of arresting fragments of the world seen like this.

Malevich remained unmoved. For him, painting remained the critical template against which artistic work in these media will be judged. Suprematism, he insists, is not simply an artistic style but a machine for artistic thinking. In other words, a research problematic. It was, in his words, "a hard, cold, unsmiling system, set in motion by philosophical thought". As such, Suprematist methods were applicable to any medium and artistic task, a position which, as I have suggested, seems to anticipate Conceptualism.

As Malevich argues in a 1927 letter to Lazlo Moholy-Nagy (1978:157), who was by then teaching photography classes in the Bauhaus, painting has to be understood not in terms of its medium specificity but as a conceptual apparatus, a problematic bound by the logic of abstraction.

> The element of Suprematist planes is different, but we are not talking about painting, but only about a method of Art, a spiritually organic means of conveying various sensations. And if it will be possible to convey the same sensations by the photographic method, then photography will become a similar technical medium such as the brush, the pencil nothing more.

Lizzitsky, Rodchenko and Moholy-Nagy were indeed to explore this possibility in their work throughout the 1920s. Each developed a distinctive photographic sensibility producing arresting images. The viewpoints chosen and visual forms and tonality employed achieve a semi-abstract mode of picturing that interrupts the reproductive propensity of photography to be simply the "pencil of nature". Each combines photographic images and abstract graphics, as with Lizzitsky's *Footballer* (Figure 7.15). Each brings a formal geometric language to bear on the photographic images, as seen in Rodchenko's *Steps* (Figure 7.16) with its strong diagonal

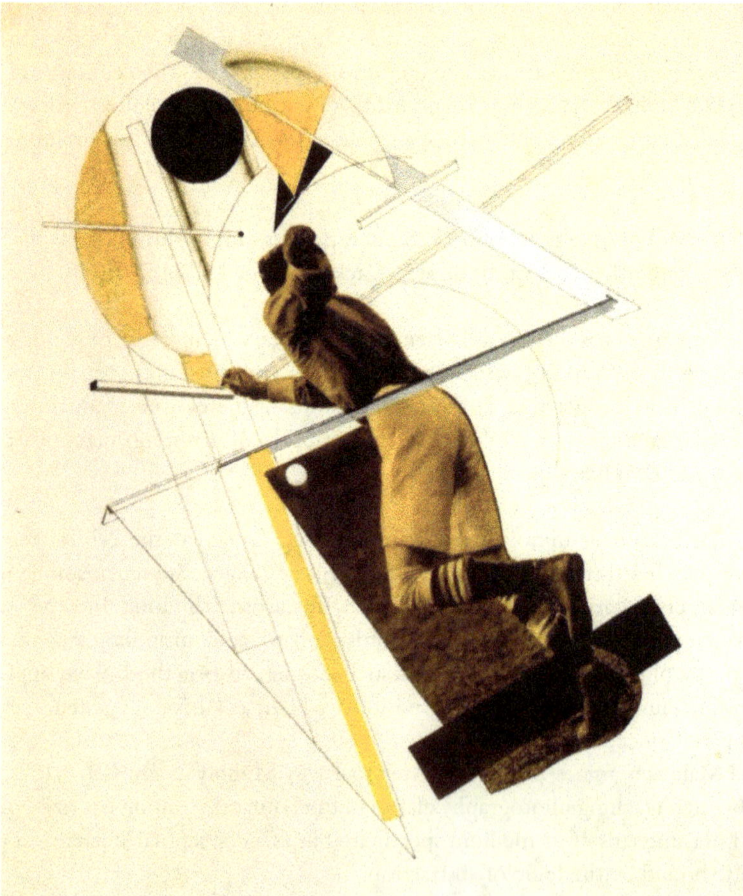

FIGURE 7.15 El Lizzitsky, *Footballer* (1921), illustration for Ehrenburg's "Six Tales with Easy Endings", photo collage, pencil, gouache, ink on cardboard, State Tretyakov Gallery, Moscow

Source: public domain

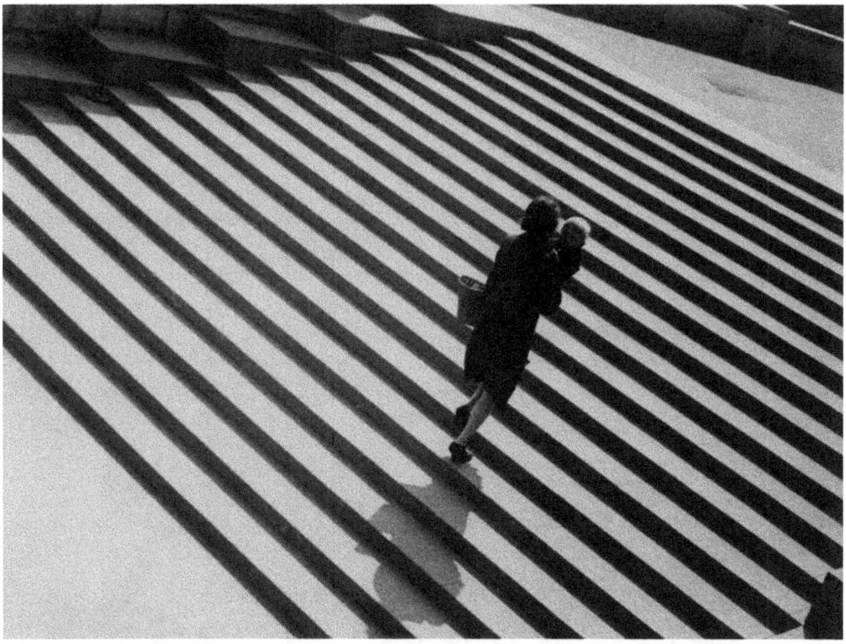

FIGURE 7.16 Aleksandr Rodchenko, *Steps* (1930), gelatin-silver print, 12.4 × 12.7 cm

Source: courtesy of the A. Rodchenko–V. Stepanova Archive Moscow

lines of shadow, or in Moholy-Nagy's circles and diagonals abstracted from his overhead images of a group of sailors at work on a dockside (Figure 7.17). In other words, as Malevich had suggested, they were able to successfully employ "a method of Art" rooted in a non-objective aesthetic to transform a lens-based medium constrained by its indexical character. Each also researched and wrote illuminatingly about photography.

These tensions between painting and photography, film and graphics were highly productive ones that brought a developed research intelligence and visual sensibility to lens and graphic practice but also challenged the limits of easel painting and a fine art aesthetic. Arguably, such tensions could be profitably explored in contemporary creative arts education with a view to both facilitating the visual and critical skills of film and photography students and the technical and professional competencies of fine art students who opt to employ lens and screen-based tools in their work. This historic debate is one that might usefully assist us in our contemporary concerns to find a better fit between fine art and the myriad of applied, design, media and interactive fields with which it co-exists in the art academy today.

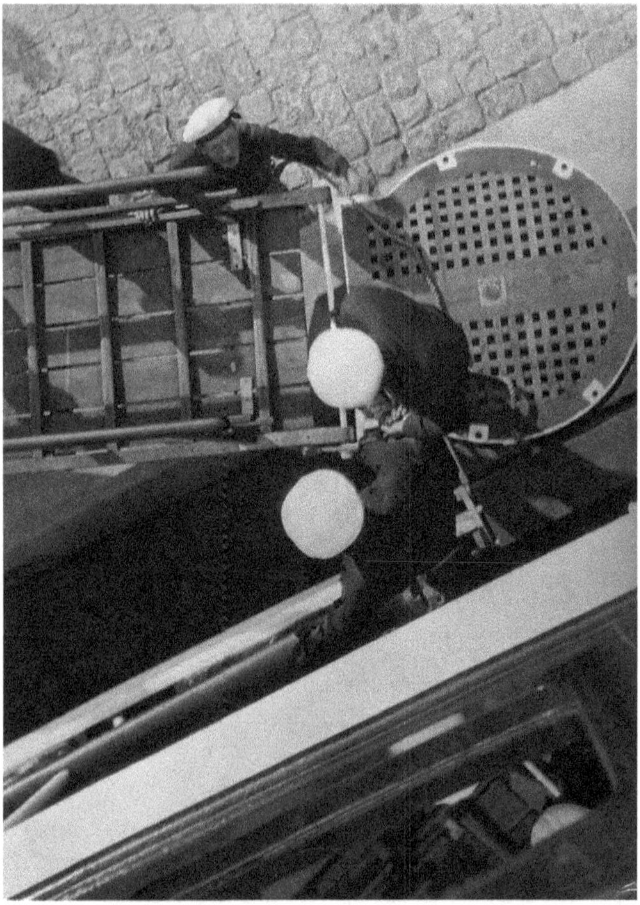

FIGURE 7.17 Lazlo Moholy-Nagy, *Sailors* (1929), Metropolitan Museum, New York

Source: public domain

7.7 The eclipse of the avant-garde

Kandinsky, politically outflanked by the Constructivists and perhaps fearful of a coming era of ideological rigidity, left Russia in late 1921 to take up a post at the Bauhaus. He would never return. He had always sought to balance theory and practice and, like Malevich, believed in a common avant-garde tradition. Moreover, he believed that the analytical momentum present in the new non-objective art was compatible with traditional forms of artistic expression concerned with the communication of emotion through painterly means. He believed his expressive approach could co-exist with the utilitarian aspirations of Constructivism and with the political responsibilities of the artist in a revolutionary society. However, against the prevailing productivist orthodoxy, he continued to believe that whatever stripped down visual language artists now employed, or whatever social and ideological purpose they followed, they should not mute their individual voice or the integrity of their vision, nor should they lose their capacity to communicate with their audience through their chosen medium.

Malevich, installed as Director of the Museum of Artistic Culture in Petrograd after 1923, was able to put up a stubborn resistance to the creeping tides of productivism and socialist realism which increasingly encroached on the avant-garde project. He had been able to develop a specialist research facility for the visual arts with state support and with an adjoining museum effectively at his disposal. He was able to hire a number of talented artists to assist him in his work, bringing a number of his former students and staff from Vitebsk to create a cadre of supportive artist–researchers. Laboratories and specialist departments were set up to facilitate this work. He secured funds for the purchase of studio and laboratory equipment to facilitate his research plans. Programmes of public lectures and seminars were arranged to report on the work of the Institute and to address live topics of debate in contemporary art practice and research. He had found a number of publication outlets for his work and that of his team, showing work at the Venice Biennale in 1924. He would have scored highly in any research assessment exercise!

In Petrograd after 1923, the "Commander" (as he was affectionately known to his colleagues) had created a citadel of Suprematism relatively untouched by the Productivist currents which were transforming INKhUK in Moscow. He had been able to create the sort of creative arts research environment that most of us would give an arm and a leg for and which makes current practice-based research initiatives look piecemeal, ill thought out and under-resourced!

However, his elevated position was short lived. By the summer of 1926, a number of opponents of Suprematism – particularly those gathered around the aggressive socialist realist agenda of AKhRR, the Association of Artists of Revolutionary Russia – were questioning the work of GINKhUK and indeed the entire Suprematist project. Malevich had already clashed with these proto-socialist realists attacking their naturalist portrayals of Russian peasants and daily Soviet life (a critique he was to extend to Eisenstein's stylized representation of the Russian worker). He was, however, careful not to pick an ideological fight (Malevich 1978: 230): "we are not enemies, but we do have different directions: at their base, they have 'representation' (mimesis), and at ours – science and life itself".

The Institute's annual show in June of 1926 provided the occasion for an AKhRR riposte. The work of Malevich and his colleagues Matyushin and Mansurov was shown in this exhibition, with Malevich (Figure 7.18) exhibiting a number of the abstract three-dimensional architectonic works of the sort that Lizzitsky had started constructing in Vitebsk. The show indicated that he had taken on board aspects of the Constructivist critique of the painterly emphasis of Suprematism. Indeed, his work was moving decisively towards an engagement with space and material and with architectural concerns which seem to sketch an imagined cityscape for Soviet urban life.

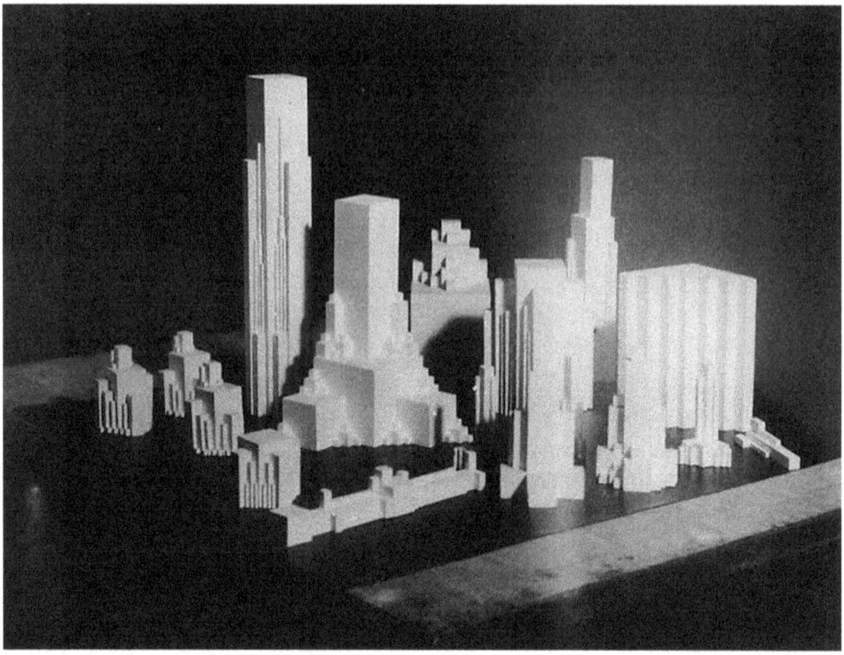

FIGURE 7.18 Kasimir Malevich, *Arkhitekton* (1926–1928), State Russian Museum, St Petersburg

Source: public domain

The show was, however, severely attacked by the party newspaper *Leningradskaya Pravda* and condemned in a searing review by a militant critic of socialist realist hue, Grigory Seryi. He did not mince his words (Catherine Douglas 1980:15).

> A monastery has taken shelter under the name of a state institution. It is inhabited by several holy crackpots who, perhaps unconsciously, are engaged in open counter-revolutionary sermonizing. And making fools of our Soviet scientific establishment. As for any artistic significance in the "work" of these monks, their creative impotence hits you right in the eye at the first glance. (. . .) The time has come to say: it is an absolute scandal that Glavnauka [the main Soviet Science Administration] hasn't yet put an end to this disgrace that it has let loose under its protection. (. . .) Now when gigantic tasks are towering before proletarian art, and when hundreds of really talented artists are going hungry, it is criminal to maintain a huge, magnificent mansion so that three crazy monks, at government expense, carry on artistic debauchery or counter-revolutionary propaganda.

The Petrograd office of the national science administration, which up to this point had backed Malevich's project, became nervous and set up a review committee to examine the work of Malevich and his team. The omens were not good. In November 1926, Malevich was dismissed from his post as Director and GINKhUK closed down. One of the first casualties of the demise of the institute was Malevich's planned publication of "Introduction to the Theory of Additional Element in Painting", a paper which would attempt to summarize the research on the development of modernist painting that he had been working on for the previous ten years. The piece would eventually be published in 1926 in Germany by the Bauhaus as a chapter of his book *The Non-Objective World*. Malevich, visiting the Bauhaus in 1927, confided to Lizzitsky that the atmosphere in Germany felt liberating after the restrictions he faced in Russia.

By then Lenin was long dead, Trotsky marginalized and the Left Opposition in disarray. Stalin had positioned himself to assume control of the Politburo and the Party. Lunacharsky, the most important ally of the avant-garde, was struggling to protect himself after Stalin moved against the Left and was in no position to champion Malevich's cause. An era of reaction was starting in the USSR. Socialist realism was becoming the order of the day. The period in which the artistic avant-garde had enjoyed unique freedoms to experiment, innovate and develop new strategies of public address with the blessing of the state was coming to an end.

7.8 Conclusion

The avant-garde has, then, bequeathed two distinct models of artistic research to us. One of these, rooted in painterly concerns, has been predominantly concerned with formal and expressive matters and the work of art as composition. It focuses on the sensate and communicative dimension of art. The other model focuses on the artwork as construction and concerns itself with the exploration of materials, form, function and social utility. It conceives of the art process as a design task concerned with the construction and testing of prototypes in an industrial society characterized by the mass production of consumer goods. The Russian and German avant-gardes struggled to assimilate these two approaches to artistic research and practice and to align the methods of formal enquiry traced by Kandinsky and Malevich with those of grounded design investigation. Most of the avant-garde in fact sought to defend a common tradition of artistic experiment and innovation, seeing a continuum between theory and practice, experiment and application, expression and technical design. And they had some success in doing this despite, on the one hand, the reservations of the purists about "applied art" and, on the other, the diatribes of the Constructivists about the redundancy of easel art and, indeed, of fine art and aesthetics in general. However, in the end a series of seismic political events – the ascent of Stalin, the failures of "socialism in one country" and the triumph of fascism in Germany – brought an abrupt end to the utopian dreams of the avant-garde.

If a hundred years later we look back at this formative period in the history of the creative arts, we do so to revisit and to repurpose the insights, aspirations and commitments present in this avant-garde adventure. Key among these is the recognition of art practice as a mode of investigation and as a form of critical intelligence to be set alongside philosophy and science. We also celebrate a body of creative achievement characterized by experiment and methodological exactitude. These avant-garde artists developed their capacity for theoretical reflection in part through their commitment to pedagogic innovation and political engagement. If this is indeed a foundational moment for artistic research, can we draw the necessary lessons for the pursuit of creative arts research today?

Notes

1 The art critic Punin went so far as to produce an equation to encapsulate the creative process S (Pi + Pii + Piii + Pπ)Y = T where S equals the sum of the formal principles governing the new art, "Y equals intuition and T equals artistic creation". While Punin argued strongly for an art criticism that would be scientific in character, he recognized the role of intuition in art practice (Bowlt 1988:171).

2 As with contemporary art practice, I am not sure how useful it is to demarcate the art of the radical avant-garde into "pure"/"fine" and applied/design disciplines. All the main figures of the movement aspired to challenge this division and were open to interdisciplinary collaborations with theatre, music and dance.

3 Following vigorous representations from the avant-garde, *Narkompros* granted artists the key role in shaping curatorial activity in the newly nationalized museums. Indeed, *Narkompros* accepted that "they were the only ones able to identify the proper qualities of an artwork and distinguish the novel features of invention" (Kokkori and Bouras 2014:164). As Christine Lodder informs us, in a two-year period after 1918 the Museum service under Rodchenko, assisted by his partner and fellow Constructivist Varvara Stepanova, acquired almost 2,000 artworks for public institutions.

4 John Gage (1999) argues that the model of the visual arts laboratory advanced by Kandinsky for INKhUK, was based on the psychology laboratory established in 1879 by Wilhem Wundt at the University of Leipzig in Germany. This makes sense given the psychological character of the research programme drawn up by Kandinsky for INKhUK.

5 It was, I guess, circumscribed by a rather mechanistic conception of how the spectator responds to the artwork. This was probably based on the reflex psychology which held sway in Russia at the time.

6 The First Conference of Museum Affairs held in February 1919 and attended by Lunacharsky identified the aims of the new Museums of Painterly Culture as being "(1) Artistic and scientific (2) Demonstrative (3) Pedagogical" (Kandinsky 1994:435), with four departments envisaged in each museum: "(1) Experimental technique, (2) Industrial art, (3) Drawings and Graphics, (4) Synthetic art" (Bowlt 1988:195).

7 As Yurkevich had argued in a piece written for the Suprematist journal in 1916 in the wake of Malevich's 0–10 Exhibition (Drutt 2003:54),

> perhaps the New Painting is not at all for easel pictures or for exhibitions. The old methods for reproducing and of viewing art are probably not suitable now. New wine demands new bottles. This is the urgent question.

8 Five Museums of the Culture of Painting were opened in Moscow, Petrograd, Nizhnii-Novgorod, Vitebsk and Kostroma, "constructed on the principle of the evolution of purely painterly and plastic forms of expression". As Lindsay and Vergo tell us (1994:435),

the largest of these was in Moscow and contained "examples of the work of the avant-garde, including Kandinsky, Malevich, Popova and Tatlin".

9 Such diagrammatic representations were at the time widely used for didactic purposes by a range of scientific disciplines in Russia. Glavnauka, the central administrative body coordinating scientific research, was highly interested in the diagrammatic representation of systems of knowledge. It organized a major exhibition and conference in 1925 to show examples of such didactic materials and to bring together scientists, artists, designers and educators to discuss these emerging audio–visual techniques. Malevich contributed a number of research charts to this exhibition and, as Kokkori and Bouras report (2014:167), his work "was considered innovative and extremely important for the demonstration of scientific methods in the study of art and the dissemination of artistic knowledge".

10 As Maria Kokkori and Alexander Bouras argue (in Achim Borchardt-Hum 2014), Malevich proposed that any museum exhibiting his work should showcase, along with the main exhibits, the contextualizing research documentation he had developed.

11 See Witkovsky (2017).

12 While Malevich produced a considerable body of writing, only a small amount of this was published during his lifetime and often with different titling and versioning. Troels Anderson's edited volumes (1978) remain the most comprehensive source of Malevich's writing translated into English.

13 The language Malevich employs to trace the development of modernist art is biological and medical, rather than sociological or semiotic. In common with other late nineteenth-century social theorists such as Herbert Spencer and Emile Durkheim, who employ organic analogies to model social and cultural systems, Malevich conceives of the modernist art movements as systems which are quasi-biological in form. As such, they are seen as prey to various viral conditions that attack them from outside, transforming them internally, much as tuberculosis or other viral conditions attack the body (Zhadova 1982:46).

14 Art History proposed to employ techniques of "scientific observation" in its analysis of cultural works. In effect, this meant providing a close reading of the forms of composition present in paintings in a particular era in order to arrive at an identification of ideal typical styles characterizing the work of specific artists, national traditions and historical periods. Art history could mutate into *Kunstwissenschaft* or the "Science of Art", and this is probably where Kandinsky and Malevich got their initial formulation of a science of painting from.

15 As Patricia Railing persuasively argues (1989), Malevich adapts the art historical concept of "style" transforming into one of "system".

16 Worringer was an enthusiastic supporter of German Expressionism. Knowledge of his work quickly circulated among the German expressionist painters of Munich. Kandinsky, David Burliuk and Nuam Gabo had studied in the Bavarian city and knew Worringer, quickly assimilating his account of abstraction. Worringer's insights into abstraction as a style were seized upon by a Russian avant-garde anxious to establish the intellectual credibility of their own practice.

17 In this paper, Foster draws upon the 1928 essay by the Russian formalist literary theorists Mikhail Bakhtin and Pavel Medvedev. *The Formal Method in Literary Scholarship* (Cambridge, MA: Harvard University Press, 1985, pp. 41–53).

18 Leporskaya began registration of Malevich's work – drawings, paintings and written documents – under his supervision in 1926 and subsequently looked after this archive after the artist's death right up to her own in 1982. See Troels Andersen (2011).

19 Malevich was to befriend a teenage Roman Jakobson (later to become one of the leading structural linguists in the world), inviting him to accompany him to a planned visit to Paris and act as a translator and "explain" Suprematism to the French. Jakobson later elaborated a theory of phonetics, which can be seen, in turn, as indebted to the sonic experiments of the Zaum poets, based as it was on "the sound as such" freed from

referential meaning. Malevich also saw a continuity between non-objective art and the concrete and non-referential character of Zaum poetry.

20 As Shklovsky argued in his essay *Art as Technique* (2016:16),

> the purpose of art is to impart the sensation of things as they are perceived and not as they are known. The technique of art is to make objects "unfamiliar", to make forms difficult, to increase the difficulty and length of perception because the process of perception is an aesthetic end in itself and must be prolonged. Art is a way of experiencing the artfulness of an object; the object is not important.

21 Malevich's credentials as a revolutionary artist were well-established by then. In 1917, following the February revolution, he had been elected Chairman of the Art Section of the Moscow Soviet and proposed the organization of a People's Academy in Moscow and Petrograd. He also proposed a foundational programme for the country's art schools. In October of that year, he was appointed Commissar for the Protection of Monuments of Antiquity and in November 1918 to the commission for the organization of a planned Museum of Artistic Culture. In early 1919, he took up an appointment at the Free State Artist Studios, later to become the Vkhutemas and in the autumn of that year moved to the Vitebsk Practical Art Institute.

22 Nina Kogan was born in 1889 in St Petersburg and died there in 1942 of starvation during the siege of Leningrad. She helped to found a painting studio for workers at the former Winter Palace (1917–1918) and later worked on the planned Museum of Painterly Culture in Petrograd (1917–1918). Introduced to Kazimir Malevich, she moved to Vitebsk in 1919, where she led the introductory course in painting and drawing. She contributed articles to the UNOVIS almanacs, decorated trams and designed theatre costumes. She moved to Moscow in 1922 where she worked at the Museum of Painterly Culture (1922–1924).

23 Of course, it can be argued that Malevich's *Black Square* anticipated such developments and that there was no absolute divide, either ideological or aesthetic, between Suprematism and Constructivism in the early 1920s – simply an evolution conditioned by the context of the revolution. Certainly, as a counter-relief the *Black Square* acknowledges the potential for transition to a broader programme of artistic innovation that would involve incorporating work with textures and materials in space and also engaging with graphic communicational techniques. The iconic black square with its distinctive photogenic quality had quickly lost its auratic, purely painterly quality and lent itself to mass reproduction. It appeared first on the cover of the 1915 and 1916 publications on Suprematism and was later used as a seal on UNOVIS publications after 1920.

24 "On Exposers. Posters" (1925), "And Painted Faces are Victorious on the Screen" (1925), "The Artist and the Cinema" (1926), "Scientific Film to Hans Richter" (1927), "Letter to Moholy-Nagy on Photography" (1927), "Cinema, Gramophone, Radio and Artistic Culture" (1928), "Pictorial Laws in Cinematic Problems" (1929). All of these pieces, bar the last, are available in the Troels Andersen collection of Malevich's writing.

25 Malevich, during a visit to Berlin, sought out Richter, the former Dadist and member of the *Blau Reiter* group, who was by then the leading avant-garde film-maker in Europe. He approached him about making a creative documentary film on Suprematism and drafted a three-page treatment for the proposed film, together with illustrative sketches. Richter later visited him in Russia to advance the project, but Malevich failed to secure support from the Soviet authorities and later attempts by Richter to produce the film in the US also failed.

PART III

Interdisciplinary dialogues

8

ART AND THE SCIENCES OF MAN

Reframing history

8.1 Introduction

In the following two chapters, we explore the interdisciplinary dynamic between the human sciences and the creative arts. I look first at art and history, focusing in particular on how the artistic imagination has contributed to recent development in what has been labelled "'postmodernist' historiography". In the following chapter, I plot some of the recent interactions between anthropology and art practice and explore the notion of an "ethnographic turn" in the visual arts.

The last hundred years has seen a dramatic growth in fields of study like psychology, sociology, economics, social and economic history, linguistics, anthropology and cultural studies. Many of these fields have aligned with various institutional practices. What, then, should we make of the relation between artistic research on the one hand and the humanities and social sciences on the other? Surely this grouping of disciplines – the study of language, literature, culture, history and society – provides the natural hinterland of artistic research? The disciplines which operate within these fields all take the realms of society and culture as their object of study. Moreover, many of these disciplines have made significant contributions to the critical study of art. Can artistic research, as it searches for its methodological auspices, adapt the research approaches pioneered in the social sciences for use in the field of creative arts research? On the other hand, the visual, sensory and emotional dimensions of human experience have become much more prominent in the humanities and social sciences over the last few years. As a result, many scholars have realized that the creative arts, with their core concerns with the expressive and sensate, can be harnessed to develop innovative research methodologies capable of exploring human experience. A dialogue has begun.

I pursue this initially via a case study in "material memory", reviewing a recent doctoral project on historical memory by Irish artist and former PhD candidate

Margaret Fitzgibbon. I then look at the research process present in two projects dealing with the Spanish Civil War, one by the German painter Pauline Zenk, the other, by documentary film-maker Tricia Ziff. This leads me to explore the field of film history by offering a case study based on my own film work. Finally, I examine the work of contemporary film-maker and artist Hito Steyerl, who like a number of other contemporary artists, such as Lubaina Himid, Mathew Buckingham, Thom Winterburn and Shirin Neshat, brings a distinctive historical sensibility and critical awareness to her conceptually driven art and film works. All of these artists wrestle with the burden of history in their work. All seek to bring a contemporary artistic sensibility to reframing the past.

8.2 Method, craft and science in history

The human science disciplines, while attentive to the advances of the natural sciences, have sought to establish their own field of study. After all, they deal not with micro-physical particles, electrical impulses, biological cells or genetic codes, but with the structuring of the lived world we inhabit and the diverse cultural practices we employ to give meaning to this. Somewhere in this mix lies the sphere of artistic activity.

In the historical, social and cultural sciences (including the creative arts), human beings are both the subject and object of enquiry. We are investigating and acting upon ourselves. And, as sentient beings, we employ tools of empathy, cultural understanding and expressive practice to "get at" historical and cultural realities. As researchers, we operate from a necessarily partial subject position grounded in specific social and political determinants. We seek to understand, but too often misunderstand, our subjects from the position of our own ideological fixations and prejudices. And this is why *reflexivity*, or an awareness of our subject position and of the limits of the methodologies and concepts we employ to understand the cultural world and the subjects we interact with and study, has assumed such importance within historical and cultural analysis. To complicate matters even further, our capacity to understand the past is itself time-bound. The concerns of the present tend to order our understanding of the past. If the past is a foreign country, it is one we navigate using our current maps. In this sense, as the Italian philosopher of history Croce argued, all history is contemporary.

Historians have at various times had to defend the integrity of their discipline against positivist arguments about its supposed lack of scientific rigour. Some of these arguments bear an uncanny resemblance to those now being employed to castigate artistic research today. However, historians have by and large weathered the storm and managed to give a fairly clear account of the methods they employ to assess evidence, to build arguments and to arrive at explanations, without having to say too much about the scientific status of their discipline. Their attention to the researcher's craft might provide a useful example for artistic research to follow as it seeks to construct its distinctive research culture based on our own field and studio practices.

At the end of the nineteenth century, the disciplines of psychology, sociology, anthropology, linguistics and literature, art history, emerged as distinctive fields of research. A fierce critical debate took place on the scientific status of history (and the social sciences). This was conducted most ferociously in the German academy.[1] The natural sciences were seen as concerned with universal laws governing the physical universe. However, positivists saw the methods of science as also applicable to social, psychological and historical and cultural phenomena. Others argued that history and the social sciences employed different modes of explanation based not on binding laws linking cause and effect but on assembling and interpreting specific constellations of historical facts, actions and cultural meanings. The law-like or *nomothetic* characteristics of the natural sciences were contrasted with the particularizing or *idiographic* character of the historical and cultural sciences concerned with specific historical events, social actions or cultural practices.

As we have seen, artistic research also deals with particulars – individual works, specific artists, processes and perspectives – rather than binding generalizations. In this sense, it is idiographic. Also, artistic criticism engages with the art of the past through the optic of current preoccupations. This applies equally to the study of Renaissance artists, those of the modernist avant-garde or indeed contemporary artists. Like history, art research deals with works, cultural practices, styles, sensibilities and representations which require interpretation, rather than events that can be causally explained by reference to universal laws. There are no general laws in art; nor are there objective valuations.

For historians, causal attribution (establishing one event as the necessary or sufficient condition[2] for the occurrence of a subsequent one) goes hand in hand with narrative strategy. Despite their twentieth-century flirtation with notions of scientific method, historians are ultimately storytellers and perhaps closer to film-makers than scientists.[3] Their research may be conceived as fact finding missions based on the location and interpretation of sources and on a critical evaluation of evidence. However, in the writing of history, such empirical considerations have to be balanced with matters of literary style and perspective in the writer's exegesis. Figurative as well as evidential elements feature in historical accounts, even if historians are not always ready to acknowledge these interactions.

In the early twentieth century, certain schools of historiography – such as the French Annales historians[4] or variants of Marxism[5] in Germany and Britain – were attracted to the logical rigour and explanatory power of the natural sciences. While historians like Marc Bloch and Emmanuel Le Roy Ladurie remained sceptical about the totalizing mission of positivism, historical research in its methods moved closer to the emergent social sciences. These new fields employed general concepts and models which seemed to offer greater explanatory purchase than the traditional focus of history on fact, event and actor.[6] Historians began to use both statistical approaches and quantitative modes of presentation and aspired to empirically grounded, law-like generalizations. History also began to engage in complex model building – for example, elaborating notions of feudalism or capitalism as time-bound formations. It used general concepts and ideal types,[7] drawn from sociological theory. All of these research strategies downplayed the role of the celebrated individual in historical process. Indeed, some sought to cast their discipline within the prevailing positivist ethos as a *protoscience*, which would evolve in time into a fully scientific form.[8] Some historians became distrustful of the storytelling dimensions of traditional history writing and indeed with the apparent imprecision of the ordinary language employed in historical narrative in comparison with the mathematical precision achievable with quantitative data. A revolution in the researching and writing of history seemed to be underway. Certainly, a dialogue with science and its ideals of objectivity, systematicity and predictive power began in earnest, particularly in the US.

Thus, analytical philosophers of science, such as Carl Hempel, a German logician concerned with identifying the formal nature of scientific explanation, turned their attention to history. Hempel argued in his classic paper "The Function of General Laws in History" (1942) that while history with its preoccupation with the unique event and role of human intervention was not yet a science and remained at the "chrysalid stage of proto-science", it could however become one by adopting objective scientific methods and explanatory criteria. Positivists like Hempel awaited a collective act on the part of historians to accept the principle of methodological unity of the sciences and to adopt the logic of scientific discovery, which they felt gave this principle its underpinning.

As we have seen, the notion of the methodological unity of the sciences weighs heavily on contemporary discussions of artistic research. University research managers often seem eager to impose a degree of compliance to scientific norms on creative art academics. Indeed, some art educators also feel that artistic research is currently in a "larval state" vis-à-vis the more established social science and humanities disciplines, whose research methods, they argue, might usefully be adopted by arts researchers.

But in the debate with logical empiricism, historians saw off the reductionism implicit in the positivist model. They were able to robustly defend their disciplinary practices. For example, William Dray (responding in 1957 to Hempel) proposed a model of historical explanation much closer to what he believed historians actually do when seeking to account for historical events. In history, the nexus between the initial conditions (C1, C2, . . . Cn) impacting upon the event or act to be explained (E) is traced not through reference to a set of laws, whether causal or probabilistic in form (L1, L2 . . . Ln), but by a judgement about whether that act was the rational thing to do in the circumstances faced by the historical actors at the time. Dray, in effect, invoked what sociologists call the "action frame of reference"[9] as the basis for historical explanation.

A historical event or a social action is best explained in reference to the set of rules and motives which guide any historical or social actor in decision-making and actions. The task of the historian, like that of the interpretative sociologist, is to render actions intelligible. The sociologist very often seeks to develop general concepts or "ideal types" to understand the systemic properties of social action and to engage in a comparative analysis of social situations to arrive at a generalized understanding. The historian is often more concerned with the specificity of an event or situation and how we might best understand this within a unique historical context or, at least, be able to assemble a number of suitable comparisons to facilitate a tentative generalization.

Historians have robustly defended the view that their research is discovery-led rather than hypothesis-driven, as positivists would have it. They tend to regard the historian's hypothesis not as the testing of a general law but rather as a rough-and-ready guide that helps to focus their research project. Such questions can provide an orientation for an enquiry. This enables the researcher to delimit their investigation as well as determining the relevance of various sorts of evidence, documents, testimony, statistical and visual sources to the historian's interpretations of events, structures or dynamics. This is also how the hypothesis functions within arts research, shaping both the processes adopted by the artist and the themes pursued in their work.

Of course, within artistic research, as with historical research, the research hypothesis is rarely encapsulated in a prior written project proposal. Often it is expressed in a preliminary sketch, or in the construction of a model, maquette, prototype, script or storyboard, as befits a visually led discipline. All of these planning and thinking devices facilitate an arts investigation and function to delimit or focus an enquiry, allowing materials and creative strategy to be tested in and through practice through processes of preproduction and execution.

Within historical research, pre-formulated hypotheses give way in the course of the historian's engagement with their sources and with drafting their account to questions of argument and narrative. As with arts research, historians can discover that they have been asking the wrong question or indeed employing the wrong methods. In historical enquiry, research questions are constantly being revisited and reframed as the investigation proceeds, as they are in the best scientific research. In artistic research, the posing of abstract research questions (often drafted in response to funding requirements) gives way to forms of understanding that emerge through investigation and innovative practice. These grounded research tasks – artists working in their studios, the historians assembling their sources, the film-maker with their treatments and scripts – are in fact the discovery process.

8.3 Material memory: exploring a family archive (case study)

One of the most fruitful areas in the dialogue between history and the creative arts has been the field of memory.[10] For example, Margaret Fitzgibbon, with a background in sculpture and site-specific work, addressed the question of familial memory and diaspora through a practice-based art investigation as part of her PhD at the National College of Art and Design in Dublin. Her work sought to appropriate a range of family artefacts to explore what Marius Kwint (1999:1) has called "material memory", a term used to allude to the way objects speak to us through the memories associated with them. Fitzgibbon re-appropriation of these family items raises concerns that have been voiced by the anthropologist Alfred Gell when he enquires about the status of native or vernacular artworks within anthropological and art discourse (1998:187), "when is a fabricated object a 'work of art' and when is it something less dignified, a mere 'artefact'?"

In Margaret's case, this was not primarily an abstract aesthetic argument about how the institutions, practices and discourses of the art world – the "art system" – in effect transform the artefact into a valued artwork.[11] Rather, it involved a personal act of remembrance. Her family archive included a collection of love letters between her parents dating from the 1950s. Her father had emigrated from Ireland to London to secure a job and housing, hoping his sweetheart would follow him. The letters involved a courtship across the Irish Sea that lasts until her mother moved to London in the early 1960s, married her father and they raised a family. As Margaret reports (Bell 2016:95), "my concern was appropriating this domestic archive to form the basis of an artwork through a process of 'memory work'". Her artistic material also includes home movies, a photo album and a surviving collection of 1960s furniture from her parents' London home, always referred to by her family as "The London Furniture". This included such items as a radiogram (seen in Figure 8.1), together with a collection of vinyl records that she and her family had listened to and which were prized consumer items charting their transition as immigrants to a world of mass consumption in 1960s England.

This radiogram was to become particularly important in her retrieval and representation of her family memories as her research came to focus on the relationship between sound and memory. As she notes (94),

> though sound and voice were already of interest to me as an artist whose background is in sculpture, they now took on a much greater significance. An attention to sound, music and "the voice" soon began to run through my entire project.

The installation can be seen at: https://youtu.be/_KjgfO6qEqo.

The project also developed an oral history element with interviews with surviving relatives, including her mother and aunts, exploring the family's experience as Irish emigrants in 1960s London. As she reports (96), "at the core of my research was an engagement with storytelling – stories handed down from generation to generation and new stories created on my PhD journey". Elements of these recorded interviews were used in the various sound tracks she fashioned and installed in the final exhibition, producing an additional documentary element to accompany the material artefacts now reframed within her artwork.

Eventually, her work took a sculptural and performative form. In one of her exhibited pieces, *Singing Letters*, a young woman performer sings/intones lines drawn from her father's love letters to her mother. The text is performed in a liturgical manner and deploys a form of the plain song chant that was popular in

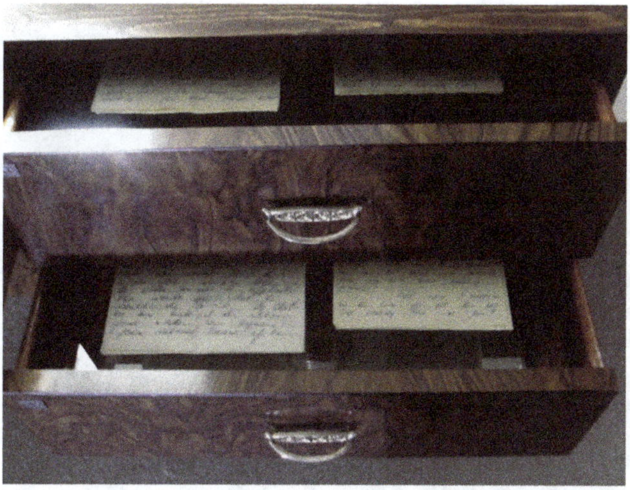

FIGURE 8.1 Margaret Fitzgibbon, *Dresser* (2011), appropriated 1950s bedroom dresser + original letters

Source: courtesy of the artist

Catholic churches in Ireland in the 1940s and 1950s. She manages to invoke the religiosity of a period when the Church sustained various moral panics about the "welfare" of young women who choose to emigrate to England.

In her research, Margaret references the work of artist–film-maker and academic Mathew Buckingham. Across a number of projects, Buckingham has explored various historiographical issues, reworking archival sources to fashion artworks that question the role of social memory and historical representation in contemporary life. He explores, for instance, traces of the native American past and US settler colonialism in the Hudson River Valley, the inception of the first English dictionary and its traces in our language use today. In the filmic installation *The Spirit and the Letter* (2011), he draws on the work of the writer and early feminist Mary Wollstonecraft and traces its significance for our own time. Excerpts from Wollstonecraft's writing are performed to camera by an actor in historical costume in a period setting. In role, she paces a large wooden floored room (Figure 8.2), seen first with her feet firmly on the floor boards and later apparently walking in a ghost like manner upside-down on the high ceiling of the room. In the installation, the film is projected on a video screen framed by an electrified "period" chandelier, reproducing the original one we see in the filmic images of the room but protruding from the floor rather than dangling from a ceiling. A mirror completes the installation, allowing the spectator to re-invert the image of the actor and turning the world upside down in an allusion to the transformative power of Wollstonecraft's vision for women (Figure 8.2). Image and reversal, past and present play off each other in a kind of visual dialectic.

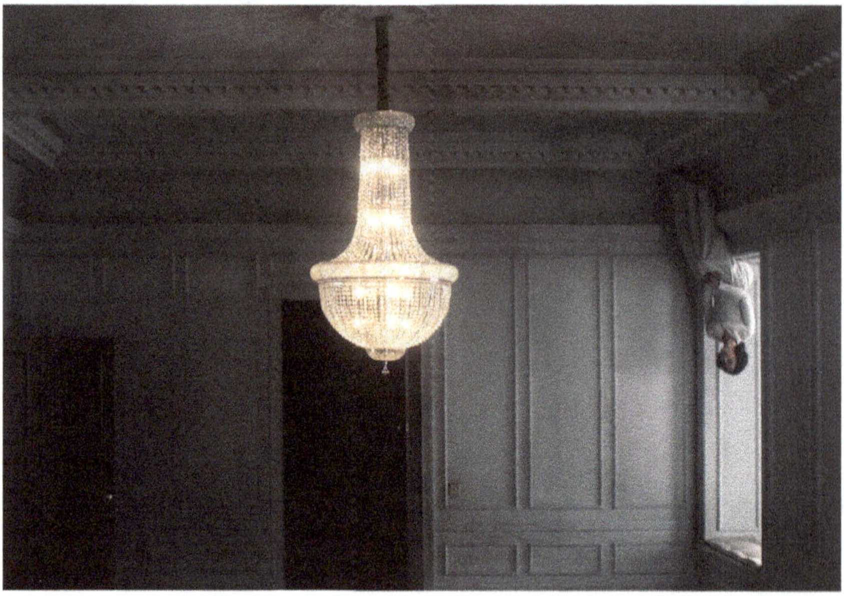

FIGURE 8.2 Mathew Buckingham, *The Spirit and the Letter* (2007), film still

Source: courtesy of the artist

Margaret Fitzgibbon's piece has a more domestic feel. It's family history. But, like Buckingham, her work fits within a tradition of "memory studies". As she notes (98), this strategy enabled her to pose a series of key research questions – "what mode of time do we encounter in the family archive? Who is represented, who is excluded, where do fiction and fact overlap? How do these artefacts operate in a 'collective memory' context?"[12] The installation offers us an embodiment of memory inscribed in objects, images and sounds and explores their role in a particular experience of diaspora.

8.4 The artwork as historiographical intervention: interpreting the Spanish Civil War (case study)

In 2015, the Musée de Paul Valéry in Sète in France presented an illuminating exhibition of the work of Catalan painter Joan Miro, *Vers l'infiniment libre. Vers l'infiniment grand* (Towards the infinitely free. Towards the infinitely great). The curators drew on Miro's extensive correspondence, diaries and interviews. These sources inform the extensive interpretative panels employed in the exhibition and provide an important context for understanding Miro's work. They show, for example, the impact of the Spanish Civil War on his painting in terms of figuration, colour and sensibility. I now examine how two contemporary artists concerned with this historical topic who have addressed the issue of memory and historical legacy in their work.

Pauline Zenk, a German artist living and working in the south of France, undertook a project, *Gravitation*, inspired by her discovery of a series of photographic images of Spanish Republican refugees who crossed the Pyrenees into France in early 1939 seeking asylum during the final part of the Spanish Civil War. The work was researched and developed as part of a residency at Les Ateliers du Plessix-Madeuc in Toulouse and shown in 2017 at the Musée régional d'art contemporain Languedoc (MRAC) in Serignan. Zenk, like a number of contemporary artists, uses the traditional media of painting and drawing (Figure 8.3) to rework original photographic images and bring out elements of a collective memory.

In an email interview, I asked her about how the project emerged and how she researched her project.

PZ: I noticed that many people in Toulouse had Spanish names and discovered that many of them were descendants of the refugees that had fled from Spain to France at the end of the Civil War. I began to research photographic material from the period initially from the internet and books on the Civil War. Then as I began to interview some of those who had been part of the exodus from Spain, I was granted access to people's family albums and collections. An important source for me was the archive of the regional newspaper *Depeche*. This had provided extensive covering of the *Retirada*, the refugee exodus in early 1939. I was struck by the fact that the paper at the time directly referred to a "migrant" or "refugee crisis" terms which, of course had a contemporary resonance.

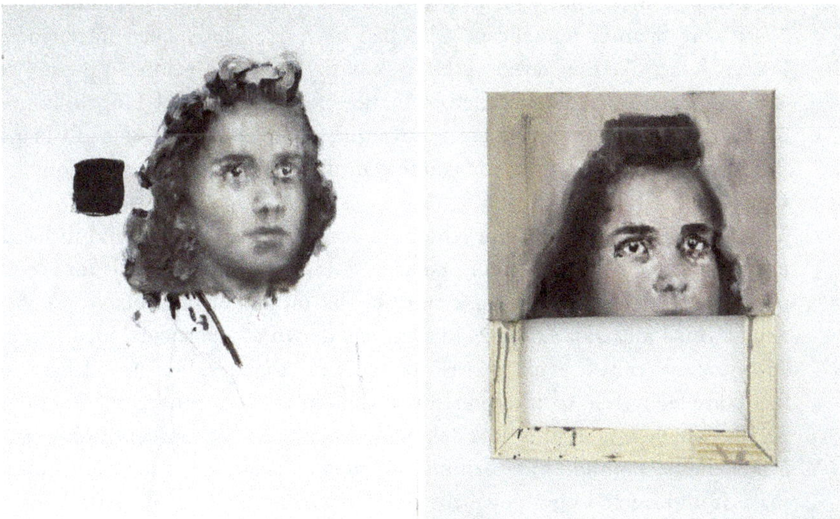

FIGURE 8.3 (left) Pauline Zenk, *Portrait of Marie* (2017), oil on canvas; (right) Pauline Zenk, *Maria coupe en deux* (2017), oil on canvas, 50 × 40 cm

Source: courtesy of the artist

DB: So, there was an oral history element to the project?

PZ: I found and interviewed quite a few people who had grandparents or parents who fled from Franco. I located two older people who had themselves experienced the exodus. One of them, Maria, now an elderly lady, I painted several times. Maria allowed me to copy the passport photograph taken when she eventually got French documentation. From this I eventually fashioned a portrait of her [Figure 8.3 left]. One remark she made to me really struck me. She told me that as a result of her flight as refugee, that she felt like she had been "cut apart". Cut apart in abandoning her home in Spain (to which she never returned) and forced to migrate to France in these circumstances. My painting of Maria would have to try and reflect this sense of loss [Figure 8.3 right].

DB: You talk about your work involving a distinctive research approach – can you describe this with regard to the Gravitation project?

PZ: I engage in factual, evidence-based historical research prior to starting my artistic interpretation, which, of course, has its own creative dynamic. My work does not claim to be an objective account of historical events and individuals. That is not my intention. But it is important for me that my artistic production is based upon a well-researched, factual basis (however complex or subconscious the connections might be that link this with my imaginative concerns).

DB: So, both types of research – the artistic and the historical – run alongside each other?

PZ: They are distinct but interconnected. For example, I discovered that many of the first prisoners sent by Vichy France to German concentration camps after 1940 were Spanish republicans who had been previously been interned in France. A fact like this stayed in the back of my mind as I painted a picture of a young republican soldier exiled to France [Figure 8.4 right]. Hopefully this makes my worked-up image more powerful, painful and expressive as I bring this historical knowledge to my painting in my brushwork and execution.

DB: Could you elaborate?

PZ: Well, both lines of investigation, the historical and creative, run hand in hand, in my case in a hybrid research practice. Indeed, it is often difficult for me to trace where the inspiration came from in this project or the source of a particular visual idea. For example, I might feel drawn to a particular image from an historical context – like Maria's passport photograph. However, I have to feel completely free to interpret this image, to alter the image, to leave out details (which would otherwise help to identify a specific moment in history). I alter the background transforming the faces or positions of the bodies as to make it impossible to pin down the exact historical moment when the source photograph was taken.

DB: So, the art process transfigures the historical moment?

PZ: Well yes. I am fascinated by how our constantly changing perception affects our sense of memory, particularly the visual dimension of this. How is this reflected in my paintings? I find images and transform them in my painting practice.

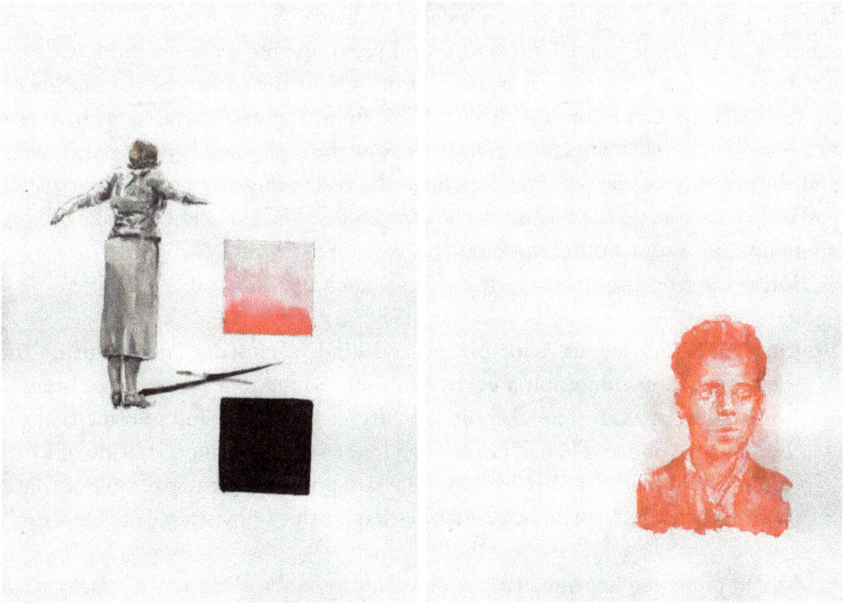

FIGURE 8.4 (left) Pauline Zenk, *A Leap of Faith* (1917), aquarelle acrylic and crayon on paper, 29.7 × 21 cm; (right) *L'homme rouge* (1917), oil on canvas, 46 × 38 cm

Source: courtesy of the artist

A historical event can become a current issue. In the case of "A Leap of Faith" [Figure 8.4 left] the topic of migration and refuge seems more topical than ever. I chose to paint the portrait of a woman standing in a coat in the Pyrenees during her crossing of the mountains as a civil war refugee. A personal portrait for some, but also a statement about a general human condition – in this case the need for escape and asylum.

DB: Can "A Leap of Faith" be read from a contemporary perspective?

PZ: The image is about making a choice – or having to make a choice between staying or going (or maybe resisting and closing your eyes). As the image shows neither choice looks promising. Depending on the dictates of history, politics and fate, this woman might be Spanish, French, German or Syrian. This feeling of taking a big leap into the unknown and feeling at a loss and alone must be there for a Syrian Woman taking her decision today to begin the refugee's journey in to an uncertain future. Burning red or pitch-black, neither option is inviting.

Pauline Zenk's work rests on and gives value to an objective record, whether captured in oral history or photography. Her research approach is, however, a hybrid one, and she believes her imaginative concerns and expressive strategies open up historical and political matters for a broader audience than a strictly factual account might. Her painting practice moves beyond photographic representation, with her brushwork making a virtue of its own limits. Often the treatment is rough, with canvases lacerated and displaced from wall to floor or partially erased with violent brush strikes, perhaps signalling the traumas of those displaced by war and terror and the fragility of the survivors' memory. In recovering elements of our recent past – such as story of the Spanish Civil War refugees – Zenk seeks to make links to contemporary issues around migration and asylum.

But, as she reports, sometimes history can catch up with you.

PZ: One encounter left its mark on me and kind of closed a circle. During the opening of my exhibition a curator introduced me to a woman who wanted to speak to me. Carmen a lady in her 80s held my hand and told me that she herself as a young child had crossed the Pyrenees with a suitcase made of a carton box and held together by cord. She said pointing to my painting of Maria with its black box apparent beside the figure: "this is my story – it is me, too!"

Trisha Ziff presented her film, *The Mexican Suitcase*, at the *Rencontres du Photographie* in Arles in 2011 in an exhibition curated by Cynthia Young and the International Centre of Photography (a trailer can be seen at https://youtu.be/9wV4yx6rXL4). The broader focus of the exhibition was a collection of photographs shot during the Spanish Civil War by Robert Capa, Gerda Taro and David "Chim" Seymour and recently rediscovered in Mexico. The negatives had been smuggled out of Spain after the fascist takeover and transported first to Paris. With the Nazi invasion of France, they had been sent to Mexico by the Mexican Ambassador to Vichy France, who had been entrusted with the collection. There they lay hidden from sight for the next seventy years. A suitcase containing the undeveloped rolls of negative was recently rediscovered in Mexico in the home of a film-maker who had been given the photographs and immediately grasped their historical significance. *The Mexican Suitcase* material (Figure 8.5 left and right) provides a unique documentary record of the Spanish Civil War. Ziff eventually brought *The Mexican Suitcase* to New York, where it now resides in the International Center for Photography (IPC).

The curational strategy in the Arles exhibition aimed not only to identify the shooting and editorial processes employed by Capa and his colleagues but also to place this work in its historical context. Ziff's film, made with the assistance of the photographer Susan Meisalas,[13] provides a particular interpretative frame to supplement our understanding of the rediscovered negatives. The IPC, for its part, produced a two-volume book with the assistance of a number of historians that aimed to explore the production of Capa's work and evaluate it within its historical context.

Each interpretative approach develops a rich historical narrative about the linkages of art, politics and fate during this troubling period of European history and, in the case of Ziff's film, seeks to trace the resonances with the present. Ziff chooses to frame the narrative of *The Mexican Suitcase* within contemporary concerns in Spain

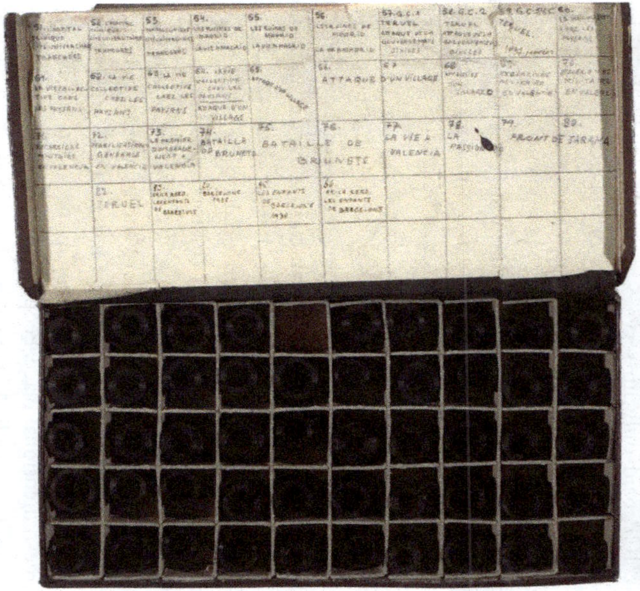

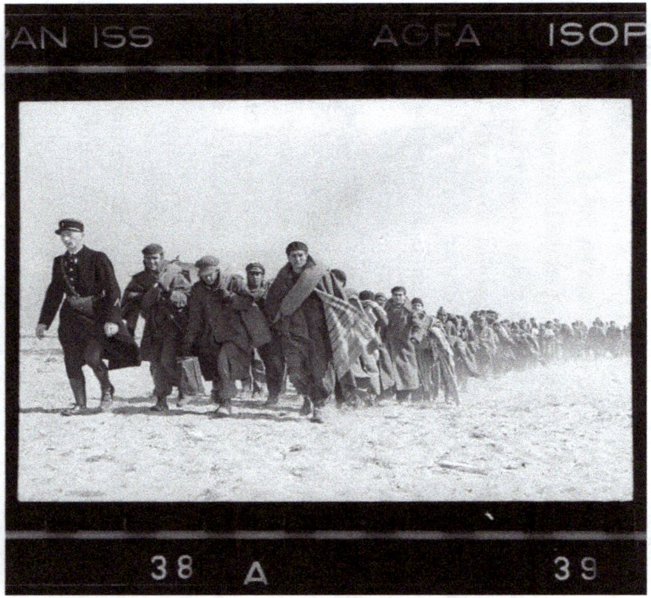

FIGURE 8.5 (left) Trisha Ziff, *The Mexican Suitcase* (2015); (right) French Internment Camp, Le Barcarés, France, March 1939

Source: (left) courtesy of the artist; (right) © Robert Capa; courtesy Magnum Photos and the International Center of Photography

with a personal and collective coming to terms with the horrors of the civil war. In the opening shot of the film, we observe a team of volunteer forensic archaeologists searching for and unearthing the remains of Republican soldiers who had been shot by the Francoists and buried in a mass grave during the war, or in its immediate aftermath. The film draws a parallel between the disappearance of the Capa photographs and their eventual reappearance and the historical occlusion about the disappeared that took place in Spain throughout the long reign of Franco, which is only now being questioned. This is not a fanciful homology. Capa's photos document the lives of the very soldiers whose grandchildren (as we see in the film) are now searching for their remains in order to honour their memory. Also, the passage of the suitcase of negatives to Mexico parallels the exodus of Republican soldiers and their families to Mexico seeking refuge after the war.

I asked Trisha Ziff how she went about researching *The Mexican Suitcase*.

TZ: I worked closely with the ICP on the historical aspect of the photographs. But what interested me was that the "suitcase" ended up in Mexico and became the symbol of all the exiles who fled Spain to Mexico. I wanted my film to address issues which remain alive today and to allow survivors from that period to give witness. My son is first generation to a Spanish exile father, so it was also about his narrative as much as the suitcase. It was a personal matter for me as well as being [a] politically relevant one. Despite adopting the name "The Mexican Suitcase" for the discovered photographs, the ICP paid only lip service to the Mexican dimension of the story. I wanted to make that a critical part of the film. So I tracked down, met and interviewed the different generations of that exile community and recorded its oral history.

DB: How in your research did you balance the historical research concerned with the Mexican Suitcase collection and its story and your creative research trajectory concerned with developing a distinctive and engaging film treatment of the topic?

TZ: I'm not a historian. I left the research work of identifying the people and the locations in the photographs to the ICP. I wanted to tell a different story and explore a different approach to history. I wanted to show the process of revealing the past, how we retrieve memory and come to see the past.

DB: Does your work as a film-maker have a curatorial aspect?

TZ: I think we make narrative choices both as curators and film-makers. Most of my films involve still images and, in a sense, the curation of these. How in a creative documentary film do you tell a story about photographic images that does not end up being a slide show? That has always been an issue for me. A film is linear, you watch it from the beginning to the end. An exhibition exists in a three-dimensional place. People move through it in different directions, stopping looking, reading, some more than others. A film is different. We expect our audience to remain engaged as the film unfolds in time through its narrative and content.

DB: How did this work in the case of your "Mexican Suitcase" film?

TZ: In the film I presented the historical images in counterpoint with the interviews. The images, together with the voices, take us to the past, so we come to feel it. I do think my experience as a curator impacts on my work as a film-maker a lot. I think in sequences of images, moments of . . . pausing . . . looking . . . reflecting.

DB: You chose to frame the film with the account of the forensic archaeologists and relatives searching or the remains of Republican soldiers who had disappeared in the civil war. Why so?

TZ: I think the question of memory is critical here. The symbol of these Spanish women literally digging for answers to the past, given the history in Spain, was much more than searching for evidence. Rather, it is a political act of defiance against a long history of fear and silence. I vividly remember the moment during the dig when they uncovered the skeleton of a Republican soldier. It reminded me to being in the darkroom and the image emerging on photographic paper . . . the revealing of an image, the finding out about the past . . . the invisible at last becoming visible.

DB: How did the film play in Spain?

TZ: Watching people in Spain visit the Mexican Suitcase exhibition was really interesting. It was not about looking at photography or art. It was as if they were looking at recovered family albums. People were searching for people in many of the photographs. No different to these women searching in the ground for their grandfathers. The process of filling in the gaps of a stolen past. For me it was a critical way of understanding the importance of the suitcase.

Ziff's work can be viewed both as an artwork in its own right and as a contribution to a politically charged research investigation.

As US historian Louis Mink reminds us (1966:44), "the very name of the discipline [history] meant originally, 'to inquire, to know, *to tell what one has learnt*'". This serves as a good description for artistic research projects like the ones we have discussed above which engage with the past but primarily through creative process. Mink argues against reducing the historian's practice to simply producing dry scientific conclusions. Historians, he observes (ibid.:38), "unlike scientists in general (. . .) must read one another's books instead of merely noting their results". The significant conclusions contained in the historian's text are indissociable from the narrative forms, expressive power and specific evidential claims present in a historical account. Historical research has no "detachable conclusions" to be circumscribed under a generalized law.

Similarly, in artistic research there are no "detachable conclusions" (even in practice-based doctoral projects). Evaluating creative arts research requires engagement with the artwork at its core and with its intrinsic language of expression. This means engaging with visual, plastic or performative forms of the art object, rather than with a text-based summary account of it. Artistic research may need to be written up, but such documentation and exegesis has to be read alongside the creative output, which, like a historical or literary work, already encodes a mode of understanding. Its meaning can be amplified by a written account but not replaced by it. Artistic research, like historiography, is not really a cumulative science, though we can certainly learn from the example of other researchers. An investigator concerned with the originality of their work cannot use the "findings", or simply "apply" the methods, of another's research and practice without effectively retracing the steps of the original researcher–artist to understand the distinctiveness of this work. Then they must then break from this mould. With the exception of the fakist, no artist simply wishes to replicate the work of another!

8.5 Imaging the past: *The Enigma of Frank Ryan* (case study)

Historians, like film-makers, purport to employ insight, intuition and empathy to understand human motives and actions and to reconstruct historical situations to render them intelligible to their readers. Because history deals with human actions and social contexts (albeit as revealed in historical documents and sources), historians have often found it useful to imaginatively recreate the world they investigate. This construction of an imagined past is even more significant for those who make "film history".

American historian Robert Rosenstone identifies film history as the screen portrayal of an historical event, a thought experiment about the past that takes on a visible and dramatic form in a movie or television programme. Through the script and its realization in the actor's performances and in the film's cinematography and visual design, the director, in effect, elaborates a series of hypotheses about the characters' motives and actions, some of which are based on historical evidence, some on imaginative constructs.

Rosenstone characterizes history films – a genre that eludes easy definition – as those that "consciously try to recreate the past" or seriously attempt "to make meaning out of the traces left to us from that vanished world" (2012:3). As he acknowledges, both academic historians and film scholars have been slow to accept his invitation to explore the potential of the history film to extend historical understanding.[14] The tendency has been to focus on film's limitations vis-à-vis the written text and to view film as a tool for the public communication of historical research or as an archival source of evidence shedding light on a period.

Film scholar Robert Burgoyne focuses on *re-enactment* as the defining genre characteristic of the history film. Re-enactment, he argues, facilitates "the act of imaginative re-creation that allows the spectator to imagine witnessing the events of the past" (2008:552). Yet, as he concedes, despite the "somatic intensity" of the history film and its capacity to deliver to an audience a sense of being present when history was being made, the viewers were "not there" and can never be. Filmic re-enactment provides not a direct window onto "how things were" – the great dream of historicism – but rather a guide as to how the past might be understood as a series of surviving fragments. Burgoyne quotes philosopher Paul Ricoeur to good effect: "re-enacting does not consist in reliving but in rethinking and rethinking already contains the critical moment that forces us to take the detour by way of the historical imagination" (553).

Let's take an example I am familiar with – my own docu-drama film, *The Enigma of Frank Ryan*, a feature-length film employing both live action and film archive made for Irish television in 2012. Ryan, a radical Irish Republican, fights in the Spanish Civil War as a member of the International Brigades. After his capture by Franco's forces in 1938, he is eventually smuggled out of jail in Burgos and transported first to occupied Paris in June 1940, then to Berlin. There he undertakes to work in a "diplomatic capacity" for German special forces, planning covert operations in Ireland, Britain's undefended flank in World War II. He is sent back to Ireland in a German U-boat but the mission is aborted after the death at sea of his travelling companion, the IRA leader Sean Russell. A key scene from the film can be seen at: https://youtu.be/b1G_c3LHdls, where Ryan returns to Berlin a broken man.

How are we to explain Ryan's apparent ideological *volte face* and course of action? Some have drawn crude generalizations: Irish armed force republicanism shades with the sort of extreme nationalism that Nazism represents, so the Nazis were welcome allies. Others adopt the "horseshoe hypothesis" invoked by US Cold War political scientists, seeing a symmetry between the Stalinist form of communism Ryan espoused and the fascist authoritarianism that he encountered in Germany.

Most historians of the period and of the Ryan case, including Fearghal McGarry, who I worked with on the project, reject these ideologically reductionist explanations, preferring to weigh the various specific factors or, in philosophical parlance, initial conditions that might have influenced Ryan's decision-making, tracing the events that led to his fateful decision to travel to Berlin to collaborate with the Nazis.

We can list these conditions. Firstly, the willingness of Irish Republicans to seek Nazi assistance to advance the cause of Irish unification and achieve full independence from Britain, a willingness captured in the motif, "my enemy's enemy is my friend". Secondly, the network of personal relationships with Abwehr II operatives that facilitated Ryan's passage to Berlin (youthful friendships with German former Celtic scholars whom he had known in 1930s Dublin). And, perhaps most significantly, the signing of the Hitler–Stalin Non-Aggression Pact in August 1939, which for someone like Ryan, close to the Communist Party in Spain, could have justified his collaboration with the Germans for Irish Republican ends. Maybe the passage to Berlin represented Ryan's only chance to escape from Franco's hellish jails and return to Ireland and its Republican struggle.

In such a case, historians would seek to identify the range of factors conditioning Ryan's actions in order to assess the weight of each factor in determining his eventual course of action. Our film reconstructs the events around his "escape" from Burgos prison in northern Spain (was he a willing participant in this?). It imagines his meeting with an Abwehr officer and former acquaintance Helmut Clissman in

Nazi-occupied Paris (did his friendship with Clissman assuage him in his decision to collaborate?). We trace his travel to Berlin, where he agrees to work with German Military Intelligence (Figure 8.6) and meets a key IRA leader (how politically compromised was he by this choice?). The filmic narrative presents an historical account in which, in the end, an Irish Republican privileges nationalist ideology over anti-fascist struggle as the likely explanation of Ryan's apparently contradictory behaviour for allying with the Nazis. Perhaps more detective work than science. Or political nous? But it relies on the historical evidence marshalled by the director working closely with a historian with a specialist expertise in this period of European history. In my video discussion with Fearghal McGarry, subsequently published in part in the historical journal *Rethinking History* (McGarry 2015), we explored the different perspectives of the historian and film-maker on evidence and interpretation (https://youtu.be/prJAog8lNNk).

FMcG: German, British and Irish intelligence reports and eyewitness accounts shed some light on Ryan's activities in Germany but these remain problematic sources. Intelligence reports focus only on Ryan's political role in Germany and reflect the views of the agency that compiled the report.

Our film uses dramatic licence to imagine Ryan's life in wartime Germany. There is, for example, no evidence that Ryan questioned German officials about the existence of concentration camps. As a committed anti-fascist, however, Ryan was aware of the brutal, anti-Semitic nature of the Nazi regime and it seems highly unlikely that he was not deeply conflicted about his presence there.

DB: Given the limited and problematic nature of sources on Ryan's life in Berlin, the film-maker has to work with scant written or oral evidence of Ryan's activity and indeed state of mind. In this context, I employ the imaginative resources of film to reconstruct a sense of what Ryan may have been experiencing as the compromised nature of his position as an erstwhile radical involved in various collaborative projects with the Nazi regime became clear to him. This means employing *re-enactment* as a creative strategy.

In scripting the section of the film dealing with Ryan's last days in Berlin I started from the assumption that it seems unlikely that Ryan, given his high-level contacts with military intelligence in Berlin, could not have been aware of the highly repressive nature of the regime he was cooperating with and indeed of the concentration camp system, although I accept that he may not have been aware of the implementation of the "Final Solution".

The key issue that arises for historians with re-enactment in the history film is accuracy. Dramatization of the past presents a gap between the historian's often more nuanced, detailed narrative, based on fidelity to contemporaneous sources, and the film-maker's concern with narrative drive, economy of storytelling and expressive potency (although historians who seek a public audience are hardly

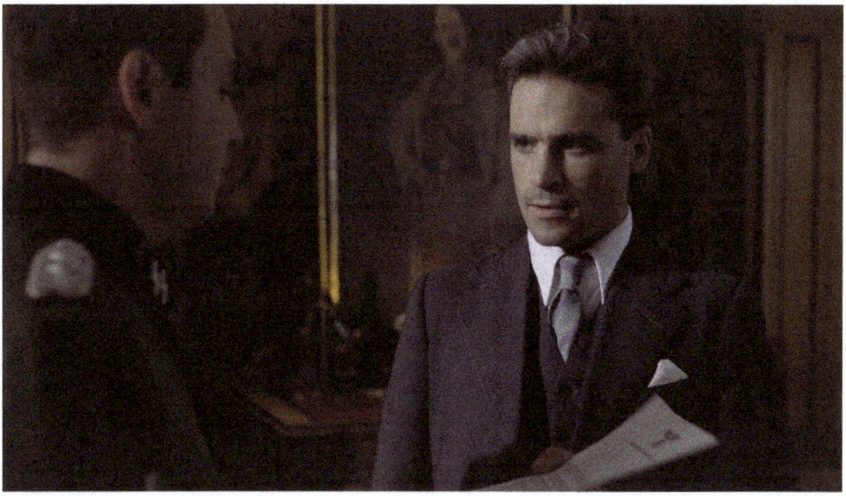

FIGURE 8.6 Desmond Bell, *The Enigma of Frank Ryan* (2013), film still: Ryan meets Abwehr II Officer, Colonel Veseenmayer, in his Berlin office

immune to these desiderata). This tension was at the core of the dialogue between historian and film-maker during the making of our film.

FMcG: The issue that I initially imagined would be most controversial was the fact that we were making a film about a figure who for many Irish Republicans is a hero. And we're focusing on the most controversial part of his life – his journey to Germany and his role there as a collaborator. Yet, one of the things I was struck by when the film was screened in front of audiences in Ireland, north and south, many of whom were from Republican and left-wing backgrounds, was that it wasn't as harshly attacked as my book had been on publication [2010]. I think this relates to key differences between a historical text, like mine, and your film.

DB: How so?

FMcG: Well, the film invites you to empathize with Ryan and as a result you come to identify with him. You adopt his point of view and share his journey. My historical text seeks to look at Ryan's activities in Berlin in a more critical way that more directly addresses moral issues about his activity. The more responsive attitude on the part of the film audience is partly because they have come to sympathize with Ryan, whereas such structures of empathy are not usually at play in a historical text.

DB: I guess the film-maker is also much more circumspect about making an authoritative judgement. I mean, the historian writing his or her book does so against other texts that are in the field and often contests these. Ryan's story is often read as emblematic of the broader relationship of Irish Republicans to the Nazi regime.

FMcG: As a historian, I'd also be uncomfortable saying that I arrive at authoritative or definitive judgements. But what I do as a historian is weigh explicitly the evidence, assess the varying alternative explanations and viewpoints and come, ultimately, to a clear judgement. In my book, I argue unambiguously that Frank Ryan, although not sympathetic to Nazism, was a collaborator.

DB: Look, I'm not saying that film-makers don't make judgements, they certainly do, but those judgements are embedded in the warp and weft of the film itself, in the development of the character and in the unfolding of the life of the film. What I think film-makers say is, yes, Ryan's sojourn in Berlin does seem highly ideologically problematic and in hindsight morally questionable. But heh, what do you think about it? As an audience, what's your judgement? In other words, we invite people to ask the question rather than feel that we were in the position to deliver an authoritative judgement, about the character and his actions.

Can film or indeed any other art form such as drama and creative writing combining fact and fiction adequately represent historical knowledge, then? This is one of the issues that practice-based modes of research in the arts have being wrestling with over the last decade. The work of directors like Trinh T. Minh-ha, whose work I discuss in the next chapter, and Julie Dash (*Daughters of the Dust*, 1992), whose films are also cited by Robert Rosenstone, can be seen as harbingers of a new postmodernist historiography. As Laura Marks argues (2000), these are historical works that have developed a methodology exploring how cinema can represent the past in ways that are distinctive and partial but potentially no less insightful than the traditional historical monograph. History film is also an intriguing variation on "counterfactual" or "alternate" history, a form of speculative and imaginative thinking about the past that historians have considered more closely in the last decade. Such films represent a form of doing history, an approach Rosenstone characterizes as a "visual form of historical thinking that cannot be judged by criteria on what is produced on the page" (2004:29).

8.6 Contemporary art and the engagement with history: Hito Steyerl's *Der Bau* and *In Free Fall*

In the work of German film-maker and artist Hito Steyerl, historical investigation, political engagement and the conceptual drive present in contemporary art come into productive alignment.

Der Bau (The Building) is a site-specific installation Steyerl produced in cooperation with the Austrian architect and academic Gabu Heindl. The exhibition's full title – *Der Bau: Unter uns, ein dekonstruktion eines Gebaeudes* (The Building: Beneath us, a deconstruction of a building) – indicates the artist's specific concern with getting below the contemporary fabric of the Bridgehead building in central Linz in Austria to provide a historical understanding of the fabrication of the building and to recover traces of the city's Nazi past (Figure 8.7). As Steyerl tells us (2010a),

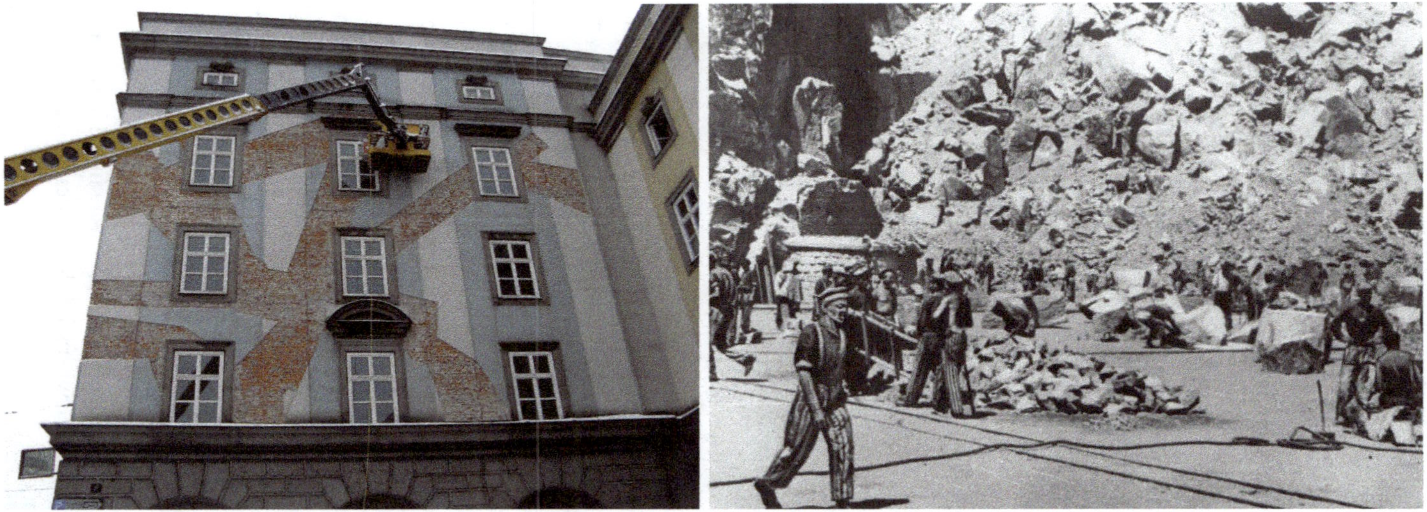

FIGURE 8.7 (left) The Bridgehead building today in central Linz, photograph by Hito Steyerl; (right) The Wiener Graben Quarry Concentration Camp (c.1941), United States Holocaust Memorial Museum

Source: (left) courtesy of the artist; (right) courtesy of Archiv der KZ-Gedenkstaette, Mauthausen

"the project deals with the construction history of a Nazi building" on Linz's main square; "it investigates its background, the stories of the people who actually built it, and also looks at the materials used in the building". It reflects her long-time interest in the perverse histories of recycled material objects.

Steyerl probes the erasures of historical memory that have taken place in post-war Germany and Austria around both countries' fascist past. As she was to discover, foreign forced labour was used to construct the Bridgehead, with some labourers drawn from the nearby Mauthausen concentration camp. Many of these would later perish there. Her historical research revealed that prisoners working in the notorious Wiener Graben Quarry located near the camp produced the building blocks for the Bridgehead building. She concludes (2010a),

> there are at least two different ways of describing this building. One and the same stone used for the building can be said to have gained its shape according to the paradigm of neo-classicist architecture, which would be the official description given to the building itself. Or it can be described as having probably been shaped by a stone mason in the concentration camp at Mauthausen, who was likely a former Spanish Republican fighter (detained and transported from occupied France). The conclusion is obvious: the same stone can be described from the point of view of a discipline, which classifies and names. But it can also be read as a trace of a suppressed conflict.

Der Bau contrasts the building and its surroundings as they appear today with archival evidence and eyewitness interviews revealing the building's grim past. Stories associated with the building and its construction by camp inmates were told in five video displays, mounted in the building's current commercial space.

Her research also revealed that the building, which today houses part of the Linz Academy of Art, including its department of artistic research, is still heated by a number of heavy iron radiators. These were most likely removed from the Mauthausen concentration camp after the war and later reinstalled in the Bridgehead building, where they have been in use ever since, including within the artistic research section of the Academy (ibid.).

> A visit with an expert confirmed that the radiators have never been exchanged in the Eastern part of the building and that, moreover, some of the radiators had already been used, when they had been installed around 1948. The make of those radiators corresponds to the few radiators seen in contemporary photos of the concentration camp at Mauthausen. Now, of course, radiators were not in use in the prisoners' barracks. They were in use in some work rooms, like the laundry room. They were in use in the prisoners' office and the prisoners' brothel, where female inmates from another concentration camp had to work.

This shocking discovery (there is, as always with Steyerl, an element of the performative in her revelations) leads her to reflect on the notion of artistic research. Seeking like many artists to exploit startling juxtapositions that bring discourse and the concrete into collision, she asks (ibid.), "but why would this very local project be relevant for a reflection about artistic research as such?" The fact that this building, replete with its submerged memories and material residues, is now used by Linz Art Academy to house a facility for artistic research is for her a telling coincidence. Her concrete focus is on the college's lack of awareness of the dreadful history of the building in which it operates. However, she argues, the case signifies a broader vacuum at the heart of academized art research. As she argues, "any investigation of the building might turn out as a sort of institutional meta-reflection on the contemporary conditions of artistic research as such". Which is to say, too often such research, encased as it is within the bureaucratic and disciplinary constraints of the academy, misses what is directly under its nose.

Steyerl senses a chilling vacuum at the heart of a model of artistic research which turns art enquiry into a variant of scientific practice. She provocatively asks (ibid.),

> what do we make of the fact that the Department for Artistic Research (its coordination office is located in *The Building*, according to the website) could soon find itself being heated by the same radiators, which were mute witnesses of the plight of female inmates in the concentration camp brothel?

As a practicing artist, Steyerl has taken aim at the sustained attempt by the academy to turn research in the creative arts into an academic field with its own set of methods, theorizations, pedagogy, institutional affiliations and income streams, in effect, treating art as akin to a corporately regularized scientific area of study. She wishes to critique this development, which she sees (ibid.) as being "complicit with new modes of production within cognitive capitalism: commodified education, creative and affective industries, administrative aesthetics, and so on". She takes her lead from Michel Foucault's concept of *discipline*. This focuses on the exercise of modern forms of power through various knowledge procedures closely connected with the organizational structures and tutelary practices of a capitalist society. Steyerl directs our attention to the way the disciplinary apparatus increasingly governs academic artistic research. This she feels often elides any consideration of the underlying forms of social and political conflict that have often been the focus of the modernist project in art (ibid.).

> A discipline is of course disciplinarian; it normalizes, generalizes and regulates; it rehearses a set of responses, and in this case, trains people to function in an environment of symbolic labour, permanent design and streamlined creativity. A discipline may be oppressive, but this is also precisely why it points to the issue it keeps under control. It indexes a suppressed, an avoided or potential conflict. (. . .) Why would one need a discipline if it wasn't to discipline somebody or something?

She contests this progressive incorporation of artistic research and practice into the capitalist knowledge economy. She also seeks to excavate a long submerged critical tradition of artistic research associated with those artists that were caught up in the struggle against fascism, or in the revolutionary upheavals in Russia in the 1920s (explored in the previous chapter). Such politically engaged art has to be read against the historical backdrop of two world wars, revolution, colonialism and the rise of fascism. As she notes (ibid.),

> if we look at artistic research from the perspective of conflict or more precisely of social struggles, a map of practices emerges that spans most of the 20th century and also most of the globe. It becomes obvious that the current debates do not fully acknowledge the legacy of the long, varied and truly international history of artistic research which has been understood in terms of an "aesthetics of resistance".

Steyerl alludes here to the work of German-born playwright, novelist, film-maker and painter Peter Weiss, still best known for his play *Marat-Sade*. His monumental *Die Ästhetik des Widerstands* (The Aesthetics of Resistance 2005) is a three-volume historical novel (1975–1981) spanning the late 1930s of Nazi rule in Germany. The volumes deal with anti-fascist resistance across Europe and addresses the precarious fate of proletarian politics. Living in Berlin in 1937, the book's unnamed narrator, together with his peers, 16- and 17-year-old working-class students, seek ways to express their hatred for a Nazi regime which had brutally crushed the trade union and socialist movement, throwing its leaders into concentration camps. The boys meet in museums and galleries. Here they explore the affinities between political resistance and art – the theme central to Weiss's novel.

Throughout the novel, Weiss explicitly uses the term "artistic research" (*künstlerische Forschung*) to refer to cultural practices such as Brecht's writing factory, set up while he was in exile in Denmark from fascism and which produced plays like *Fear and Misery of the Third Reich*. Weiss also points to the constructivist practices of writing and image making employing formalist and factographic approaches in post-revolutionary Soviet Union, which influenced Brecht. In particular, Weiss identifies the dramatic and documentary work of the Russian writer Sergei Tretyakov, a figure who Steyerl also often turns to in her work on the "biography of objects" (Tretyakov 1929/2006).

Weiss seeks to establish a genealogy of aesthetic research and practice related to the history of emancipatory struggles throughout the twentieth century. For Weiss, the radical modernist project, dramatically eclipsed in Europe by the rise of fascism, is only intelligible in terms of its relationship to the upheavals, traumas and utopian sentiments which mark European history between the two world wars. Steyerl seeks to build upon that radical tradition in an aesthetic grounded in political critique and struggle. Over the last decade, she has combined her work as an independent film-maker and artist (she was recently voted in a newspaper poll as one of the ten most influential artists in the world) with life as an academic with a penchant for penetrating, critical writing. Her practice has evolved from being primarily a film-based artist, reflecting her film-school training in Munich as a documentarist and subsequent work as a production assistant and researcher for German film-maker Wim Wenders, to engaging in a more fluid mix of research projects, lectures, performances, publications and installations that have given her work a global reach, making it all but impossible to pigeon-hole her.

Steyerl's written essays and lectures do not provide an exegesis of her films but are performative works in their own right, offering an amplification of themes pursued in her creative work and allowing more abstract vectors of speculation and critique to emerge.[15] For instance, in 2010 she made the film *In Free Fall* and published a paper the following year with the same title, based on a previously given lecture (2010b) in which she explores a range of ideas about contemporary visuality.[16]

For Steyerl, "free fall" serves as an extended metaphor of our current historical condition. It directly alludes to stock market crashes past and present. More generally, the term alludes to an existential state we all are living through in a world seemingly out of control. "Disorganized capitalism" has led to the collapse of economic security for many and to the absence of any meaningful narrative within which people can make sense of social change. As she notes (2011),

> many contemporary philosophers have pointed out that the present moment is distinguished by a prevailing condition of groundlessness. We cannot assume any stable ground on which to base metaphysical claims or foundational political myths. At best, we are faced with temporary, contingent, and partial attempts at grounding. But if there is no stable ground available for our social lives and philosophical aspirations, the consequence must be a permanent, or at least intermittent state of free fall for subjects and objects alike.

Her 2011 essay also explores the collapse of traditional forms of visual perspective, whether in the verticality and visual flattening present in contemporary drone and satellite photography, a development we have seen precociously anticipated in the non-objective art of Kasimir Malevich. In turn, this emergent vertical perspective can, she believes, be read politically as an expression of the forms of class hierarchy and distribution of power characteristic of contemporary capitalism.

> The view from above is a perfect metonymy for a more general verticalization of class relations in the context of an intensified class war from above – seen through the lenses and on the screens of military, entertainment, and information industries. It is a proxy perspective that projects delusions of stability, safety, and extreme mastery onto a backdrop of expanded 3D sovereignty. But if the new views from above recreate societies as free-falling urban abysses and splintered terrains of occupation, surveyed aerially and policed

> bio-politically, they may also – as linear perspective did – carry the seeds of their own demise within them.

In the essay, as in her film, she plots the links between visuality and power within contemporary capitalism. However, Steyerl, despite the speculative and indeed polemical character of her writing (something she shares with a number of contemporary artists), remains attentive to the craft of film-making. It can be argued that her film work represents a more focused and coherent exercise in an aesthetics of resistance than her essays, which are often characterized by gestural and performative features rather than by any great analytical clarity, evidential reach or political import. Maybe the discipline of film practice encourages her to make her arguments in a much more concrete manner.

Her thirty-minute film *In Free Fall* can be viewed as an experimental exercise in contemporary history. It consists of three elements titled *Before the Crash*, *After the Crash* and *Crash*. The term "crash", like that of free fall, alludes to the periodic crises in capitalism and, in particular, to the great stock market crash of 1929. It also references the air disaster as a staple narrative of recent Hollywood cinema. In her film, these two meanings come together around a particular geographical site, research investigation and creative strategy.

In an interview conducted after the release of the film (2010d), Steyerl describes its creative origins.

> I first started shooting the film after I saw a picture of an airplane graveyard in Mojave in California. During each financial crisis planes get stored at this graveyard as its unprofitable to operate them, but they get parked or stored. The older ones get dismantled for parts or scrapped for aluminium. And, because it's very close to Hollywood, there is also a fourth way of dealing with the planes, which is to use them in Hollywood movies to stage crashes and explosions.

The film focuses on the life cycle of the Boeing 4X-JYI jet manufactured for TWA airlines, with its historic association with Howard Hughes, the entrepreneur and film director. The plane is later bought by the Israeli Air Force and figures in the Israeli defence forces raid to free the Entebbe hostages in 1976, before being taken out of service and scrapped or used as a film prop.

Her treatment of the topic mixes documentary concerns with imaginative one in her recycling narrative (ibid.):

> So, I had this initial fantasy that a plane would be stored there because of the financial crisis then blown up for a Hollywood film, the debris would be collected, shipped to China where it would be converted into recycled aluminium, then DVDs would be made using that aluminium and the pirated version of the film in which the plane is blown up initially would be printed

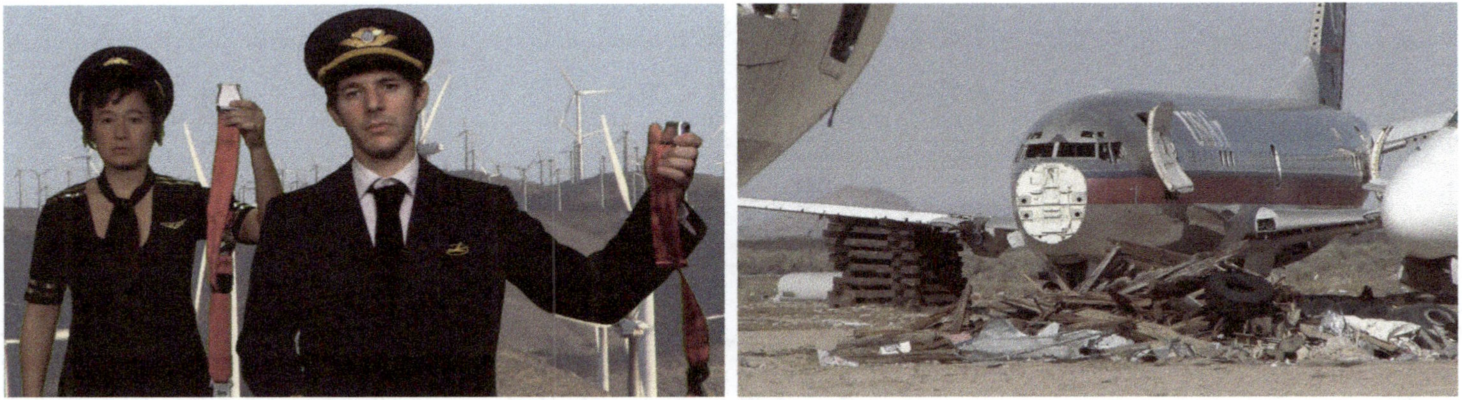

FIGURE 8.8 Hito Steyerl, *In Free Fall* (2010), video HDV, 32′, single channel, sound, colour, film stills

Source: courtesy of the artist and Andrew Kreps Gallery, New York

> on these DVDs, the film of the plane being blown up would be printed on the remnants of the plane itself.

She draws on a wide range of found footage scraps to construct her film – low-resolution clips depicting plane crashes, snippets from television science programmes about recycling, the plane crash and explosion from 1994 movie *Speed*, a jaunty sound track of disco hits. Images of Klaus Kinski, Charles Bronson and Robert Loggia appear among the rapidly cut, scratch video sequences. This found footage, much of it dragged in degraded form from YouTube, is intercut with her own pristine high definition video footage of the Mojave scrapyard and the Mojave Air and Space Centre, shot by a cameraman who, in turn, contributes to the film as an interviewee describing the disastrous impact of the 2008 crisis on his own life and who is seen alongside Steyerl in some shots (Figure 8.8 left). She conducts documentary-style interviews – some real and some involving actors.

Asked about the relationship between her writing/theorizing/researching activities and her creative film and imaging work, Steyerl acknowledges the different registers and crafts involved (2010d):

> I call myself a film-maker and a writer. I was trained as a film-maker. . . . I became a writer by accident maybe! But of course, there are huge differences. I always compare it with being a carpenter and a weaver at the same time. These are hugely different crafts. You work with different tools and you produce different things with different means. I end usually working on one phenomenon sometimes with the camera, at other times with the computer keyboard and the pen. The modes of production and also the results are completely different.

Hito Steyerl's methods are then both immersive and historico-critical. Her research and creative practices deal with contemporary history, whether the submerged traces of Nazism in Austria or the recent history of capitalist crisis and aggressive restructuring read in the biography of a Pan Am jet. Her methods, however, are those of the artist, not the academic historian. And, as we have seen, it is from her field investigations and creative experiments within specific projects that various theoretical trajectories begin to emerge.

8.7 Conclusion

As art educators seek to trace a methodological trajectory for creative arts and media research, it may be profitable to attend to the arguments made by historians about the autonomy and distinctiveness of their discipline. This sort of grounded focus on sources, evidence and argument seems applicable to our own nascent field of research. In addition, across a range of case studies and critical arguments, we have surveyed some of the synergies between history and the creative arts currently being explored by both artists and historians. The field of memory has been an important site of methodological encounter between history, the creative arts and a range of other disciplines, as has the concern with film and "post-memory", which focuses on the materiality of the image and its role as a visual or sonic trace of the past. The creative arts, whether film, photography, installation and site-specific work, performance and drama, have their own distinctive way of doing history and engaging with the past.

Notes

1 The forms of positivism which held sway in France and Britain had perhaps less purchase in Germany where neo-Kantian philosophy and hermeneutic scholarship led both philosophers and historians to seek to establish the scientific basis of the historical and cultural disciplines on a basis which did not simply seek to ape the natural sciences.

2 For philosophers, *x* is the necessary cause of *y* if every incidence of *y* is preceded by *x* as an event, though the occurrence of *x* may not always lead to *y* as other factors may combine with *x* to lead to the outcome *y*. On the other hand, *x* is a sufficient cause of *y* if the occurrence of *x*, in and of itself, always leads to the event *y*. However, another cause *z* may the cause incidents of *y*. Accordingly, the presence of *y* does not imply the presence of *x*. In historiography, causal attribution seems more straightforward. It usually rests on a notion of sufficiency and a juristic notion of "beyond reasonable doubt" in causal attribution.

3 Hayden White (1987) is perhaps the most prominent philosopher of history who has revealed the range of figurative devices historians employ in their accounts of the past and who has emphasized the artificial, fictive and anti-realist nature of all historical narrative.

4 Jacques Rancière, in his *Names of History* (1994), offers us a meditation on historical discourse and, in an extended discussion of the French historian Michelet, reveals the importance of narrative and voice in all history writing.

5 It has to be said that the fiercest debates within Marxist historiography in Britain centred on the attempts by certain theoretically motivated historians (such as Perry Anderson 1980) to advance the notion of Marxism as a science, a position advocated by the philosopher Louis Althusser (2005). In turn, this tendency was vigorously contested by a humanist wing of Marxist historiography, closely identified with Edward Thompson (1978).

6 Certain Annales historians developed an interest in the history of mentalities. This aimed to describe and dissect the forms of belief, knowledge and classification employed by people in a specific historical period to make sense of the often stratified world they occupied. Such *histoire des mentalités* produced studies on reading, attitudes to childhood, popular religion, class sentiments and, in many ways, drew on structuralist anthropology's reading of culture.

7 The concept of an "ideal type", as used by the German social theorist Max Weber (1949), is an analytical construct – for example his generalized conception of bureaucracy – that serves as a template to explore similarities as well as deviations in concrete cases under sociological analysis. While using such concepts, historians place more emphasis on the distinctiveness of each historical case.

8 Indeed, sociology already seemed to provide an exemplar of such an evolution. As developed in France by Emile Durkheim, it seemed possible to study a phenomenon as deeply personal as suicide by employing social statistics and models of social structure in order to reveal patterned regularities in social life (1897/2002).

9 The action frame of reference is a sociological approach often identified with Max Weber and systematized by the Harvard sociologist Talcott Parsons (1951/1991). Social behaviour is comprehended by locating patterned social actions within the normative frameworks which condition the individual motives and behaviours of actors.

10 That said, traditional historians, with their privileging of material, written and statistical sources, have been wary of the evidential status of the forms of oral testimony, which rest on personal and collective memory. On the other hand, sociologists, anthropologists, cultural and media studies scholars and increasingly creative artists have been drawn to the field of memory studies and have made their distinctive contributions to this, often pushing the boundaries of historiographical analysis and how we think about memory and its constitutive role with regard to identity.

11 Alfred Gell (1998:187) rejects the idea that an artefact can be distinguished from an artwork purely on aesthetic criteria by "having certain qualities of visual appealingness or beauty". Rather, as he notes, "a work of art may not be 'beautiful' or even interesting to look at, but will be a work of art if it is interpreted in the light of a system of ideas that is founded within an art-historical tradition".

12 These issues are explored in a 60,000-word critical exegesis that seeks to contextualize her practice.

13 So, for example, the academics examining Margaret Fitzgibbon's PhD work had to closely attend to her exhibited installation and not just assess the written dissertation.

14 Robert Rosenstone (2012) reports a striking example of this, namely the *American Historical Review*'s May 2006 decision to end its policy of reviewing individual films, in part acknowledging its reviewers' lack of expertise and interest in the medium of film. As the editor, Robert A. Schneider noted (in the May issue) that "when historians review films, they usually write about what they know about – accuracy, verisimilitude, and pedagogical usefulness. These are not inconsiderable as commentary, but it is a far cry from what we expect from them in a book review". See: www.historians.org/publications-and-directories/perspectives-on-history/may-2006/on-film-reviews-in-the-ahr.

15 As Sven Lütticken (2014:2) observes of Steyerl's use of the essay form,

> to essay is to try, to attempt. The essay is a form of doubt – a format in which one can explore doubtful theses. While the essay is at root a literary genre, in the twentieth it leapt into new media. (. . .) Steyerl's written essays are not explanations of her films. Even when they share a title, as with *In Free Fall* (a 2010 film and 2011 article), the relation is one of mutual complementation and contradiction.

16 An example of such a lecture uploaded to Vimeo can be seen at: https://vimeo.com/81109235.

9

ARTISTIC RESEARCH AND THE HUMAN SCIENCES

Staging anthropology

9.1 Introduction

Today, culture is no longer seen as the esoteric pursuit of artists but as the lived experience of everyone. The advent of visual culture as a field of enquiry has moved the study of the visual beyond art history's traditional concerns with painting, sculpture, architecture and design towards an engagement with the domain of popular culture and the media. This approach has sought to plot the various modes of cultural representation at play within these spheres and has employed various semiotic and ethnographic approaches to do this. This understanding of mass mediated culture has inevitability drawn the creative arts closer to a range of social scientific and critical humanities concerns.[1]

Much contemporary visual and plastic art is no longer primarily concerned with transformations in visual language and style or with formal characteristics, as modernist art historically has been. The autonomous status of the work of art has been destabilized and the boundary between artefact and artwork, between document and construct, fact and fiction has been harder to establish. Conceptualism seeks to challenge conventional patterns of thinking about art practice and its relation to the world, while relational approaches seek to elicit new kinds of responses from audiences. We now have a focus on the socially constructed character of "artness" and on the interactions of the artist with a public, often in site-specific locations. These developments have led contemporary art to engage with a number of adjoining fields of practice, such as social activism and community development, and with a range of social sciences, such as urban studies, media studies, human geography and, of course, social anthropology.[2]

In this chapter, I examine the dynamic between art and anthropology. I explore a number of case studies which hopefully illuminate this relation from a research point of view. I look in particular at the work of a number of established artists and film-makers, such as Jean Rouch, Susan Hiller and Trinh T. Minh-ha, who have engaged with anthropology. This "classic" work, much of it produced over thirty years ago, continues to set the terms of the debate between anthropology and art. Here we have our contemporary "masters" and, most notably, "mistresses" of practice-based arts research. In it, we see a "working through" of a range of issues such as cultural representation and difference and an engagement with the ethics of appropriation and collaboration. I argue that this capacity on the part of artists to address a range of methodological, ethical and political issues through the specific challenges encountered within their work has played an important role in the fashioning of a paradigm of practice-based creative arts research. I also explore some examples of recent doctoral work which seem to operate within this ambit.

9.2 An ethnographic turn in the visual arts?

Some art theorists, like Hal Foster, anticipating perhaps the emergence of a relational focus within contemporary art, have sought to identify an "ethnographic turn" in the visual arts (1996a). He talks about a "quasi-anthropological paradigm" being adopted within the visual arts. Within this, the artist identifies themselves as a researcher into culture.

Similarly, art historians such as W. J. T. Mitchell (1996), responding to the advent of visual culture as a field of study, have identified an anthropological sensibility at the core of the new art history. Correspondingly, anthropologists such as George Marcus and Fred Myers have acknowledged that "art has come to occupy a space long associated with anthropology, becoming one of the main sites for tracking, representing and performing the effects of difference in contemporary life" (1995:1).

Today, a range of interesting collaborative research between anthropologists and artists is emerging. Anthropologists, particularly those interested in sensory and emotional matters, are experimenting with a range of art practices in their investigations.[3] On the other hand, artists, especially those committed to socially engaged practice, have been attracted to ethnography as a research method. Certainly, the critical dialogue between visual anthropology and contemporary art, particularly around ethnographic film, photography and curatorial matters, has provided one of the clearest exemplars to date of a coherent, practice-based research activity in the creative arts that has generated both creative and knowledge outcomes of some significance.

Over the last number of years, social anthropology has been particularly attentive to the complex relationships that obtain between a researcher and their subjects. This reflexive dimension gives anthropology a particular relevance to artists concerned with collaborative and research-led practices and indeed with their own subjectivity. As we have seen, such reflexivity necessarily entails the artist–researcher exploring their *subject position* (their own gender, class and ideological location) and the *research tools* (methodologies and epistemological assumptions) employed to understand others. It involves a sensitivity to the cultural and material differences which may exist between themselves and those whose cooperation they seek in their art and media activity. As Foster notes (1995a:305), for contemporary artists, whether professionals or students, "it is the self-critique of anthropology that renders it so attractive, for this critical ethnography invites a reflexivity at the centre even as it preserves a romanticism of the margins". As he wryly observes, many artists often seem to find other people's culture more interesting than their own and distant political struggles more vibrant than domestic challenges. Ethnographic enthusiasm, the pursuit of the exotic and various forms of displaced idealism, often seem to go hand in hand.

However, Foster has less to say about how the creative arts and in particular lens-based practices have played a not insignificant role in encouraging the discipline of anthropology to embrace a reflexive approach in its ethnographic research practices.[4] As we shall see, it has often been the employment of the still and moving camera within anthropological fieldwork and the issues this raises with regard to the representation of one's subjects that has encouraged the development of such reflexivity within recent anthropology.

Foster argues that an anthropological frame of reference now serves "as a lingua franca in artistic practice and critical discourse alike" (1995a:305), providing a contextualizing and intellectually legitimating role for art practice. In the past, art history with its canons and narratives of tradition offered this, but as we have seen, with modernism the cultural hierarchies and value systems on which art history rested have collapsed. As we have seen, the art process itself is now conceived as a field of cultural investigation and as a sort of social research. Artists conceive of themselves as researchers entering a field of culture to pursue an investigation and subsequently report on this via the work they make. The art involved is seen as a question of process as much as the business of product. Artists seek to both respect and decode the "otherness" of the social subjects they encounter. They develop art strategies and fashion works through their engagement with the cultural artefacts, discourses and lives of their "subjects", navigating the culture of the other with some degree of awareness of the ethical issues likely to arise around representation and participation.

Ethnography is quite literally the practice of writing about a social group and its culture – usually in the context of a structured field investigation. With its origins in the discipline of social anthropology, it is usually understood as the systematic study of people and cultures in which the researcher observes and reports – in writing or graphically – upon a social group, culture or subculture. Ethnography relies greatly on up-close, personal experience and cultural engagement. Participation, rather than just observation, is one of the keys to this process and, hence, perhaps the attractiveness of this research approach to those interested in relational approaches to art making.

Ethnography has been associated with the practice of *participant observation* (a methodology which involves the researcher gaining an intimate familiarity with the language, culture and mental habits of a group by participating over a sustained period in their rituals and forms of life). Increasingly, the notion of ethnography now seems to operate well beyond the field of anthropology. As a method, its use has spread into fields like educational studies (ethnography of the classroom), design (ethnography of user-centred design),[5] sociology (ethnographies of youth culture and metropolitan ethnic subcultures) and film studies (ethnographies of the audience or of the film production as a subcultural world). So, art practice and research is by no means unique in its enthusiasm for this approach. Clearly, ethnographic description can be captured not only in written forms of documentation but also by means of photographic and film practice and, indeed, by employing any process of artistic representation mobilized to explore and portray a field of cultural relations. Indeed, as early as the mid-1970s, leading anthropologist Margaret Mead, an enthusiastic advocate of the use of both still and moving image photography within anthropological research, bemoaned the fact that (Mead 1975:4)

> all over the world, on every continent and island, in the hidden recesses of every industrial city as well as in the hidden valleys that can only be reached by helicopter, precious, totally irreplaceable, and forever irreproducible behaviours are disappearing, while departments of anthropology continue to send fieldworkers out with no equipment beyond a pencil and a notebook.

However, once the camera became a key tool of anthropological fieldwork, a new catalyst was introduced into the ethnographic mix and new creative aspirations embraced by social researchers.

9.3 The tarnished legacy of anthropology

Of course, the use of photography and film within anthropology has a long history coterminous with the discipline itself. Indeed, this less than happy history has been the focus of a number of artists and film-makers who have sought to explore issues of cultural representation, power and voice within a contemporary postcolonial context.

To get a measure of just how deeply compromised the discipline of anthropology has been historically, consider one of the first uses of the camera for research purposes within the emerging discipline. In 1891, Trinity College Dublin established an anthropometric laboratory under the direction of the anthropologists Dr C. R. Browne and Professor Alfred Cort Haddon (modelled on the lab set up by the eugenicist Francis Galton in London). Haddon, who had initially trained as a zoologist, took up a post in Dublin after conducting ethnographic research on the aboriginal peoples of Torres Straits. This work was later published in the volume *Head-Hunters, Black, White and Brown* (1901).

Seeking to apply social evolutionary theory to Irish ethnicity, Haddon and Brown undertook extensive fieldwork on the Aran Islands off the coast of Galway, an Irish-speaking archipelago that the two believed was inhabited by natives largely untouched by metropolitan Anglo-Saxon culture.[6] Their study might therefore offer empirical evidence about the racial characteristics of the Irish in comparison with the "Anglo-Saxon". The ethnographic survey of the islands and their inhabitants was conducted between 1891 and 1893 and employed photographic documentation alongside cranial measurement and the recording of the eye colour, skin pigmentation and general physiognomy of the islanders (as seen in Figures 9.1 and 9.2). This anthropometric activity was underpinned by eugenic thinking and represented an attempt, as the researchers put it, "to unravel the tangled skein of the so-called Irish race". As Annie Combes observes (1991:205),

> the same rhetoric of extinction and preservation, once applied by academic anthropologists to specifically colonized races, as a means of validating anthropology's expansion, was now systematically applied to certain communities within the British Isles.

And as the image from the survey shown in Figure 9.1 indicates, Haddon and Brown's fieldwork seems to have involved the use of the Royal Irish Constabulary (RIC), the country's police force, to assist the anthropologists in administering the anthropometric research process. This was a paramilitary force which had been used in the past, and would be in the future, to enforce evictions on the Aran Islands.

In 1898, a mere three years after the invention of the Lumière Cinématographe, Haddon returned to the Torres Straits and employed a 35mm Newman and Guardia movie camera to produce one of the world's first ethnographic films. This footage depicts various islanders in traditional fire making, displaying costumes used in rituals and participating in dance and ceremonies (http://aso.gov.au/titles/historical/torres-strait-islanders/clip2/).

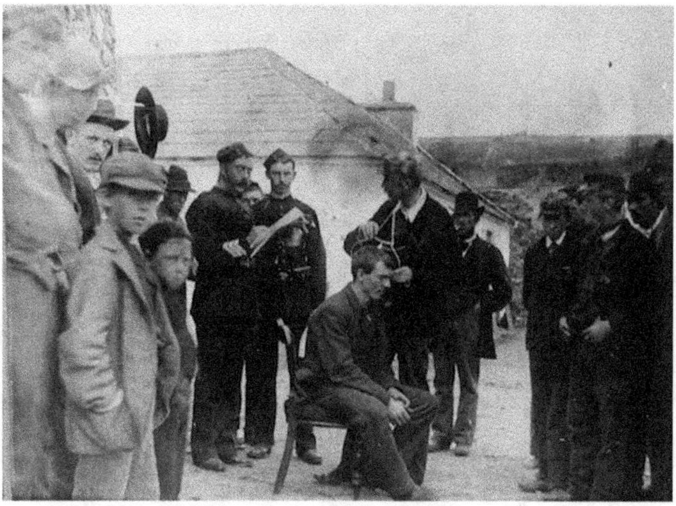

FIGURE 9.1 Trinity College Dublin (TCD) researchers conduct cranial measurement and photographic documentation of an Aran islander c.1892, National Library of Ireland

Source: public domain

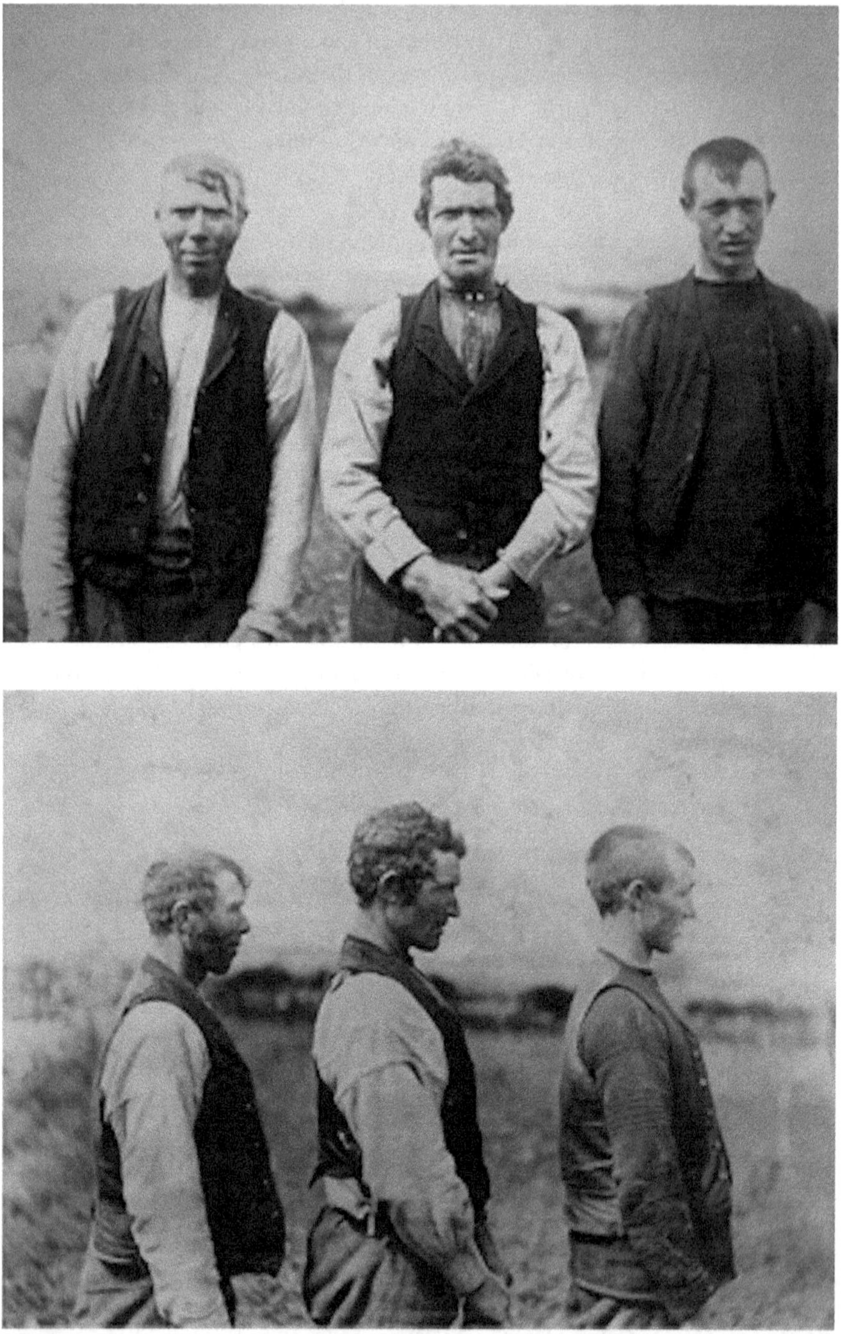

FIGURE 9.2 Photographic portraits taken during the TCD survey: (top) full frontal and (bottom) side view of three Aran islanders

Source: public domain

Haddon was to become an advisory curator at the Horniman Museum in London, an institution concerned with "illustrating natural history and the arts and handicrafts of various peoples of the world" (Horniman 2019). He may have played a role in the 1908 Franco-British Colonial Exhibition in London. This presented a "reconstructed" Irish village to the public on the White City site alongside Scottish, Somali, Dahomeyan and Senegalese hamlets, each with its compliment of "natives" involved in handicraft and other "traditional" activities (Rains 2011). Certainly, Haddon's ethnography of the Aran Islands, together with his role in the British Ethnographic Survey and long-term interest in the surviving folk culture of the west of Ireland, made him a recognized authority on Irish peasant life, and the design of the village (captured in a postcard image in Figure 9.3 left) seems to resemble settlements on the islands.

Such images, with their blend of faux ethnological themes and picturesque imagery, rapidly became deployed in the promotion of tourism in Ireland, later generating in the hands of the John Hinde company a distinctive genre of picture postcards (Figure 9.3 right). Needless to say, contemporary Irish artists, such as Sean Hillen (Figure 9.4) and John Kindness (Figure 9.5) have had a field day playing with such iconography. Hillen's use of photomontage techniques, indebted to artists like John Heartfield and Peter Kennard, leads to an interrogation of photography as evidence (the use of the medium privileged in ethnographic appropriations of photography). Photography is reframed by these artists within an approach concerned with the politics of representation. I asked Sean about what inspired his *Collecting Meteorites at Knowth* picture, aware of the parallel poster work by John Kindness, designed as a public art project for billboard display in a rundown inner-city area in Dublin.

SH: It's true (as I usually tell) that I knew of John Kindness's witty pastiche of Hinde's card, where as someone put it "the baskets were piled high with unidentifiable suspicious packages and the donkey garlanded with syringes". However, in my mind the structure of "Collecting Meteorites" is simply stolen from one of the fifteenth century plates of "Les Très Riches Heures du Duc de Berry".

DB: But it was part of a broader series of photomontages which juxtapose touristic representations of Ireland with a range of other found images which erupt apocalyptically within the Irish landscape?

SH: Yes. The Irelantis project marked my transition to Dublin after 10 or 12 years in London. I was consciously trying to leave behind the direct reference to the Northern Ireland conflict in my work. Irelantis involved a new "enquiry". An itch that it's addictive to scratch. Irelantis as a re-imagined postmodern Ireland is epic and yet familiar, offering scope for a critical take on spectacle and hyperbole through a reworking of these iconic postcard images.

I asked him about the process employed in his work.

SH: The physical process of gathering the elements for the composition was important and essentially is a playful one involving locating and searching through large amounts of source material for pieces that have a particular visual quality, as well as those which simply "caught my fancy" – a photo of a large meteorite that skimmed the earth in the late 80s, an artist's impression of the Sun from the front of a *Scientific American*, then the layers of observatories.

It's a very liberating method that suited my general purpose. In the case of most of my collages I really don't "intellectually" know what I'm specifically looking for until I find it.

DB: But it's not all serendipity?

SH: I will actually hold up two images and move them around to see how they might combine; a critical thing is that I've trained my eyes to align and focus separately; so that I can overlap and move around both images in my "internal" field of view and to evaluate how they might be cut together to some advantage.

There can be little doubt that the widespread deployment of lens-based media within social anthropology has raised a range of acute issues for the discipline around representation, reflexivity, collaboration and, in particular, the mediation of ethnographic accounts by visual tropes. Most of these concerns first became manifest within ethnographic film. In the pioneering work of French ethnographic film-maker Jean Rouch, we witness a searing critique of colonial anthropology. In its place, we are offered a participative model of ethnographic film-making. His approach challenged the practices of both anthropology and cinema. Not surprisingly, this model has had particular purchase within recent practice-based arts research.

In his 1970 film *Petit à Petit* (Little by Little), Rouch travels to Paris with the three West African characters who had figured in his earlier film *Jaguar* (1954–1967) to conduct a sort of "reverse ethnography". The film playfully turns the tables on the entire enterprise of colonial anthropology. In *Jaguar*, Rouch had told the story of three young men who set out from rural Niger on the road to Accra, in what is now Ghana. They hope to find fame and fortune and eventually return home as heroes. They are following the well-trodden path of the migrant worker from rural West Africa in search of work in developing urban areas. In Accra, their final destination, the spirit of recent political independence and Pan-Africanism is strong. The culture they encounter is stridently modern. The film has been described as a sort of road movie combining elements of documentary realism with fictive constructions that enable the three protagonists, Damouré Zika, Lam Ibrahima and Illo Gaoudel, in effect to play themselves, and to tell their story in a manner that always seems more than simple documentary testimony.

In *Petit à Petit*, Rouch continues with this strategy of *ethno-fiction*. Damouré, Lam and Illo now play the roles of a delegation of African businessmen who travel to France, purportedly to research how people live in high-rise apartment blocks, with a view to building one themselves in Niger. The intrepid travellers step out onto the streets of Paris to conduct their own ethnography of metropolitan life.

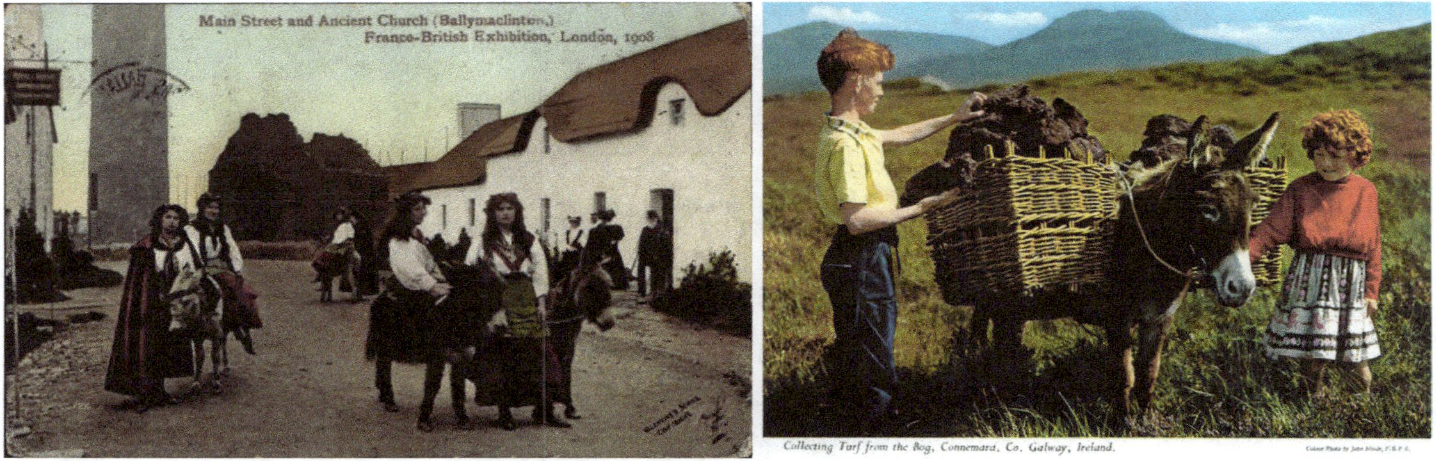

FIGURE 9.3 (left) Postcard of the Irish Village at the 1908 Franco-British Exhibition; (right) John Hinde postcard (1965–1970)

Source: (left) public domain; (right) © John Hinde Archive

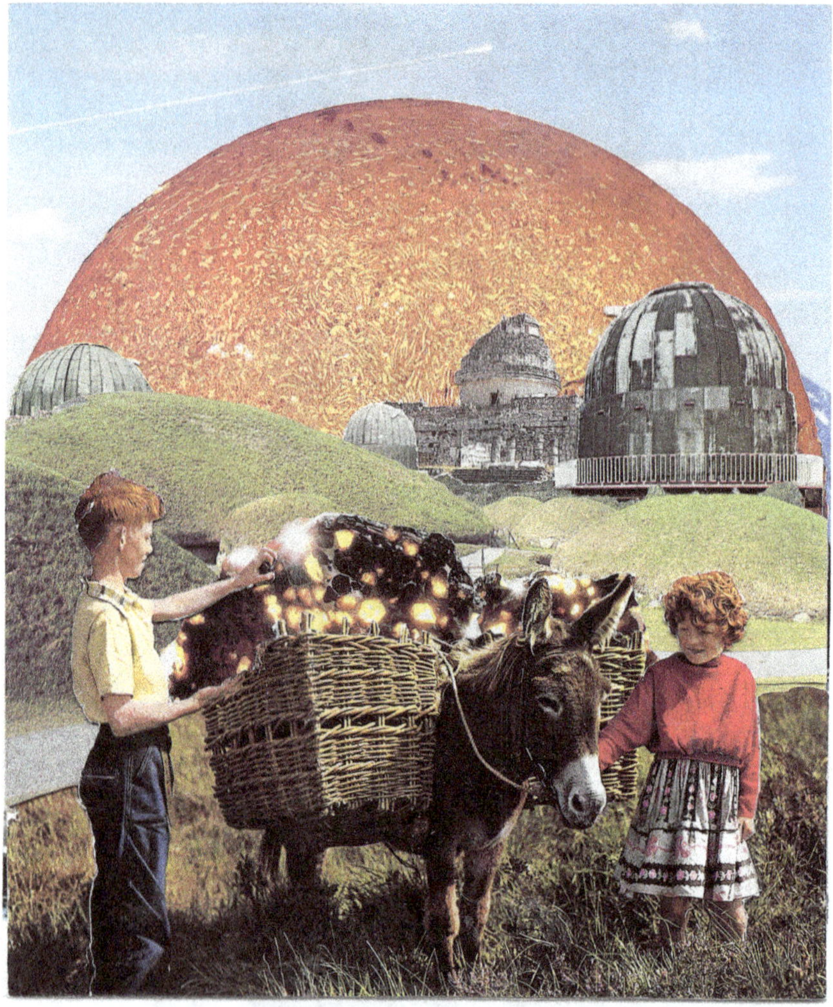

FIGURE 9.4 Sean Hillen, *Collecting Meteorites at Knowth* (1996)

Source: courtesy of the artist

In one scene, Rouch has Damouré Zika measure the heads of Parisian passers-by with callipers on the steps of the Musée de l'Homme (France's leading museum of ethnology). In other scenes, he inspects a girl's teeth and queries a young man about his habits (Figure 9.6). This *tour de force* playfully reverses the traditional asymmetrical power positions of anthropologist and native we have glimpsed in Haddon and Brown's anthropometric activities on the Aran islands (Rouch's debunking scene can be seen at: www.youtube.com/watch?v=IxnbYV4-su8). This is not ethnography. It's more fun than that. However, Rouch's films certainly represent a

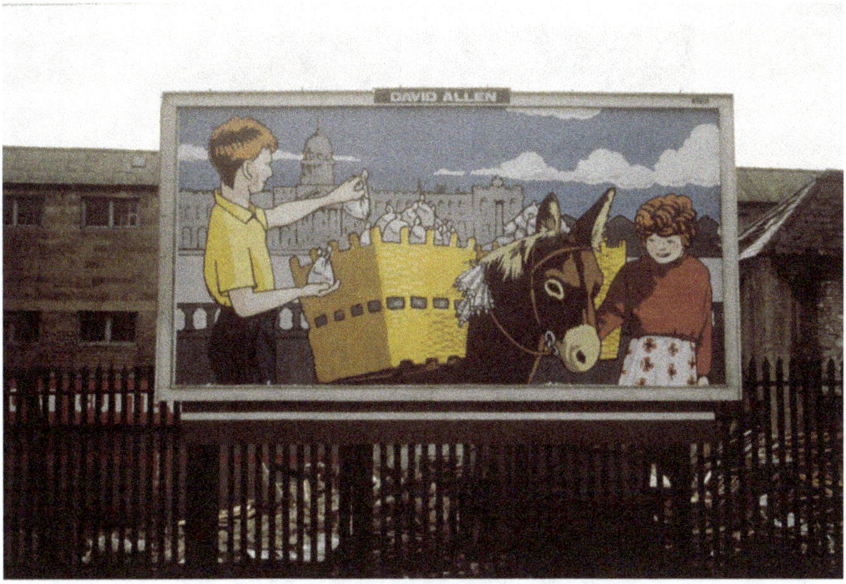

FIGURE 9.5 John Kindness, *The Dealers* (1987), hand painted poster display, Dublin

Source: courtesy of the artist

critical interrogation of traditional ethnographic methods and suggest imaginative new directions for anthropologists to follow.

The Colombian artist Alberto Baraya has recently reprised Rouch's trope. In his exhibition *Expedition Berlin, Herbarium of Artificial Plants*, shown at the 8th Berlin Biennale in 2014, he visits the Naturkunde Museum in Berlin armed with a set of anthropometrical measuring callipers and is photographed having his head measured by a number of visitors of non-European appearance to the Dinosaur Gallery (for a clip, see: https://vimeo.com/98546192). He had already experimented with this motif in his 2013 piece *Machu Pichu Expedition*, where we see a photographic image of him as the intrepid traveller clutching a bunch of plastic artificial plants and having his head calibrated by a Quechuan woman close to this well-trodden Inca tourist site (Figure 9.7).

The main focus of Baraya's work over the last ten years has been a deconstruction of the taxonomies employed by nineteenth-century scientists to classify botanical specimens "discovered" in Latin America (focusing in particular on the Spanish Royal Botanic expedition of Jose Celestino Mutis). He approaches these taxonomies as a set of colonial configurations of nature through which relations of power and economic domination can be read (interest in such taxonomies has been widespread among contemporary artists). His specific artistic strategy is to "research" a collection of artificial plants. These are not native but "Made in China" – as mass manufactured decorations. Many of these artificial plants or flowers, made

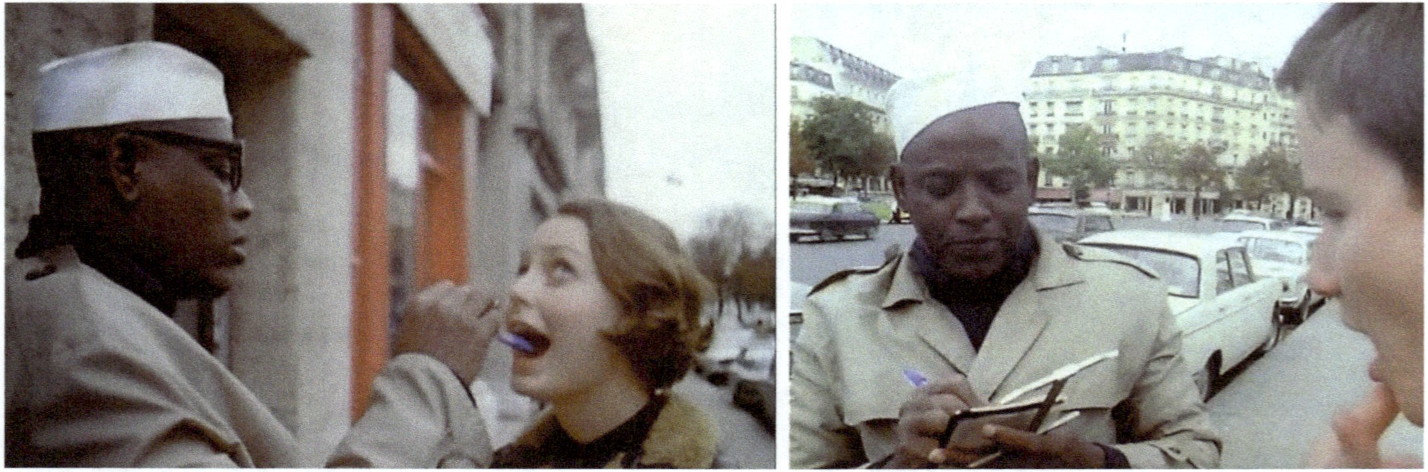

FIGURE 9.6 Jean Rouch, *Petit à Petit* (1970), film stills: Damouré Zika conducts his own ethnographic research on the streets of Paris

Source: courtesy of Icarus Pictures

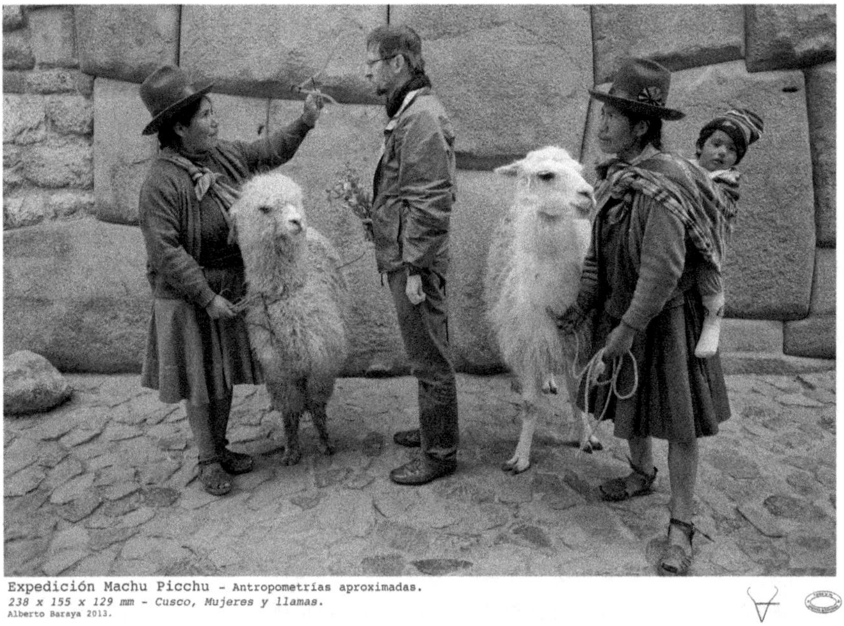

FIGURE 9.7 Alberto Baraya, *Machu Pichu Expedition* (2013)

Source: courtesy of the artist

from plastic, cloth or paper specimens, have been stolen from restaurants, lifted from waiting-rooms and funeral parlours or pocketed at someone's house. In effect, he re-enacts the act of plunder associated with the collection activity of the great historical scientific expeditions of the nineteenth century. As he freely admits (Roca 2007),

> by picking up some plastic flowers on the street, I behave like the scientists that Western education expects us to become. By changing the goals of this simple task, I resist this "destiny". In that moment all assumptions are put into question, even History.

He then mounts and classifies his specimens in a meticulous manner like a nineteenth century botanical collection, complete with a detailed card index system, providing in effect an ironic commentary on our fascination with the exotic and the taxonomies of colonial science. Baraya is concerned, in particular, with deconstructing the figure of the *viajero*, the traveller and scientific observer, whether botanist or anthropologist, that developed forms of knowledge to ideologically underpin a colonial system of expropriation. His *Herbarium of Artificial Plants* (seen as exhibited in the Berlin Bienalle in Figure 9.8) parodies the empirical objectivity of a botanical naturalist, just as his expedition images lampoon the anthropometric activities of the anthropologist.

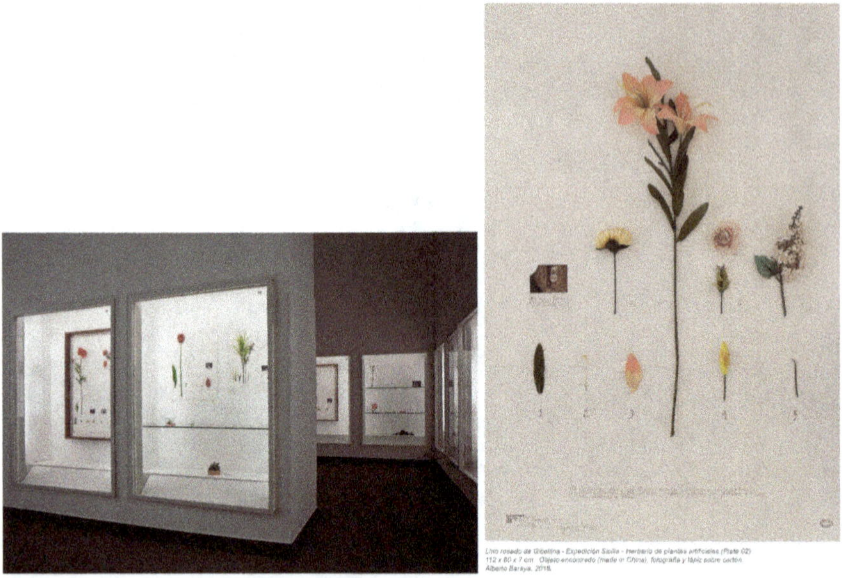

FIGURE 9.8 (left) Alberto Baraya, *Herbarium of Artificial Plants*, installation view, Berlin Bienalle, 2016; (right) Close-up of a "specimen"

Source: courtesy of the artist

As he communicated to me in an email correspondence,

> in my work I explore a range of scientific methods, not to directly advance my art practice, but as source material in developing an art insight. My practice contests the "imperial" forms of knowledge present in positivist and colonial science offering a different way of knowing. So, the main issue is not to question the results of scientific research, but to expose the mechanisms of manipulation that lie behind the promotion of a certain idea of progress and to reveal the political use of certain forms of scientific knowledge.

I asked Alberto about the relation between his interest in classic botanical taxonomy and anthropometric research, both of which are parodied in his work.

> If at first glance they study different categories, they probably share a common root. And this can be traced by studying the different faces of their own knowledge. The Spanish artist Rogelio López Cuenca identifies the role of these two types of scientific practice in the service of the economic and political conquest of the Americas. Probably I am more interested in the conceptual aspects of this phenomenon rather than engaging in a militant anti-imperialist gesture. In general, the great botanical or ethnographic

> expeditions presented themselves as projects for human progress. However, at the end of the day, the results are not always used for praiseworthy purposes. My figure of the ironical scientific researcher seeks to reveal this darker side of science.

Jean Rouch's West African ethnographic film work also reveals the darker side of the human sciences. Films such as *Jaguar*, *Les Maîtres Fous* (1955) and *Moi, un Noir* (1959) were not widely known in the Anglo-Saxon world in the 1960s and 1970s as a tradition of visual anthropology got underway. However, as we have seen, *Chronique d'un été* made both film buffs and anthropologists sit up and take notice.[7]

As visual anthropologist Faye Ginsburg observes (2005:111), Rouch's practice "was far in advance of contemporary rethinking of both anthropology and film making". Rouch was developing an entirely new kind of ethnographic and documentary film practice that blurred the boundaries between: producer and subject; fiction and "reality"; Europe and Africa; the practical and the poetic; the mundane and the magical.

For example, in *Jaguar* Rouch shot most of the film on a small 16mm film camera without sound. At this stage, he did not have available to him the synchronized sound recording equipment that he was later to use with such effect in *Chronique d'un été*. But in Rouch's hands, even this technical limitation could be exploited for creative and critical ends. Rouch showed his participants – and by now film collaborators – the rushes for the film, encouraging them to improvise a commentary reacting to what they saw on screen and recalled from their journey. Robert Flaherty had employed a similar strategy some 30 years earlier. In his *Man of Aran*, he recorded the islanders' reactions to his rough-cut to create a sound track dubbed onto certain parts of the film, such as in the famous shark fishing scene with its shouts and exhortations. Flaherty had to transport his participants to a studio in London to achieve these sound effects (the islanders were supervised by Seamus Delargy, the director of the recently established Irish Folk Lore Commission). Rouch was able to record the sound track for *Jaguar* in the field, later mixing it with a musical track provided by a number of Niger musicians for the final dub. Again, the playful, performative and vernacular character of the commentary produced by Rouch's collaborators departs from the authoritative, voice of God narration used in the travel documentary and quasi-ethnographic films of the time.

This collaborative mode of film-making, as well as producing an entertaining movie, also facilitated a mode of anthropological understanding of a particular cultural and socio-economic phenomenon – migratory journeys in West Africa and the transactions between tradition and modernity which occur within the experience of such migrants. The project in effect produced a new form of ethnographic knowledge released through the film-making process. However, such film-making is not about providing a visual illustration of anthropological data, or as Jay Ruby awkwardly calls it, "a pictorial expression of anthropologically constructed knowledge" (1998). Rather, it involves a process of enquiry in which the ethnographic understanding emerges through the collaborative processes of producing a film. It is, in other words, through the very grain of the film-making that an anthropological understanding emerges.

As Jean-Paul Colleyn, Rouch's collaborator in the ethnographic laboratory of the École des Haute Études en Science Sociales in Paris, suggests (2005:115),

> before having read Mikhail Bakhtin, Rouch and his friends had invented dialogical films. With no need for James Clifford's, Bill Nichol's, or Clifford Geertz's theoretical reflections, he had invented reflexive, performative and ironic films.

It is for this reason that Rouch has been heralded as the originator of a particularly rich tradition of practice-based research in the moving image. And, indeed, the area of ethnographic and documentary film has been one of the fields in which critical theory and creative practice have had perhaps their most productive alignment. Within the ethnographic film we glimpse the emergence of a grounded and rigorous example of practice-based artistic research.

9.4 Lens-based ethnography: documenting Dublin's docklands (case study)

A number of contemporary creative arts researchers have taken inspiration from Rouch's work. For example, Moira Sweeney in her Dublin Institute of Technology PhD project (2016) sought to explore the shifting occupational and communal identities in Dublin's docklands. Sweeney employed her still and video cameras to document her interactions with her subjects – dockland workers who have witnessed the transformation of their working lives wrought by the containerization of the port, casualization of dock employment and decline of the north inner-city dockside as a vibrant neighbourhood. She describes (2016:54) her interdisciplinary research approach as involving "the convergence of at least three distinct methodological and professional tributaries: the imagination and production skills of a television documentary maker (she has previously worked in broadcasting), an inquisitive geographical imagination, and an emerging interest in modes of self-reflective enquiry" (which she reports finding in visual anthropology).

A significant influence on her research was the work of photographer, film-maker and critical theorist Allan Sekula. His 1995 photographic essay *Fish Story* revisits documentary traditions of photographic representation in order to explore the subject matter of the sea and shipping and the circulation of commodities within maritime space. He addresses complex economic forces which are by their very nature difficult to visually represent. One of the key issues explored in Sekula's photo-essay (and subsequent in the 2010 film *The Forgotten Space*, made together with veteran film scholar Noel Burch; Figure 9.9) is the connection between containerized cargo movement and the growing internationalization of the world

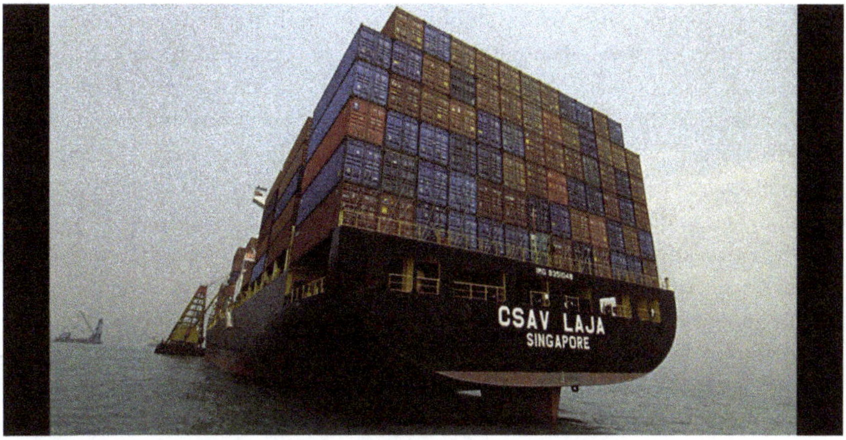

FIGURE 9.9 Still from the film *The Forgotten Space* (2010), 110 mins, colour, by Allan Sekula and Noel Burch

Source: courtesy of Noel Burch and Doc-Eye Film

industrial economy. He seeks to trace how these systems of circulation impact on the restructuring of the occupational life of ports.

Sweeney found in *Fish Story* an example of a creative and critical practice she could follow. Like Sekula, she seeks a visual storytelling strategy that might best address this marine political economy.[8] She worked with a number of Dublin dock workers over a two-year period, observing, interviewing and photographing them at work, adopting what she describes as a "visual ethnography approach" in her investigation. Among other things, this meant acknowledging that she as a film-maker and her subjects existed in a shared and imaginative space constructed in part through the actions of her camera. She invokes the notion of the *ciné-trance* employed by Jean Rouch. For Rouch, as we have seen, the film-maker, far from adopting a distanced and objective stance when recording the activities of their subjects, enters into an intuitive and ecstatic experience of sensory immersion as they film. "For me", Rouch admits, "the only way to film is to walk with the camera, taking it where it is most effective and improvising another type of ballet with it, trying to make it as alive as the people it is filming" (2003:38).

Rouch's camera does not seek to provide an objective record of the cultural practice filmed. Rather, through its provocative entry into the social world studied, camera and film-maker play a part in the reconstruction of that world. As Jean-Paul Colleyn (op. cit.) reminds us, Rouch

> never sought to be the unnoticed observer, the "invisible witness", favoured in US observational documentaries of the 1960s. He hated the metaphor of a filmmaker as "a fly on the wall": His camera drove straight into the centre of the action, changing it, and provoking reaction.

This *ciné-provocation* seeks to be illuminating for both researcher and their subjects, producing a range of insights into cultural relations which could be shared by researcher and subject alike. In Rouch's hands, the ethnographic film process underscores the fabricated nature of the ethnographic encounter and the collaborative character of all anthropological research. Perhaps for this reason, his approach has found such favour with a generation of young artist–researchers seeking a critical model to guide their practice.

The rituals of working life that Moira Sweeney filmed in Ireland were those of manual workers on a Dublin dockside rather the trance activities of the Songhay-Zarma in equatorial Niger that Rouch was concerned with capturing. However, the common feature is the involvement of each film-maker in a form of cultural exchange that is seen as collaborative in character. Sweeney adopts the highly mobile use of the camera by the researcher–film-maker in relation to the subjects filmed that Rouch has described as a "strange kind of choreography". This hand-held use of the camera a visual prosthetic facilitates the entry of the researcher into the events they encounter in the field. As the use of the term "choreography" suggests, the subjects often actively collaborate with the film-maker to facilitate the portrayal of a scene. This means that

> realities are co-constructed and that meanings always change as context of interpretation change, continually revealed and modified in numerous ways. Provoking, catalysing, questioning, and filming are simply strategies for unleashing that revealing process. (. . .) people respond by revealing themselves, and meanings emerge in that revelation.
>
> *(Feld 2005:16)*

After closely observing the work rituals of the stevedores, dockers and labourers over several months, Sweeney's filming begins to focus on a study of Amy, one of the few women working on the docks (Figure 9.10 left). The film-maker seeks to produce a moving image portrait of Amy, seen at work loading and unloading the docked cargo ships. In a sequence shot in the tiny cabin of the crane she operates, Amy senses the difficulty that Moira is having steadying and focusing the camera in the enclosed space. In an accommodating gesture, Amy instinctively shifts her position to facilitate a film-maker struggling to capture a key shot – a small gesture but one that signals the collaborative and performative character of such documentary work. Her work finally culminated in the exhibition of two multiscreen installations sited within a large dockside warehouse (Figure 9.10 right).

Her completed films were projected in this space and her participants invited to be her first audience. Like many documentary film-makers who have moved out of broadcast and film exhibition, she chose to experiment with an installation approach, believing that this "has the capacity to intervene in sites and construct interactive spaces in relation to audiences that employ sensory means that break the mould of the white cube gallery" (Sweeney 2016:63).

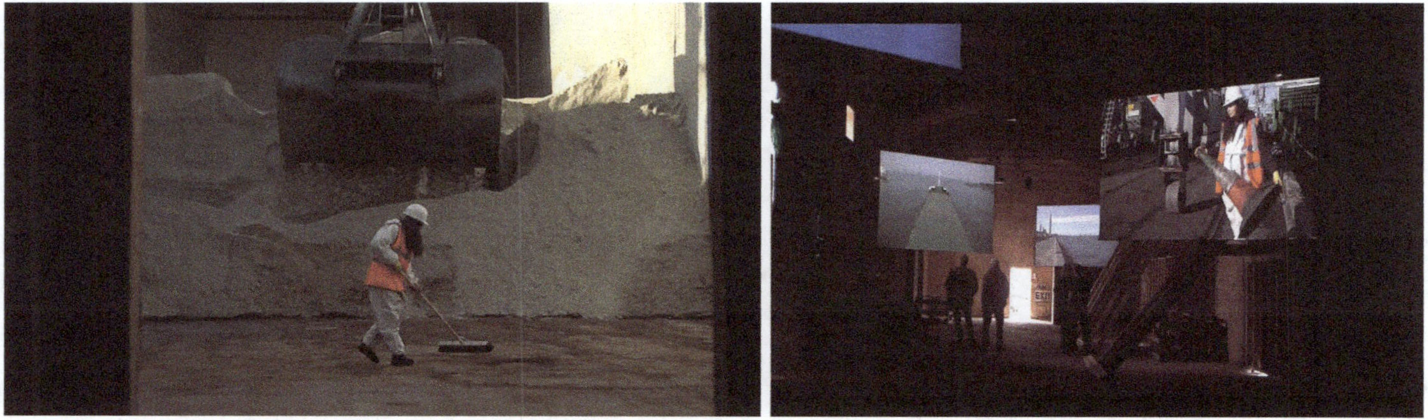

FIGURE 9.10 (left) Moira Sweeney, *Rhythms of a Port* (2014), screen shot: "Amy cleaning the hold"; (right) installation view, 5 Digital HD screens in a dockland warehouse

Source: courtesy of the artist

Schneider and Wright have identified a generation of lens-based practitioners, like Sweeney, who work like anthropologists but exhibit as artists. These researchers employ what they refer to (2013:19) as an "observational aesthetic", producing film and photographic works "exhibited in gallery or site-specific settings which facilitate the use of multiple screens allowing for different kinds of physical encounter with the work".

This method of presenting ethnographic research moves beyond social scientific reportage, a discourse which often employs objectivist language and scientific tropes that distance the research from the subjects whose collaboration facilitated its outcomes. Sweeney felt that her chosen mode of exhibition "was both in keeping with the embodied approach adopted in the filming and responsive to the specifics of the site of my research and to my participants" (Sweeney op. cit.: 63). As a researcher, she was picking up on a well-established tradition of artists seeking to explore the challenge of site-specific work and of developing new modes of display conducive to the collaborative and dialogical character of their practices.[9]

Anthropologist, curator and historian of photography Elizabeth Edwards has sought to more broadly engage with this recharged photo-documentary dynamic. She urges anthropologists to acknowledge the art of photography and not simply to see the camera as a field tool for documentation but also to acknowledge and learn from the expressive, formal and conceptual dimensions of recent photographic practice. Her concern is to encourage the imaginative and dialogical, rather than strictly scientific, character of anthropology. As she admits (1992:56), her approach draws mainly

> from discourses outside the accepted canons of visual anthropology: from contemporary photographic criticism, the dilemmas of photographic practice and the ambiguous spaces of postmodern documentary practice. (. . .) these discourses constitute precisely that set of "counter rules" which might enable still photography to constitute an active and individual voice within the discipline, premised on the expressive rather than the realist.

This notion of *counter rules*, derived from Feyerabend's suggestion that science needs challenges from outside its normal paradigm to renew itself as an empirical domain, is a useful one. Art practice provides such counter rules, disturbing the "imperial" authority of anthropology. Conversely, anthropology also can serve as a set of counter rules which might inform contemporary art practice. As Schneider and Wright suggest (2013:12), "anthropology has much to teach artists about the kinds of negotiations and culturally relative inflections of terms such as *participation* and *collaboration*, particularly in colonial and postcolonial contexts". For example, those who celebrate the relational character of art activity have been slow to elaborate the methodological auspices of such "socially engaged" work and might usefully attend to how over the last twenty years anthropology has sought to elaborate an appropriate methodological and ethical response to the challenge of conducting collaborative fieldwork and adequately representing its subjects.

9.5 Susan Hiller: a coming to terms with anthropological method

Another established artist who has been highly influential with regard to current creative arts research practice is Susan Hiller, sadly no longer with us. Hiller trained as an anthropologist and conducted field research before developing her visual arts practice, and this experience has undoubtedly shaped her work as an arts practitioner and researcher in various ways. She reports that as an undergraduate student in the early 1960s at a New England liberal arts college, she had a keen interest in art practice but opted to study anthropology as a major. Many of the early figures in the discipline were women, such as Ruth Benedict and Margaret Mead, and Hiller tells us that in anthropology she found "role models of adventurous women and a sense of pioneering" which appealed to her as an enquiring teenager (1996:17). Indeed, she quickly found that anthropology was a better fit with her artistic interests than the stodgy art history curriculum on offer. This largely focused on the Renaissance and dealt exclusively with canonical male artists. Anthropology seemed a better fit for an aspiring artist driven by intellectual curiosity and a feminist awakening.

She also reports her first encounter with an anthropological collection discovered hidden away in a college basement (1991:1):

> the look and the sense of "the primitive", artefacts to be touched and smelled, works of ancient or exotic origin, objects of mystery; plus, an idea of intellectual community, of scholars united in the decipherment of a puzzle and the quest for enlightenment on the fundamental question, "What is the nature of man".

Later, she describes how as a postgraduate anthropology student she was "so intensely moved by images I saw during a slide lecture on African art that I decided to become an artist" (ibid.). This fascination with ethnological artefacts, their display and the retrieval of their original meaning has become an abiding interest throughout her artistic career.

Conducting anthropological fieldwork was another matter. Hiller has reported in a number of publications (1991, 1996) the discomfort she experienced when, as part of her doctoral studies in anthropology, she undertook ethnographic fieldwork in Mexico, Guatemala and Belize. This unease centred on what she felt was the problematic role of the participant observer in such research. The ethnographer enters and participates in the cultural world of the subjects they are studying. However, they also effect a certain distance from their informants with a view to achieving a degree of objectivity in their analysis.

Hiller reports that she found the bifurcation of self within anthropological fieldwork disconcerting, as "participation was clearly meant as a kind of formal pose to be adopted for a limited period" (1991:2). Once the data was amassed in the field, like colonial booty it was taken back to a metropolitan centre for analysis and domestic consumption. The participants in the research who had shared their lives and culture with the researcher, if not forgotten, were certainly marginalized within the scientific undertaking.[10] The search for an objective stance seemed to lead to an alienation from her subjects. As she notes (1996:104),

> this alienation is already encoded by the split term participant/observer, which implies that in some way I have never been able to understand, the object of enquiry is separate from the enquiring subject. From my perspective, this is a clearly false, ideological split, that creates a hierarchy which locates embodied knowledge and the contradictions of lived experience "below" the abstractions of overarching theory.

As she recalls (1991:2),

> I promised myself happily to abandon the writing of a doctoral thesis whose objectification of lived events was destined to become another complicit thread woven into the fabric of evidence that would help anthropology become a "science". In contrast I felt art was, above all, irrational, mysterious, numinous. (. . .) I would relinquish factuality for fantasy.

This inner conflict seems to have propelled her towards a life as an artist in order, as she puts it, to "find a way to be inside all my activities". She also came to believe that art practice could facilitate a more collaborative basis for working with the subjects she was engaging with. Her experience of ethnographic research had given her a certain hypersensitivity to issues around representation and considerable unease with regard to what she perceived as the hierarchical ordering of theory and practice within the discipline of anthropology. She reports her unease with anthropological formulations that (1996:18)

> tend to abstract meanings and generalized applications, assertions, of the superiority or precedence of theory over experience, as though theories were formulated and uttered somewhere outside the constraints of time, place, subjectivity and the body, and are therefore free of prejudices and mistakes.

Of course, Hiller was not the only 1960s anthropology graduate student uneasy with traditional ethnographic theory and practice and eager to critique the positivist and imperialist underpinnings of their discipline.[11] The rise of anti-imperialist movements in post-war Africa and South-East Asia, and mass student opposition in the US to the Vietnam War after 1967, transformed a discipline which historically had been complicit with colonialist endeavour.

In the minds of a generation of radicalized students, objectivist social anthropology was fundamentally flawed, as was indeed the entire positivist project for the social sciences. A new era of reflexive, critical and collaborative ethnography was emerging within anthropology. This drew upon: Marxist political economy and the analysis of the complex transactions between the first and third worlds; feminist critiques of patriarchy and expansion of notions of the political; symbolic interactionist and phenomenological approaches to understanding culture, identity and subjectivity; deconstructionist theories which plotted the linkage between knowledge and power.

One suspects that had Hiller stuck around the world of anthropology[12] she would have found a much more congenial intellectual context than the one that conditioned her own formation as a field researcher. Instead, Hiller opted to develop her work and sharpen her visual interests and critical skills within the folds of creative arts practice. However, her desire as a fieldworker for "connection, empathy, identification" was in time facilitated by her art practice with its collaborative possibilities. However, as she was to later admit (1991:2),

> repudiating an objectifying discourse (anthropology) in favour of a subjectifying discourse (art) does not even begin to resolve the extraordinary lived contradictions of merely being a subject in a culture that (. . .) does not allow a "synthesis between ideology and poetry".

The significance of Susan Hiller's work for an emergent paradigm of practice-based creative arts research surely lies in her awareness of the challenge of balancing the subjective and objective vectors of critical and creative practice within any art research methodology and informed way of working.

Moreover, she is fully aware that modernist art is as compromised as anthropology in its dealing with the culture of colonized peoples (1991:2):

> While anthropology tries to turn the people, who are its subject-matter into objects, and these objects into "theory", art tries to turn the objects made by those people into subject-matter, and eventually into "style". Both practices maintain, intact, the basic European picture of the world as a hierarchy with "ourselves" at the top.

Over many years, Hiller has been highly critical of the ways European artists and curators have in their own work appropriated the forms of visual design present in various non-European cultural artefacts in an unacknowledged way. Such cultural product is often labelled as "primitive art". Both anthropology and modernist art "have history" in this regard. In the sphere of culture, both anthropology and modernist art have traded in the exotic and primitive in a manner which has uncomfortable parallels to colonialism's economic exploitation of native resources via rapacious extractive industries and through unequal exchanges of trade. As she notes (1996:4),

> for many years the data of ethnography have been source material for Western artists. At the very beginning of modernism, African art profoundly modified the formal ideas of artists like Picasso, Derain and Matisse, who adopted sculptural or pictorial conventions from exotic cultures without any

> knowledge of the original meaning of the material . . . it was Picasso and Derain who discovered that African art could prove a source of inspiration to European artists in search of new techniques, in something of the same way that Africa itself could prove a source of wealth to Europeans in search of new markets and materials.

As she fully admits, artists, like anthropologists, deal with the stuff of culture – with representation, transaction and power. They are, in part, cultural anthropologists. And, like anthropologists, they need to develop critical methods adequate to their role as cultural producers and critics (1996:24) and become "experts in their own cultures".

> Artists. . . *modify* their culture while learning from it. . . . Artists *perpetuate* their culture by using certain aspects of it. . . . Artists *change* their culture by emphasizing certain aspects of it, aspects perhaps previously ignored. The artist's version may show hidden or suppressed potentials . . . by revealing the extent to which shared conceptual models are inadequate because they exclude or deny some part of reality. Artists everywhere operate skilfully within the very socio-cultural contexts that formed them. . . . They are experts in their own cultures something that most anthropologists ignore and of which most artists are unaware.

Hiller's 1976–1977 project *Fragments* shown at the Museum of Modern Art in Oxford and also at the Hayward Gallery in London (Figure 9.11), is seen as one of her more anthropologically influenced projects. It provides a useful exemplar of her research and critical practice. The artist invites us to engage with what seems at first sight to be an antiquarian exhibition. She presents us with a collection of traditionally decorated Pueblo pottery shards. As an assembly, the piece initially suggests a museological structure. It seems to be offering us an interpretation of a set of archaeological remains or anthropological artefacts.

These ceramic fragments were acquired from Pueblo woman potters. In a subsequent interview (1996), she tells us she actually collected the pottery shards from "the surface of a mound which had been a rubbish dump for Native Americans". She subsequently painted a sample of three hundred of these ceramic pieces in "earth colours", a creative strategy which she sees as analogous to the decorative practices of the Pueblo potters. These artisans, she tells us, "get their ideas for pottery designs from broken pieces of old pottery". Her project follows a strategy of appropriation and reworking of found fragments. However, in its respect for its sources and in its attentiveness to the original context within which such work was made and given meaning, it differs in significant ways from classic modernist art strategies of appropriation. The latter opportunistically refashions everyday artefacts (domestic or colonial) and integrates these into the artwork via processes of *bricolage* and reproduction which often fail to acknowledge the original sources of the material utilized.

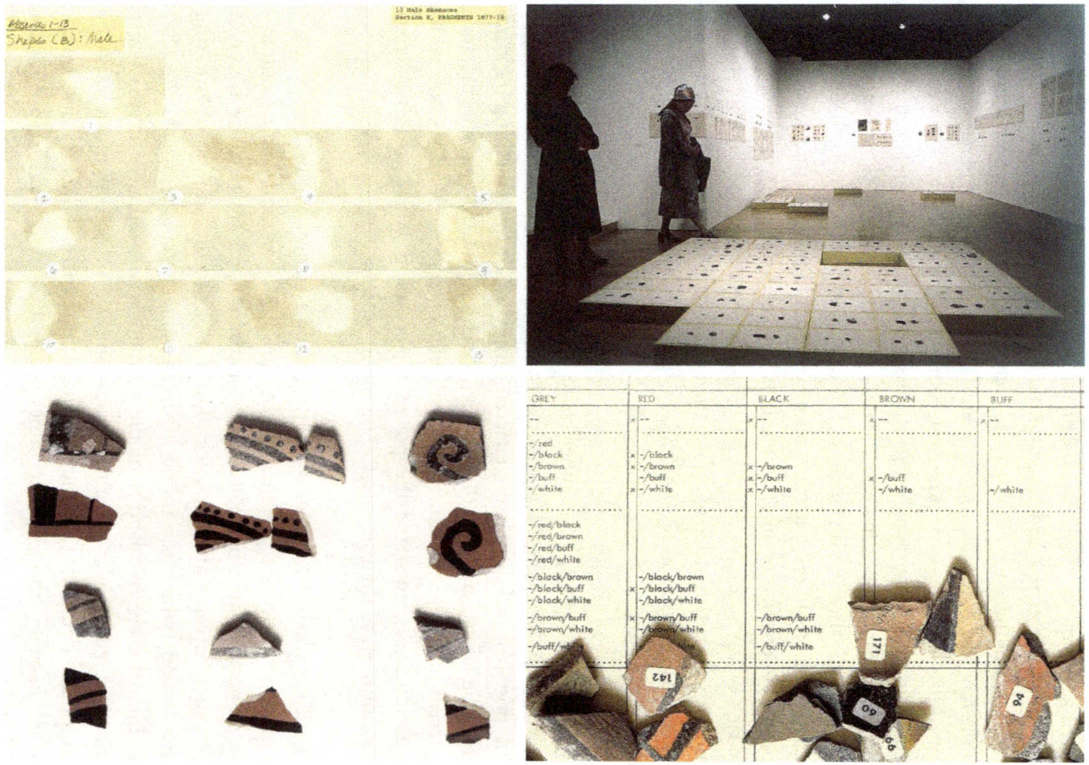

FIGURE 9.11 Susan Hiller, *Fragments* (1978), installation, Hayward Gallery, London

Source: © Susan Hiller; courtesy of Lisson Gallery

In the exhibition, the shards are arranged – some of them numbered – in a formal pattern. This initially appears to accord to archaeological conventions of display. An introductory section to the exhibition tells us this pottery is the product a 2,000-year craft tradition and that the decoration of the pottery involves the use of ancient mythological symbols combined with the potters' personal interpretations, often inspired by their dreams.

However, while aspects of the material practice of these Pueblo potters are reported in the accompanying charts and diagrams, gouache drawings and photographs, no substantial ethnographic or contextualizing detail is given with regard to any fieldwork conducted. But maybe this is the point. The work, like that of Alberto Baraya, is conceptualist in form, not ethnological. She is rejecting, or least problematizing, an ethnographic viewpoint which drives a wedge between researcher and subject, metropolitan and native, between etic and emic, art and craft, theory and practice.

Yet, despite the conceptualist thrust in her work, the installation has been viewed by critics – aware of Hiller's background in anthropology – as some sort of fieldwork report or scientific documentation. This is a somewhat strange interpretation given the lack of any substantial documentation of the ethnological auspices of the collection. Moreover, the ordering of the material is hard to read as a coherent taxonomy. As Denise Robinson (2004:100) succinctly argues, this again is to miss the point of the work:

> Hiller's work opposes the inheritance of the 19th Century roots in the museum which forces its objects into a taxonomy of compliance, yet there is no post-colonial duty being enacted simply to suture something back into place; the very nature of the fragments refuses this. After gathering these materials and objects through a specifically charged archival system, which in its initiation fantasized the power of recording, Hiller creates a hyperconsciousness of their new status – a living breach.

Hiller's work is perhaps best seen as an artistic interrogation of the discourses and institutionalized forms of representation of "primitive" art and "exotic cultures" that figure in both western museology and anthropological practice. Far from conforming to such interpretative practices, Hiller seeks to dismantle this "othering" of a native craft in which the mode of display and interpretation has the effect of diminishing the creative labour of the women who produced these artefacts, thereby marginalizing any account they might give of the object's meaning.

Lucy Lippard identifies a distinctive ethical horizon in Hiller's work. Indeed, Hiller herself has used the term "an ethics of appropriation" with regard to how artists and anthropologists might source and acknowledge their material. Lippard attributes this sensibility to Hiller's familiarity with a range of the critical debates within social anthropology around representation of the "other" and the ethico-political obligations of the researcher. As Lippard notes (in Hiller 1996:12),

> Hiller does not idealize "otherness" and has never exploited the surface imagery or displaced symbols of other cultures, as so many artists did in the

> 1970s and 1980s. She brought with her from anthropology a methodology rather than any specific theory. When she uses cultural artefacts, from potsherd to postcards, she does not project meanings onto them but retains their "idiosyncratic nature" which affects the way they are received and illuminates their significance in relationship to our own culture.

Over the last decades, a debate has raged within the creative arts about the ethics of representation, appropriation and collaboration, with a significant body of critical writing on Primitivism and the culture of modernism emerging (see Barkan and Bush 1995). However, Hiller is not simply reproducing such debates. Rather, she is "working through" her own practice as an enquiring, reflective artist and providing a critique of those models of anthropology (and of art practice and curatorship) which fetishize native art and diminish the agency of the native producer. Again, I would suggest that this is the import of her work for a current modelling of artistic research. Perhaps it is also the best response to Hal Foster's diatribes on the unthinking adoption of ethnographic tropes by contemporary artists. Hiller's work provides an important exemplar of an artistic research which is informed by ethnographic concerns but which, in turn, offers a critique of anthropological theory and practice.

9.6 Trinh T. Minh-ha: speaking nearby

The work of film-maker Trinh T. Minh-ha can also be read as a critique of traditional ethnographic tropes and neo-colonial forms of representation. It advances a notion of experimental film practice as a research journey. Born in Hanoi in 1952 and brought up in South Vietnam during the war, Minh-ha emigrated to the US in 1970. There she studied music and literature. She later visited Paris as an exchange student, where she developed an interest in ethnomusicology and in film, eventually finding herself teaching composition and musicology in French-speaking West Africa. As she tells us with regard to her first film (1992:112),

> I made *Reassemblage* after having lived in Senegal for three years (1977–80) and taught music at the Institut National des Arts in Dakar; in other words, after having time and time been made aware of the hegemony of anthropological discourse in every attempt by both local outsiders and insiders to identify the culture observed.

Reassemblage, shot and edited over several years and released in 1982, is an experimental documentary film portraying the lives of women in various rural communities in Senegal. The published synopsis for the film provides a basic but apposite description of the film.

> Women are the focus but not the object of Trinh T. Minh-ha's influential first film, a complex visual study of the women of rural Senegal. Through a complicity of interaction between film and spectator, *Reassemblage* reflects

> on documentary filmmaking and the ethnographic representation of cultures.

Minh-ha has identified the film as a concerted attempt on her part to break from a certain anthropological framing of the postcolonial subject. As she notes (1992:112),

> you don't have to be a connoisseur of film to be aware of the problems that permeate anthropology . . . whether you look at a film, attend a side show, listen to a lecture, witness the fieldwork by either an expert anthropologist, or any person subjected to the authority of anthropological discourse, the problems of subject and of power relationship are all there. They saturate the entire field of anthropological activity.

However, as she freely admits, freeing herself from "the authority of anthropological discourse" in her research was a struggle that could only be resolved through the warp and the weft of her film practice, shot by shot, edit by edit (ibid.:113).

> While I was filming, for example, I realized that I often proceeded in conformity with anthropological preoccupations and the challenge was to depart from these without merely resorting to self-censorship.
>
> There were a number of things I did not want to reproduce in my work: the kind of omniscience that pervades many films, not just through the way the narration is being told, but more generally, in their structure, editing and cinematography, as well as in the effacement of the filmmakers, or the invisibility of their politics of non-location.

This meant questioning the established codes of ethnographic and documentary film-making. It involved rethinking the use of sound within documentary film (via experiments with silence and the use of non-synchronous combinations of sound and image). It entailed considering the framing of her subjects (abandoning the privileged viewpoint of the objective observer for the intimacy of the immersive encounter). Similarly, the placement and movement of the camera were addressed in a rupturing of the established cinematographic conventions and notions of visual truth associated with the pan, the long shot and the close-up within documentary production. In the edit, she on occasion opted for a disjunctive juxtaposition of shots (employing jump cuts and awkward transpositions between different visual textures). All of these strategies sought to reveal the artifice of filmic conventions and to acknowledge the presence of the film-maker, so often occluded in traditional documentary practice.

As the title of the film suggests, the film is a reassembling and working through of a range of documentary materials. Thus, while the film might be described on first viewing as an ethnographic film on Senegal representing an exotic culture within an anthropological framework, it is clearly more than that. As we seek to make sense of the fragmented visual and spoken text presented to us – characterized by the use of close-ups of Minh-ha's subjects, jump cuts, ellipses and repetitions in the narration – we quickly realize that in formal terms the film is qualitatively different from any traditional observational film.

Towards the start of the film she tells us, in a phrase which has become a watchword for her nuanced approach to ethnographic engagement, that she does not intend to speak about the Senegalese communities she is filming, "just speak nearby". That is to say, she wishes to distance herself from the forms of objectivism that have characterized the ethnographic gaze. In conventional documentary films, particularly educational, travelogue and ethnographic ones, the film or television spectator is encouraged to occupy the position of a voyeur with regard to the cultural scene being portrayed, a viewpoint that mirrors the objectivist gaze and dominant position assumed by the film-maker with regard to the subjects being filmed. In *Reassemblage*, Minh-ha seeks to develop a radically different approach to her subjects by employing an indirect language of representation which, as she explains in an interview (Chen 1992:87),

> does not objectify, point to an object as if it is distant from the speaking subject or absent from the speaking place. A speaking that reflects on itself and can come very close to a subject without, however, seizing or claiming it. A speaking in brief, whose closures are only moments of transition – these are forms of indirectness well understood by anyone in tune with poetic language.

The film is comprised of a rich and complex tapestry of images and sounds shot by the film-maker at various locations. It has a complex, multifaceted sound track comprised of synchronous and non-synchronous sound elements. On occasion, these are used in a realist manner and sometimes in an experimental one, as in her periodic use of silence over montages of images. The 40-minute piece (which has been removed from YouTube) has no clear filmic narrative, story climax or structured closure. Nor has it a binding audio commentary providing an explication of what we are seeing on screen. There is minimal use of titling and graphics as framing devices. However, throughout the film we hear a first-person voice-over – that of Trinh T. Minh-ha herself. She speaks in a quiet, reflective, even poetic manner. This narration sets the political context for the film, as in her opening lines, "scarcely twenty years were enough to make two billion people define themselves as underdeveloped". It also addresses (time code 26.44) her own subject position and gaze as a film-maker and the ethical import of this. Over observational footage of teenage girls washing clothes in the Senegal River, she comments,

> I come with the idea that I would seize the unusual by catching the person unawares. There are better ways to steal I guess. With the other's consent. After seeing me labouring with the camera, women invite me to their place and ask me to film them.

The film, while concerning itself with the broader politics of development, also seeks to deconstruct documentary form in order to explore the part played by cultural representation in securing western visions of the postcolonial world. At one key point in the film, her reflections on the ethnographic film process seem to come together. Over a long shot of canoes being paddled on the river (28.37), followed by footage of donkeys being watered, she admits that "for many of us the best way to be neutral and objective is to copy reality meticulously".

The camera, she acknowledges, facilitates this mimicry (1992:105). We see a medium shot of a child playing with a teapot outside a village house (30.48). Then the camera moves in closer to the subject. In her narration, she acknowledges the intrusion of the photographic gaze.

> Stressing the observer's objectivity
> Circles round the object of curiosity
> Different views from different angles
> The a, b, c . . . of photography

The film falls silent (31.46), as it does at various points in its duration. She offers us a series of affectionate film portraits of Bassari women and children (Figure 9.12) who bashfully return the gaze of the camera. As she comments, "what I see is life looking at me. I am looking through a circle in a circle of looks". But as she wistfully concludes (33.12) with regard to the traditional use of the documentary camera, "creativity and objectivity seem to run in conflict. The eager observer collects samples and has no time to reflect upon the media used".

As Minh-ha notes, anthropology attempts to treat film and, in particular, the ethnographic film, as a methodological appendage of a scientific discipline. Film and photography are often primarily viewed as forms of "data collection" concerned with providing evidence for anthropological claims rather than modalities of investigation and meaning in their own right. They are treated as recording instruments rather than as disciplines which are part of a larger family of creative practices alongside poetry, music and the visual arts. Trinh T. Minh-ha, with her background in music and propensity to treat language in a poetic and figurative manner, identifies her work within this broader field rather than with ethnographic film practice *per se*. She identifies a realist drive within anthropological discourse which conditions how field researchers approach film. Anthropology, she argues, employs "a neutralized language that strips off all its singularity to become nature's exact unmisted reflection" (1989:53). This "positivist dream" was present at the birth of photography and is the basis of anthropology's attraction to film and photography as possible research tools.

Of course, this naturalism is also present in much documentary film-making. Factual film privileges the indexical quality of the photographic image and the objectivity of a documentary gaze which aspires to directly capture reality through the lens. Minh-ha questions this naïve epistemology and seeks to reveal its ideological auspices. Too often for the anthropologist, film becomes, as she puts it

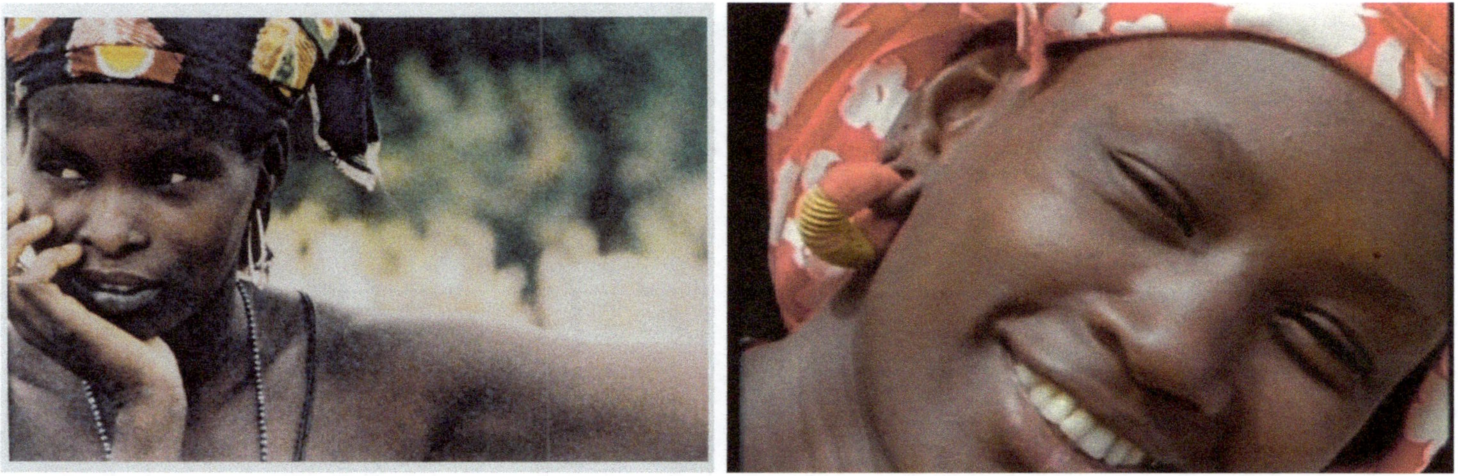

FIGURE 9.12 (left) Still from the film *Naked Spaces* (1995), 135 mins, 16mm, colour, West Africa, by Trinh T. Minh-ha; (right) Still from the film *Reassemblage* (1982), 40 mins, 16mm, colour, Senegal, by Trinh T. Minh-ha

Source: courtesy of and copyrighted by Moongift Films

(1992:264), "a mere vehicle for the transmission of knowledge" and, indeed, for forms of knowledge production characterized by scientistic pretension and reification of the culture of the Other. The ethnographic or documentary film can become

> a closure: you wrap up a story – some information, an analysis for the viewer – instead of keeping the filmmaking alive as a mode of knowing. Working always involves a desire to modify one's consciousness or the limits of one's thinking, and to shift from one place to a better one.... One is quickly bored with one's work where this generative quality is absent.

She attempts instead to solicit a new seeing on the part of the audience. This is a film, like many experimental films, essentially constructed in postproduction around the artistic vision, creative research strategies and critical sensibilities of the film-maker. Its research content lies not so much in its fieldwork programme (Minh-ha has been criticized for drawing her material from a number of Senegal communities and conflating these in her portrayal of the lives of the women filmed) but in its visual experimentation and in the harnessing of the results of this to her critical aims of deconstructing the language of ethnographic reportage.

However, she is reluctant to have this work pigeonholed as an experimental film despite the character of its cinematography, picture assembly and sound editing (1992:114):

> So, while the techniques are not surprising to avant-garde filmmakers, the film still does not quite belong to that world of filmmaking. It differs perhaps because it exposes its politics of representation instead of seeking to transcend representation in favour of visionary presence and spontaneity which often constitute the prime criteria for what the avant-garde considers art.

As she explains,

> "experimental" for me is not a genre nor an approach to film-making. It is, in a way, the process of unmaking ready-mades, or more commonly put, of making visible what remains invisible (ideologically, cinematically) to many, including to oneself (. . .) If "experimental" is a constant questioning of the relationship between the film-maker and film-making, then it cannot be separated from the material, whether one chooses to call this material documentary or fiction.

It is perhaps no coincidence that Minh-ha is a musician, and the film can be read as a composition constructed in the experimental practices of the editing suite. She was eventually to publish a script for *Reassemblage* in her book *Framer Framed* (1992), a text which also contains a series of interviews providing an idea of her working methods as a film-maker. Her script, she tells us, was not written prior to production of the film. It is, in effect, a transcript of the completed film that reports the editorial experiments and decisions made in the cutting room and sound studio rather than a shooting script drafted to provide a template or recipe for the production. As she says (1992:225–226),

> I would say the strategies used are almost the same for the images, the music and the voice over . . . this film brought together a number of activities whose end-point was not known to me in advance. Any motive that I had before starting work on the film necessarily dissolved as I went along, so that the creating consisted not so much in inventing something new as in rediscovering the links within and between images, sounds and words.

Her scripts are, then,

> tools that one works with rather than texts that one tries to conform to. It is important to keep in mind that the script is no more than a kind of skeleton. It is like a dead form that the film leaves behind once it is completed.
>
> *(Spangler 1993:2)*

The script is both a heuristic guiding an investigation and an account of a creative process.

In artistic research more generally, experiment with form, content and voice is often the key research activity. The generative or creative strategies explored in a project provide for a distinctive "mode of knowing". It's not clear that many anthropologists recognize that photographers, film-makers and artists who engage with cultural difference may be employing a completely different research stance to that found in their own discipline. Indeed, some anthropologists have questioned what the research significance of a film like *Reassemblage* could be. They say they have no evidence that it was based on any systematic fieldwork or demonstrable ethnographic data – a judgement which rather misses the point.[13]

Certainly, such work does not have as its main focus the methodological agenda or substantive concerns of anthropology.[14] Rather, it pursues other expressive and critical goals emerging out of the process of "making work" and asserts the primacy of practice over theoretical articulation. Moreover, Minh-ha has herself acknowledged that her critique of anthropology and the development of a mode of reflexivity within her practice are not original and that her approach is built on an ongoing critical tradition within cultural studies – "many have done it before and many are doing it now" (Chen 1992:85).

Similarly, Susan Hiller has been reluctant to elaborate a generalized critique of anthropology, preferring to demonstrate in her practice another way of working through cultural "stuff".[15] Trinh T. Minh-ha, like Hiller, is also quite clear that her focus as an artist is making work, not talking up products or achieving "scientific breakthroughs".

While both artists write critically and accept that writing involves encounters with other disciplines such as anthropology, they are primarily concerned with working through their own distinctive art practices and with broader matters of representation and power. Their work goes beyond the methodological debates within anthropology and engages with questions of film and artistic form, modes of address, the politics of post-coloniality and feminism and, in Minh-ha's case, a particular concern with the voicing of the experience of women from the Third World. As she concludes (Chen 1992:85)

> if the tools are dealt with only so as to further the production of anthropological knowledge, or to find a better solution for anthropology as a discipline, then what is achieved is either a refinement in the pseudo-science of appropriating Otherness or a mere stir within the same frame.

9.7 The auto-ethnographic impulse in artistic research

How might such thinking around what is sometimes called the "crisis in representation" impact on contemporary ethnographic approaches within the creative arts? Of course, ethnographic research does not necessarily involve fieldwork in distant locations and with "exotic" cultures. Ethnography as a research method is equally applicable to engagements within an artist's own culture. Indeed, a researcher can choose to engage with their own personal sphere, social practices and cultural sensibilities through what has become known as *auto-ethnography*, and many artists have exercised this choice in their research. Auto-ethnography is a research practice which emerged out of qualitative approaches within interactive sociology. The researcher focuses on their own life experiences, addressing the emotional and haptic sphere often ignored by traditional social science. From a methodological point of view, "their over-riding concern is to shrink the distance between the experiencing subjects and their accounts of lived experience" (Ellis and Flaherty 1992:4). Through their documentation of their field experiences and, on occasion, through role playing (there is inevitably a performative character to all ethnographic work), they seek to address broader social and cultural questions. These can be addressed in a body of writing or, increasingly, through art practice. By means of various strategies of display and self-reflection, the researcher seeks to connect their autobiographical story to wider cultural, political and social meanings.

Thus, for example, Carol Rambo Ronai explores her experience of being an exotic dancer in a Florida strip club. As she tells us in a short paper on her PhD research (1992:104),

> the data presented here comes from field notes I took during and after my experiences of dancing. They blend events from several nights into a typical "night in the life of a dancer". The participant role is never clearly separated from that of being a dancer, a wife, or the other roles I enact. These materials

> will give the reader a perspective on becoming a dancer/researcher, writing about the experience, and the impact of multiple identities on my "self" as a participant observer.

Auto-ethnography differs from traditional anthropological ethnography in that it embraces and foregrounds the researcher's subjectivity and emotional reactions to situations rather than attempting to limit this, as in many "objective" approaches within empirical social research. Ronai gives us a flavour (106) of her anxieties as she enters the site of her fieldwork, the dark misogynistic world of the strip joint. She passes a number of leering male customers on her way into the bar.

> Regardless of my cognitive desire to be the intrepid sociologist, braving new frontiers. Going where most woman dare not tread, my dread of the coming night imposes it presence on my reality in the form of a tightness in my chest that constricts my breathing. The tightness is paralysis. (. . .) Get a grip, damn it it's only a fucking bar, I mentally slap myself, finding my desire for dependence both attractive and repulsive. I get a charge of machismic satisfaction from not giving in to these emotions. To divert my attention, I again see my enthusiasm for my research, but it not there. The closer I get to the door the more I don't want to dance.

It's often unclear if the outcome of such research activity is social scientific in character or an exercise in experientially based creative writing.[16] Of course, auto-ethnographers largely reject a notion of social research as the pursuit of detached, objective knowledge. The outcome of such research inevitably seems to become "clothed in exclusionary scientific discourse" (Ellingson and Ellis 2008:450). Instead, they often embrace a range of creative strategies in their work. Accordingly, this approach has had considerable attractiveness to artists engaged in practice-based and reflective research.[17]

Auto-ethnography can then be read as one possible critical response to the crisis experienced within the social sciences and humanities around the representation of the other and the alienating effects of traditional field observation methods on both researchers and their subjects. By focusing on the lived world of the researcher themselves and on the forms of narrative we deploy to make sense of our experience, auto-ethnography can arrive at novel forms of interpretation which acknowledge the role of emotions, subjectivity, performance and storytelling in social life. Moreover, auto-ethnographers have been open to exploring a range of methods drawn from the creative arts and in particular from drama and performance studies, storytelling and creative writing, photography and documentary film-making.

PhD student Ruby Wallis, with a background in photography, experimented with various lens-based approaches in her investigation of the "new age" settlement of Cooloorta in the west of Ireland. As part of her doctoral research, she plotted her complex relation with its landscape and residents across a series of photographic projects. In one of these, titled *Autowalks*, she employed a hand-held

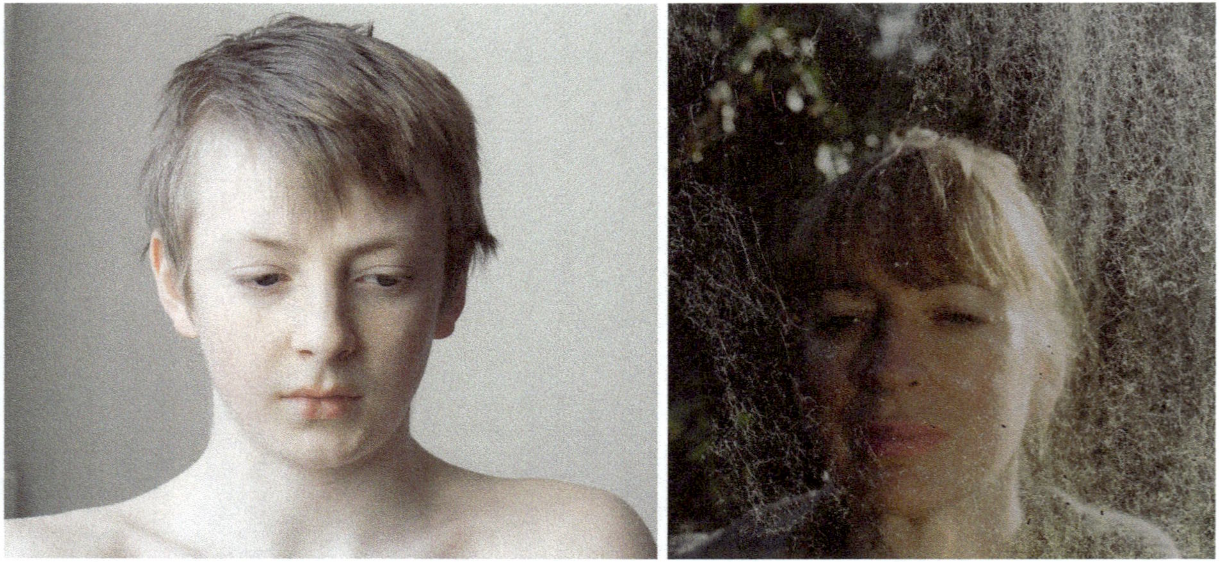

FIGURE 9.13 (left) Ruby Wallis, *My Brother Django* (2014), digital photograph; (right) *Dido* (2014), medium format photograph

Source: courtesy of the artist

video camera which she took with her in a series of walks through the hamlet. In this perambulation, she encountered a number of the residents either by accident or by arrangement. She had initially employed a medium format Bronica camera to photograph Cooloorta. This is the sort of equipment used by documentary photographers to deliver, as she puts it (Bell 2016:42), the "contained and compositionally framed images which photography privileges". She now opted to use a small portable video camera which "seemed to be an ideal way to represent place from an embodied position and within the auto-ethnographic framework I began to employ in my research". The resulting footage, shot hand-held in the act of walking, produced a more fluid approach, with a distinctive first-person, diaristic feel. She also cites the influence of Jean Rouch in her evolving methodology, observing that for Rouch, "the camera is often seen as a prosthetic extension of the sensing body of the film-maker", with ethnography always conducted "in the first person".

A second project, *Moving Stills*, took the form of a series of video portrayals of Cooloorta residents, some of whom were her relatives (her son is seen in Figure 9.13 left). These were stylized portraits employing the long take and fixed camera position and later slowed down in postproduction in a series of experiments with the role of temporality within photographic portraiture. In her work, traditional ethnographic concerns with mapping a cultural habitus are combined with formal preoccupations with the use of different imaging approaches and the impact of these on the representation of an interactive situation. The camera, of course, is an extension of the ethnographic gaze and as such serves to distance the researcher from their subjects. Thus, Wallis's project does not simply document Cooloorta and its residents but also alludes to the dynamic of intimacy and distance that all photography involves. Like the anthropologist, the creative arts researcher has to navigate a delicate path between subjective experience and objective analysis.

This navigation takes a particularly acute form in what has been called *native ethnography*. In this approach, the researcher is investigating a community or culture which they are, or have been, a member of themselves. Wallis has written illuminatingly in her dissertation about some of the problems of conducting a native ethnography. She recounts how in picturing subjects with whom she had had an intimate connection she comes to reframe her relationship to the community. As she assumes the researcher role, she questions her own identity as a former Cooloorta "native". Gradually over the duration of her research, her sense of the distance between herself and her participants seemed to grow. She was an insider who had by choice left the communal fold to become an outsider. Her role as a researcher seemed to amplify her problematic existential status. Her project sought ways of giving visual expression to this interactional complexity.[18] As she concludes, "I was no longer exploring the cluster of dwellings and my complex relation to this community, but perhaps thinking through my own state of chosen nomadism" (46).[19]

Art educationalists have in general been attracted to ethnographic models of research in so far as these seem to provide a methodological anchorage for practice-based arts projects which engage with a cultural field or social site or, most particularly, a range of personal experiences. Certainly, within the art college there has been a vogue on the part of some students for confessional testimony clothed in a form of pseudo-ethnographic reportage. This has been particularly marked in photographic and film work. Tutors appreciate the grounded focus within anthropology on techniques of field reportage, triangulation of interpretation and ethical reflection, which can act an as antidote to student self-preoccupation and narcissism.

Art students, of course, are well used to documenting their process and to drafting proposals and statements which either lay out the auspices of their work or seek to place it in relation to cognate work and evaluate it within various critical parameters. The techniques of fieldwork note taking are in many ways quite similar to the traditional artists' work of studio and project documentation. Instruction in each discipline hopes to encourage forms of field or studio reportage that is both descriptive and analytical and which facilitates reflexive practice. Many artists and art educators have found in visual anthropology and its fieldwork methods and reflections a model which artistic research might borrow from and adapt, and on occasion critique; an impressive range of text books are now available that detail a range of visual anthropological methods which might guide creative arts students in their research.[20]

That said, the employment of a notion of ethnography within much art and design research often seems like a figurative appropriation of anthropological terminology. Many PhD students, while increasingly mindful of anthropological perspectives and methods, want to read and draw their methodological inspiration more widely and often consult a range of philosophical and critical positions (existential and social phenomenology, history of visual culture, film and photographic theory). Nor does "art ethnography" always employ the empirical and critical rigour found in anthropological fieldwork. After all, its outcome is an artwork, not an academic monograph. Indeed, as we have seen, it has been artists operating within a tradition of experimental film and conceptual art making – traditions which historically have overlapped with documentary film-making in significant ways – that have made the most significant critical contributions to the development of ethnographic film and reflexive cultural research.

Notes

1 From pop art on, the predominant focus within contemporary art has been navigation of the cultural world of mass media images, appropriating, reworking and ironicizing the outputs of our systems of mass communication. There has been only the most meagre acknowledgement of the critical and theoretical challenges this might entail. The issue has in effect been outsourced to fields like media and cultural studies.

2 Some anthropologists have lately sought a measure of fusion between art and anthropology. For example, Arnd Schneider and Christopher Wright (2013) have, across a range of publications (2005, 2010, 2013), sought to plot the interactions between contemporary anthropology and art practice. They identify a focus within each discipline on collaboration, creative and representational strategy and methodological innovation. This, they

believe, is creating the basis for the emergence of a hybrid approach to exploring cultural patterning that employs artistic means guided by anthropological perspectives. Some have christened this hybrid form of research and practice "art ethnography".

3 As Schneider and Wright note, a notion of ethnography as a form of conceptual art has been proposed by a group of anthropologists from Cambridge who draw upon a range of historical sources from the early twentieth-century Russian avant-garde to 1960s conceptualism in art (Wilson and Ssorin-Chaikov 2012).

4 This is somewhat strange, as Foster in his writing on modernist visual arts practice and its relation to the evolution of art historical method has traced how the reflexivity embraced by avant-garde artists and concern with the material means of expression transferred itself to art history. A similar process seems to have occurred in the synergies between surrealism and anthropology.

5 To give but one example, a recent collection of research papers on fashion design, *Fashion Studies: Research Methods, Site and Practices*, edited by Heike Jesse (2016), contains a section on "Exploring Fashion Practices Through Ethnography", with papers on such topics as "Ethnographic Entanglements: Memory and Narrative in the Global Fashion Industry" or "Urban Fieldnotes: An Auto-ethnography of Street Style Blogging". Indeed, ethnographic approaches have assumed a core position within design research, where such observational and interpretative strategies seem particularly apposite to "user centred" design approaches (Milton and Rodgers 2013).

6 In actual fact, the Aran Islands, while geographically peripheral in relation to metropolitan Britain, were from the mid-nineteenth century integrated into the trading and migratory relations that characterized Ireland's relation to Great Britain within the Victorian capitalist economy. Many islanders were migratory labourers working at various times of the year in the farms of lowland Scotland and northern England; others became assimilated as waged labour in the expanding centres of industrial production. Nor were they unfamiliar with ethnologists. William Wilde (father of Oscar) had organized a field trip to the islands for the delegates of British Society for the Advancement of Science in 1857.

7 As the stock of the *Nouvelle Vague* rose, so did Rouch's reputation as an innovator who straddled the fields of film and anthropology. The term *cinéma vérite* had been invoked by Edgar Morin to describe the new filmic approach, very much in homage to Dziga Vertov's work. Rouch himself had reservations about the name, "fearing that it was tainted by the pretension of an absolutist notion of truth" that he felt both traditional anthropology and American direct cinema toyed with (Feld 2005:14).

8 In a similar manner, Juan Orrantia, in his project *Entrada Prohibida* (*Forbidden Entry*) dealing with the ports of Mozambique, in particular the Indian Ocean port of Beira, finds inspiration in Sekula's work.

9 As Schneider and Wright observe (2013:19), multiscreen exhibition facilitates the presentation of multiple narratives to an audience. It can

> bring about a simultaneous shift in the relation of the work to truth. It draws attention to the multiplicity of truth in a way that even a single-screen, linear narrative that contains many voices does not. The installation allows for a literal overlapping of narratives.

10 Indeed, traditional anthropological method inevitably seems to drift from an emic viewpoint and collaborative sensibility concerned with comprehending the lived world and thinking of the local subjects studied by the ethnographer, to an etic perspective and an objectivizing tendency which seeks to place an interpretative grid over "native" experience and mentalities in order to arrive at a "scientific" understanding of a culture.

11 As she recalls (1996:45), at this time in the US anthropologists

> were being told that as social scientists our work (. . .) was value-free, it was objective, and it was meant to exemplify pure academic research values. Well, the Vietnam War in a very practical sense showed everybody in the social sciences that this wasn't so,

because the data of anthropology were being used by government agencies to subvert village life in South-East Asia.

12 For example, as a 1960s American graduate student in anthropology, the film work of someone like Jean Rouch was not available to her, and the sub-discipline of visual anthropology was still in its infancy.

13 *Reassemblage* had a difficult reception in some quarters. Anthropologist Jay Ruby judged Trinh T. Minh-ha's films (2000:288) to be "uninspired derivatives of 1960s US experimental film", finding her criticisms of documentary film and anthropology "uninformed by the tradition of self-criticism easily located within both fields". He also questions her competence to assess anthropological method. This is a strange argument, as Minh-ha makes no claims to professional skills in the discipline, except perhaps in the specialist area of ethnomusicology. She speaks and writes as a film-maker and cultural activist, not as a social scientist.

14 Jean Rouch was always much closer to the institutions and practices of anthropology than either Susan Hiller or Trinh T. Minh-ha. He completed a PhD dissertation in the subject on his work in Niger and usually described himself as an anthropologist and film-maker, rather than as an artist. Indeed, Minh-ha has been critical of Rouch's work, in part responding to the unease of many African film-makers with Rouch's approach to African culture and in part reflecting her belief that he had not fully broken from the ideological constraints of anthropology.

15 As Susan Hiller argues (1991:6), "theoretical consistency – even when the theory is innovative or radical – only continue[s] to prop up the idea that everything is known and/or knowable, one of the more geriatric assumptions of the European world-view".

16 That said, many anthropologists have been uncomfortable with the performative character and lack of empirical constraint characteristic of some auto-ethnographic research. On the other hand, artists often take their lead from figures like Sophie Calle who playfully fuse documentary-style work with a performance art aesthetic in a manner that not only cleverly combines fact and fantasy but questions these epistemological distinctions.

17 Hal Foster (1995a:304) remains sceptical. Commenting on the new-found interest with auto-ethnographic approaches on the part of artists, he observes that "such self-othering easily passes into self-absorption, in which the project of "ethnographic self-fashioning" becomes the practice of philosophical narcissism".

18 In the final practice element of her research, *Turlough Swim*, Wallis produced a three-minute video piece based on the activity of swimming in a lake adjacent to the hamlet. She attached a portable flip camera to her head just above eye level, then entered the lake. This facilitated a point-of-view style of filming (inevitability the *kino eye* of Dziga Vertov is invoked). The spectator shares the line of sight experienced by the film-maker and can immerse themselves in the haptic dimension of the swim (for a clip, see: www.rubywallis.com/turlough-swim.html).

19 Elements of all three sub-projects which made up her PhD research were presented in an installation exhibited in the gallery of the Burren College of Art (situated relatively close to Cooloorta), which the examiners of her PhD visited and evaluated.

20 These range from general texts on the use of visual methodologies within social research, such as Marcus Banks's *Visual Methods in Social Research* (2001) and Gillian Rose's *Visual Methodologies* (2001); to texts dealing more specifically with the methods of visual anthropology, like Sarah Pink's *Doing Visual Ethnography* (2013); to practical handbooks on making ethnographic films, such as Ilisa Barbash and Lucien Taylor's now classic *Cross-Cultural Filmmaking* (1997); and, as we have seen, texts such as that of Schneider and Wright (2013) or Bell (2016), which offer examples of art projects and PhD research shaped by ethnographic considerations.

10

ARTISTIC RESEARCH AND CRITICAL THEORY

10.1 Introduction

Finally, then, how should we conceive of the relation between theory and practice in the creative and media arts? How might we address the disjunction that seems to have arisen between the two within current creative arts education and research? In this concluding chapter, I look at the processes of theorization at play in creative arts research. As we have seen, research in the creative arts and media disciplines emerges from practice and seeks to inform that practice. That is its cognitive and professional interest. But what of the deployment of theory within arts research and practice? What form does this take? How successful have we been in our theoretical endeavours in forging a productive relationship with creative practice? Do we as artists and media makers have the right kind of theory?

Now, the notion of theory deployed in art educational institutions and media schools is, as I have suggested, a highly distinctive one. It is not a theory of practice based on inductive reasoning or abduction from the results of artistic activity. Nor is it based on generic theories of creativity informed by psychological experiment and analysis. It is not primarily rooted in an analysis of visuality, either as a symbolic system or set of perceptual processes.[1] Nor is it a theory of artistic technique, still less one of aesthetic value. In fact, somewhat paradoxically, it is really only incidentally interested in creative arts activity *qua* practice.

Put bluntly, the relationship between theory and practice within the creative arts has become characterized by a certain dyspraxis.[2] "Theory" – whether literary, cultural or filmic – has effectively become a conversation among a coterie of academic professionals, and one without either much explanatory utility or political saliency. So, we may have to look beyond current constellations of theory and postmodernist sophistry to find a suitable model for critically informed research practice in the creative arts. Yet a *theory of practice*,[3] or to be more precise, a set of theories related to the different creative art practices, seems to be exactly what we need at present to undergird arts research.

In this chapter, I also look at a number of recent creative arts projects which seem to operate within a critical paradigm and which employ a disparate range of research practices. These are largely of a lens- and screen-based character and fall between documentary report and fictive construction. They seem to indicate the capacity of contemporary art to engage with contemporary political realities – when it chooses to do so. These projects employ innovative creative strategies to "get at" this world. However, the generalizations arrived at are provisional and specific. The theoretical insights gained do not displace the primary creative imperative which motivates the work. Nor do they stand in for it. As I have suggested, one of the tasks of artistic research is to draw out the insights present in such work, opening it up to a series of critical interlocutions and methodological queries which might both clarify its auspices and trace its import for further work in the field.

10.2 Theory in the creative arts and media

Art and media theory (including film theory) is significantly different in form to those types of theory found in the natural sciences, or indeed in those social sciences, which search for objective and binding universal laws governing human behaviour. However, it seems to have certain affinities with those modes of historical and anthropological investigation which focus on questions of meaning and interpretation, and on issues of social criticism and political and cultural action.

At the same time, we have to continually remind ourselves that artistic research is discovery-led rather than hypothesis-driven. The forms of understanding that emerge from such research activity are intimately bound up in creative endeavours and professional practices. To invoke Deleuze and Guattari's terms, artistic research, if it is a science, is a "nomadic" rather than "royal" one. The core task of the artist as researcher remains that of making and mapping work, bringing artistic intelligence to the research process and research expertise and intellectual openness to the development of their creative work. However, specific knowledge outcomes and a measure of theoretical generality can be abstracted from the business of making work. This, I have suggested, is best achieved through acts of reflection and via critical interrogation of specific projects, often in interlocution with theorists from other disciplines.

So, how do we generally use the term "theory" within the creative arts and media subject domain? Perhaps the key notion is that creative work can be illuminated and partially understood by placing in within the context of various critical discourses and social dynamics and in relation to other work produced within a field or around a theme. Over the last three or four decades, a notion of critical theory, with its origins in social and political analysis but significantly refashioned within literary and cultural studies, has become perhaps the leading way of conceiving of the role of theory within creative arts and media practice. Students, as we have seen, are expected to "critically contextualise" their practice work in relation to not only cognate artwork but in reference to bodies of theory. Art and media academies host departments of contextual, critical or visual cultural studies, tasked with teaching such theory and with supporting students in making sense of their work.

The critical sources presented to students have often been embedded in complex theoretical architectures – structuralism, poststructuralism, deconstructionism, feminism, post-colonialism. Indeed, it is not always clear if the majority of creative arts students, or indeed their tutors, fully grasp the theoretical pedigree of these approaches. Nor is it apparent that the broad-brush strokes provided by the theory on offer can really animate the creative practice for which it is meant to be providing a critical context.

There has been some synthesis. Cultural studies and its sub-variant, visual culture, has been identified as providing the contextualizing resources for practice-based arts research. However, as I have argued, this approach often treats culture as a form of symbolic discourse and can be unsympathetic to the vicissitudes of practice. Individual cultural items – books, films, photographs, artworks – are treated as texts and interpreted as such. This methodological disposition reflects the profound "linguistic turn" that took place in the humanities and social sciences from the end of the 1960s. The perspectives and methods of structuralist linguistics, then deconstructionist reading, spread across these fields, offering sophisticated procedures for analysing cultural relations, media messages and communication flows and for "reading" visual art and film works. In turn, the expressive powers of the artist–director often seem to be treated as secondary in relation to the interpretative activities of the audience–consumer. Yet any theory of practice has to address the expressive character of the creative arts and the core role of the artist as a producer of meaning, albeit in communication with an audience within a cultural ecology. This, for example, was the theoretical task Kandinsky and Malevich wrestled with.

As we have seen, a number of art movements, such as art and language,[4] have sought to combine their commitment to view art in conceptual – rather than material or retinal – terms, with an interest in language. For if art were indeed conceptual in nature, then it was closer in form to a language than to a set of perceptual cues or aesthetic values. Indeed, it could be argued that in its essence art *is* language.

Conceptualism, as Benjamin Buchloh put it (1990:107), entailed "a mapping of the linguistic model onto the perceptual model". As he points out, this move both destabilized the status of the artwork and the parallel notions of the artist as original creator. It also shifted our understanding of the spectator as the receiver of the perceptual stimuli generated by the artist – the communicative model that had animated modernism and which is exemplified in the work of Kandinsky. Conceptualism acknowledged the key poststructuralist motif of the "death of the author" and amplified the role of the audience in the constitution of the meaning of the artwork.

Conceptualism also wrestled with issues around the social responsibility of art. Many artists accepted the challenges involved in aligning their calling with a critique of the inequalities of capitalist society and the concomitant commoditization of culture. They foresaw a neo-liberal ideology that was to initiate an assault on public culture and which aimed to assimilate contemporary art – whatever its radical gestures – into an "economy of enrichment" (Boltanski and Esquerre 2016). However, at the level of theory, the intellectual fascination was with a semiotic modelling of cultural practice and perceptual experience. And this inevitably led to a theoreticism within art writing and art education, as it did within the field that was to later emerge as cultural studies.

This new formulation of theory tended to separate the study of culture from that of political economy and to break from classic Marxist concerns with understanding the dynamics of class-based politics. It now looked beyond class affiliations for the basis of collective resistance to a capitalist order undergoing a series of crises with the end of the long post-war boom. These "rituals of resistance", as Stuart Hall, the doyen of British cultural studies, christened them (Hall and Jefferson 1975/2006), might be found in a youth generational revolt; in feminist, ecological and other social movements; in the negotiation of the quotidian; or in an emerging politics of identity. Radical politics had moved beyond a vanguard.

From a political perspective, this new mutation of critical theory also represented (and acknowledged) the fragmentation and weakening of both the traditional social democratic left and its communist alternatives, and the emergence of a new form of politics. This now centred on subjectivity, everyday life and consumption activities, and on notions of identity and entitlement. In these "Hard Times", the original emancipatory interest of critical theory, with its commitment to radical social transformation, seemed to give way to a more diffuse interest in the esoteric business of theory itself. Theory was no longer burdened with the challenge of establishing a concrete relationship to practice via modes of political activism and cultural intervention.

10.3 Paradigms of the theory–practice relation

However, to make real sense of the notion of theorization within artistic research and its relation to creative arts practice, we may need to step away from the specifics of creative arts activity and examine a number of paradigms which have governed our thinking about the relation of theory and practice across the broader field of the social sciences, and the arts and humanities.

Some 40 years ago, the analytical philosopher Brian Fay provided us with a summarizing account of the various methodological approaches that have shaped the social sciences and humanities and the constitution of their knowledge. In his short text *Social Theory and Political Practice* (republished in 2015), Fay explores the question of the relation of theory to practice and how this is posed within these paradigms.[5]

10.3.1 Positivism

The first paradigm Fay identifies is the *positivist* one. As we have seen, within a positivist modelling of science the relationship of theory to practice is envisaged as reflecting the symmetry between explanation and prediction. If one can explain a natural phenomenon by reference to causal laws, then one can predict the occurrence of such phenomenon. Moreover, such systematic knowledge facilitates technological intervention designed to engineer favourable material outcomes. So, for example, we build bridges, originate new medical treatments and develop a range of industrial products on the basis of our scientific knowledge. Natural science, based on covering laws verified by empirical data, can make predictions and successful interventions in the physical world, advance technological progress and transform our lives. It can also deliver nuclear oblivion and ecological disaster.

This model is a powerful one in so far as it advances an instrumental relationship between theory and practice closely related both to the dynamics of industrial capitalism and the self-image of the west. Moreover, as we have seen, positivism envisages a unity of scientific method and asserts that its model has applicability across both the natural and social and cultural sciences. Human behaviour is seen as subject to the same sort of universal laws as physical phenomena. As Fay argues (1975:19),

> just as the natural sciences have provided men with a certain kind of knowledge by which they can control their natural environment, thereby making it more hospitable and productive, so also the knowledge gained from social science will enable men to control their social environment, thereby making it more harmonious and congruent with the needs and wants of its members.

Positivist social science, particularly as pursued in the US, generates a model of politics and governance as a "technological fix". It seeks to provide the statistical data, computer models and forms of systemic analyses of social, political and economic behaviours which might provide an effective "value free" and "evidence-led" guide to the policy making of a cadre of "experts".

So, for instance, public expenditure on the arts can be justified on an "objective cost–benefit" basis by economic research which demonstrates in a "value free" manner the beneficial impact of such spending on the economic development of a city or region. In practice, however, the social, historical and cultural sciences remain concerned with questions of value and meaning and embroiled in sectional interests. Who, for example, will benefit from public expenditure on the arts when this leads to a stimulation of the property market, the gentrification of urban areas and the marginalization of working-class families from the housing market? Is arts and cultural investment primarily about property values or about communal well-being? Both the posing of social and economic problems and the solutions advanced by "experts" seem to be mired in various unacknowledged ideological positions reflecting economic and political interests. Moreover, the social sciences, and in particular economics, seem to have severely limited predictive power (for example, economists seemed completely unable to foresee the 2008 crash, despite the warning signs).

Most social scientists recognize that policy making has not been primarily guided by evidence-based procedures but has rather been conditioned by sectional interests and subject to the naked deployment of corporate power. Accordingly, they identify a critical dimension to research in the social sciences and the humanities. Of necessity, social theory must explore the conflicts, contradictions, inequalities and social exclusions that characterize contemporary society and develop the reflexive tools to do so. Art practice faces similar challenges that emerge out of its intimate relationship with the capitalist marketplace and its status as public culture.

The positivist model has not had much real direct purchase on creative arts research to date – save as a discourse of metric-driven bureaucratic accountability employed by the academy and research funders, and perhaps also in the specific area of evaluating public policy for the arts and culture. Nor, despite the aspirations of positivist philosophers championing the unity of scientific method, has it served to provide a methodological bridgehead for interdisciplinary work between the creative arts and sciences. As we have seen, such collaborations rarely involve any serious methodological dialogue. Each subject area tends to remain true to its own disciplinary considerations and constituency, even if it is open to peering over the fence into the back yard of the other.

In certain specific areas of art–science research collaboration – such as evaluating the incorporation of art activity into medical provision (particularly hospital treatment for the elderly or psychiatric patients) – artists have usefully collaborated in research projects with medical researchers. These projects have employed a range of empirical methods to generate data that explores the efficacy of patient participation in art processes with regard to their general health and well-being (a clear statistical relationship has been established).[6]

But such research is very much the exception in relation to interdisciplinary encounters between the creative arts and science. Whatever theory is within the creative arts, it is not primarily based on an empirical investigative process concerned with testing hypotheses about arts activity, or the behaviour of artists and their audiences with a view to arriving at general laws. As we have seen, some art educators have sought to model the creative arts PhD on such a paradigm, believing this to represent a useful way of designing and monitoring the practice-based research process. Yet, whatever the pedagogic value of such an approach, it remains the case that few, if any, creative arts PhD research projects arrive at corroborated and potentially replicable empirical results that might facilitate the building of a body of theory that might, in turn, inform the practices of other artists. Art simply does not work that way.

10.3.2 The interpretative paradigm

Fay identifies a second paradigm of social science research, the *interpretative*. This focuses on providing an understanding of human action and has its own conception of the linkage between theory and practice. Such action differs from other observable physical behaviours (whether human or inanimate) in so far as it is rule-governed and determined by social practices, cultural relations and historical conditioning. As Fay argues, a large part of the vocabulary of social science is comprised of "action concepts". We explain the action of any group of social actors by positioning this within the rule-bound context that renders it both intelligible and, to some extent, predictable.[7]

In a sense, both making art and making sense of art rely on such action concepts. Artists make statements about their artistic intentions and the aims of their projects. Their projects often involve interactions with others. Audiences attribute motives to artists as producers and ascribe meaning to their work. They employ various types of accounts to make sense of the work they confront. Critics, curators and connoisseurs offer their own intellectual, professional and market assessments of artworks and artists. The art world can be envisaged as constituting an *art system* (Halsall 2008:67), understood as the totality of these social interactions. The art system is comprised of the production of the artwork; its exhibition in a gallery and exchange in a market; and its reception by various audiences, together with the accompanying artistic discourses which facilitate these relations.[8]

So, the interpretative social scientist does not regard intentions, including artistic intentions, as private mental events. Nor do they treat the artwork as a unique expressive entity. S/he does not peer into the head of their subjects to magically divine the inner springs of their actions, nor simply rely on intuition and empathy to ascribe motives to actors or meaning to an artwork. So, no theory of artistic expression is to be found here. Rather, social scientists seek to contextualize human action within the social and cultural structures that condition them at the level of meaning (Abel 1948). Through an understanding of the set of assumptions, definitions and conceptions which structure the world for actors in certain definite ways, the interpretative social scientist comes to understand the world as socially constructed and individual motives as embedded in societal narratives. So, we understand art practices not as private acts of expression issuing from mental states but as socially constructed and discursively conditioned symbolic actions best understood by reference to artistic conventions and traditions which channel individual acts of artistic creation into partly intelligible patterns. Within the interpretative paradigm, the focus is on the normative and discursive factors conditioning art actions rather than on the artist's expressive activity and reported intentions – however interesting it may be to know what the artist thinks they are doing with regard to the work they undertake.[9]

One of the virtues of the action frame of reference (and of employing an anthropological approach with regard to artistic research) is that it focuses on social practices and the construction of meaning through interactions with others rather than simply on the play of discursive structures that fascinate many cultural theorists. Accordingly, it seems to be in a position to provide an illumination of artistic practice in terms likely to be recognizable by art practitioners.

It can, for instance, employ ethnographic means to provide a grounded understanding of the conduct of artistic life, including exploring how artists develop, organize and deliver their projects; how they interact with their sponsors, audiences and critics; and the discursive strategies they employ to facilitate their "art actions". As we have seen, anthropologists have joined with practicing artists to develop modes of research which combine the traditional ethnographic approaches of the anthropologist with the expanded toolkit of sensory-based understanding that art practice can facilitate. Indeed, as we have seen, artists can themselves adopt a reflective stance with regard to their own activity, conducting their research via an auto-ethnographic practice which can illuminate the auspices of their own work.

It is also possible, I think, to trace an intellectual lineage between a model of interpretative social science – particularly as expressed in phenomenological accounts of the social world (Schutz 1967; Berger and Luckman 1967) – and the sort of "relational aesthetics" which have been so influential of late within the art world. Both approaches emphasize the socially constructed nature of reality. Both see social situations as fashioned by those who participate in them via their symbolic interactions. The core premise of the relational approach as sketched by curator Nicholas Bourriaud is a sociological one. Contemporary art, he argues, is concerned with "the realm of human interactions and its social context rather than the assertion of an independent and private symbolic space" (2002:14). Through its engagement with its audience,[10] art explores the socially constructed nature of the world we inhabit. It also anticipates social and cultural arrangements which challenge the forms of commodification and bureaucratic rationalization that characterize late capitalism.

Art, in Bourriaud's words (16), "is the place that produces a specific sociability" by producing new sites and forms of conviviality and communication in and through the art process. These sites may be "interstitial", dwarfed by the monolithic structures of capitalist modernity, much as Rirkrit Tiravanija's fragile bamboo structure of his 2018 piece *Bamboo* (seen earlier in Figure 3.7) is dwarfed by the multistorey, corporate skyline of downtown Singapore. However, these sites model forms of dialogue and participation and sketch a social imaginary which supersedes the commodity form while challenging the fetishism of technology.

That said, as critics of relational aesthetics have argued[11] (in a manner parallel to those social theorists who have reservations about the quietest character of the interpretative–interactionist paradigm), such artwork cannot supersede the social and political antagonisms that are a feature of capitalist society. Indeed one of the responsibilities of public culture is to acknowledge these antagonisms, whether the privileges of the few ranged against the rights of the many, private aggrandizement alongside public squalor, technological advance beset by ecological destruction, or the collapse of work and economic security. Simply elaborating a "microtopia" of convivial dialogue within the gallery or field site around an art "event" is clearly not enough in terms of art discharging its social and political responsibilities and represents a serious failure of criticality within art theorizing. For it remains the case that individuals' freedom of action is constrained by social conditions and political structures over which they have no control. Indeed, most social interactions are not the result of consciously chosen outcomes, as envisaged in relational art, but are rather determined by the social relations of production and exchange characteristic of capitalism, structures which often seem as binding as the laws of gravity.[12]

10.3.3 The critical paradigm

Fay identifies a third model of social science – a critical paradigm with its own envisaged relation to practice. It is with regard to this paradigm that much discussion of theory within the creative arts takes place. Critical theorists seek to distinguish their aims, methodological and theoretical premises and forms of explanation from standard understandings in both the natural and the social sciences. For them, social inquiry combines philosophical criticism with social science and cultural methods. Moreover, it aims not only to facilitate objective explanation of social relations but also to recognize human agency and to promote enlightened social and historical change. Such an approach permits the critical enterprise to be *practical*. Theory is practical in a distinctively politico-moral sense. As Fay argues (1975:97),

> for it to be so its theories must not simply explain the sources and nature of discontent experienced by the social actors, but also must demonstrate how it is that such discontent can be eliminated by removing, in some specified way, the structural contradictions which underlie it.

The goal is to produce the forms of knowledge that permit the members of a society to understand and challenge the structures of domination and oppressive stasis under which they live. In other words, critical theory has an "emancipatory interest". It seeks to not only understand the world as currently constituted but to transform this in an enlightened direction via political and cultural intervention. As Marx, still in 1845 a young Hegelian,[13] cryptically put it in his eleventh thesis on Feuerbach, "philosophers have only *interpreted* the world in various ways; the point is to *change* it" (Marx 1968:30).

A critical social science may start from the concrete experience, aspirations and cultural sensibilities of those who find themselves marginalized within capitalist society and excluded from its system of rewards, but it will inevitably seek to trace the structural conditions, political dynamics and forms of cultural and ideological hegemony which underpin that social inequality. Indeed, the distinguishing feature of critical theory is that it is first and foremost a *critique of capitalism as a social and economic system* rather than merely a method for reading a text or symbolic structure. In other words, it rests on a materialist premise.

Fay acknowledges (1975:93) that "the major philosophical analysis of the concept of critical theory is to be found in the writings of the Frankfurt school". Indeed, in the context of the current welter of postmodern, postcolonial, feminist and queer theory resonating within the arts and humanities, it is worth reminding ourselves that critical theory has its intellectual origins[14] in a body of social theory developed by Marxist intellectuals in 1930s Germany, in the context of a profound economic crisis and the rise of fascism (a period that seems to have some resemblance to the current era we are living through). Figures such as Theodor Adorno, Max Horkheimer and Walter Benjamin, associated with the Institute for Social Research in Frankfurt, sought to reshape social theory, philosophy and political economy in order to address the acute crises shaking European society in the 1920s and 1930s.

Subsequently critical theory was developed as a sociological and philosophical approach by Adorno, Horkheimer and Herbert Marcuse to understanding mass consumer capitalism and its political order in post-war America. Their analysis began to significantly depart from classical Marxism in its downplaying of the role of political economy within the critique of capitalism. At the same time in Germany, the second-generation Frankfurt School scholar Habermas, a figure like Adorno and Horkheimer haunted by his country's recent experience of fascism and war, sought to bring social critique together with a theory of communication, with a view to arriving at an enlightened social democratic settlement. His analysis drew on a range of philosophical traditions – Marxist, Pragmaticist and Analytical Philosophy – to achieve this.

A body of philosophically inflected, and historically grounded, sociological and political theory emerged within the social sciences and humanities, concerned with mapping the mutations of culture, power and subjectivity within advanced capitalist societies.

However, this body of theory, freed from any relationship to party or mass movement, became academicized. Cut adrift from any form of socialist politics, it moved away from a specific critique of capitalism to a more diffuse concern with issues of representation, identity and difference. Literary and cultural concerns replaced economic, political and ideological ones.

How influential has this body of theory been with contemporary art practice and media theory? Certainly, there are very few practicing artists who attended an art school in the last twenty years who are unaware of the broad tenets of critical theory, though the form in which they encountered such thinking will have varied depending on the interests of their lecturers and the prevailing intellectual fashions which periodically sweep through the art and media sector. Moreover, in areas like film and photography, and more generally within the broader fields of media and communication studies, the critical paradigm, particularly its culturalist variant, has been highly influential. On the other hand, it's not always obvious what purchase, *if any*, such theorization has on the creative practice activities of the majority of art and media students. Indeed, it's fairly clear that students often gain more from working through exemplars of critically informed practice than they do from raw encounters with dense bodies of demanding theoretical debates conducted in a political vacuum.

10.4 Exemplars of contemporary critical arts practice

So, let's now look at some examples of contemporary, critically informed practice. We need to bear in mind that few artists directly reference bodies of critical theory in their work, preferring to exercise their criticality primarily in and through their practice and in the context of their real-world engagements.

10.4.1 Neil Beloufa's game playing

The Franco–Algerian artist Neil Beloufa has recently exhibited at MoMA in New York (2014), at the ICA in London (2014) and at the Venice Biennale (2013). He works mainly, but not exclusively, with film and video, though he often employs an installation practice to give his work an immersive dimension. Like Hito Steyerl, Beloufa addresses the crisis-torn nature of global capitalism and the various ideological narratives developed to rationalize this economic and political order.

Beloufa's installations combine projected films, three-dimensional machine-like sculptural elements, paintings and drawings (an installation example can be seen in Figure 10.1). These elements often seem to clash within the gallery, offering the viewer an immersive but hard to read experience.[15] As Sandra Patron – the curator of Beloufa's exhibition *Sustainable Development*, mounted in summer of 2017 at MRAC in Serignan – comments, "fully immersed in the artist's installations, the viewers do not know if they have walked into some dystopian future or into the heart of our contemporary neuroses" (MRAC 2017). Beloufa uses the exhibition space intelligently to confront the viewer with a range of visceral experiences that are initially uncomfortable for those habituated to the stereotypical and anodyne images of the conventional mass media. As Patron observes (ibid.) of the installation of his work in her own white-walled gallery,

> in a space with no light which looks like a sanitised airport lounge, the artworks form a landscape playing on antagonisms: a high-tech bar seemingly welcoming us for a drink turns out to be a border control desk controlling and excluding visitors, a group of sofas seemingly inviting us to rest suggests strange high-tech beds for accommodating detained refugees; Iranian political tags on the wall become Orientalist decorative elements shorn of their transgressive message.

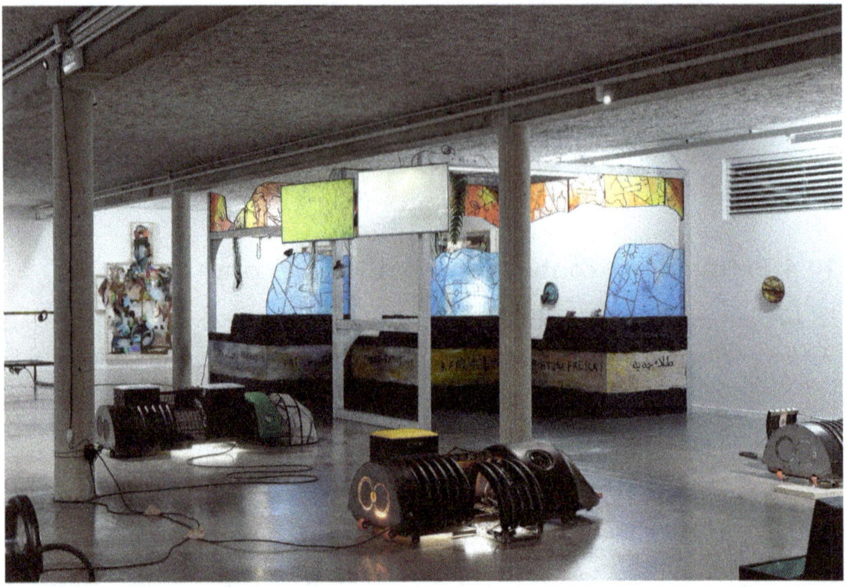

FIGURE 10.1 Neil Beloufa, *Sustainable Development* (2017), installation view, Regional Museum of Contemporary Art, Serignan (MRAC)

Source: courtesy of the artist

In short, Beloufa plays with various antagonisms: the factual and the fictive; the original and the stereotyped; hard and soft power within capitalism; the material body and its virtual appearances; neo-liberal violence and utopian longings; hegemony and resistance; domination and emancipation.

At the core of the *Sustainable Development* exhibition is Beloufa's film *World Domination*. This was originally a commission from the Museum de Tokyo in Paris in 2012. Since then, it has been employed in a number of his installations (Figure 10.2).[16] For this 28-minute film, Beloufa recruited a number of non-actors from a variety of ethnic and national backgrounds – Caucasian French, white North American, Latin American, Asian and African – and asked them to perform for his camera in a game scenario. Each individual was asked to act as a notional representative of their nation and present to the group a social or political problem affecting their country, with a view to canvassing opinions and arriving at political decisions about how such a problem might be solved. The participants are filmed gathered together around a conference table like a political cabinet or as delegates in an international diplomatic meeting. The rub was that each "politician" was instructed that the solution to their pressing problem could only be found in waging war on one of the countries in the world. As Beloufa relates in a 2015 web interview,

> it was an experiment . . . it's normal people that I found on the street or in bars in Paris. Basically, nothing is written, what was interesting to me is to

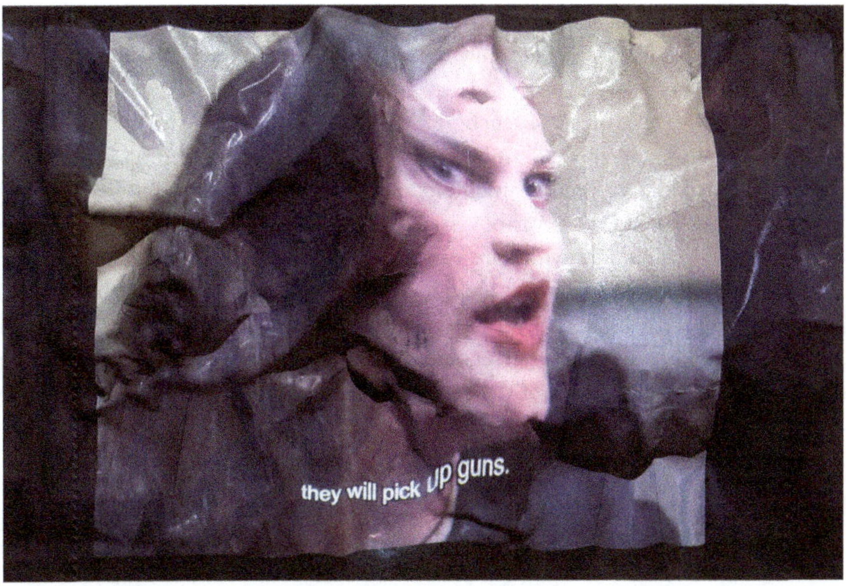

FIGURE 10.2 Neil Beloufa, *World Domination* (2012), shown at Artesmundi 7, Cardiff 2012

Source: courtesy of the artist

> bring a G8 type conversation, or geo-political conversations [to the screen]. These [meetings] are highly mediatized but we only have pictures of the conference room and then a summary [communique]. At the same time geo-politics has become like trash tv . . . in a demagogic, populist way?
>
> The rules of the game were a bit perverse. But then many political leaders – like Margaret Thatcher – [faced] with a local problem use war, solving a national problem with an external enemy, creating an economy of war. So, I gave this rule to [my participants] and I was curious what they would say. Everything was coming out naturally through that game rule. It became alright to be a bit racist and tough. And that rhetorical game becomes a borderline crossed, a fascist, violent, ironic thing [emerges]. But it was more of an experiment.

Beloufa's film is shot entirely within the confines of the "conference room" in a series of mid-shots of the group in discussion, interspersed with close-ups of the presenting "politicians" and cutaways of reaction shots from the other participants to the person holding the floor (a clip from the film can be seen at https://vimeo.com/131175653).

As the discussion develops, the participants move from the reasoned presentation of the problems besetting their polities to an increasing identification of other countries and their nationals as the source of their domestic difficulties. With some inevitability, they eventually arrive at proposals for waging war on their competitors/enemies as a solution to their national problems. The film plays as a comedy, but a pretty black one. This *reductio ad absurdum* sounds a chilling note for contemporary viewers currently witnessing the resurgence of populist nationalism and warmongering.

For example, one of the French "representatives", speaking in the measured tones of a senior civil servant, employs the language of macroeconomics to identify the source of the high levels of unemployment affecting France. This, he argues, is the product of the actions of various Asian nations who are manipulating exchange rates and flooding markets with state-subsidized goods produced with cheap labour, thereby destroying the French industrial economy (arguments made by Donald Trump with regard to the US's trading relations with China and by Marie Le Pen and the Front National in France). As diagnosis slips into policy prescription, a fellow French participant from the business world proposes a military solution to restore "proper terms of trade". This theme is seized upon by a third French representative present, a senior politician with his own political agenda.

French "business representative":	There are many Asians who are a cheaper workforce than our citizens. We know that war is controversial. But war is also pragmatic. We must attack Asia!
Senior French "politician":	Because war justifies a lot of things, you can pass many laws that aren't very social. So the nation will focus on fighting the enemy.
French "economist":	I suggest using force to face these Asiatic countries. This is the only way to create once again a dynamic situation in our countries.

And so the dialogue spins into a deadly diatribe.

Beloufa often projects his filmic images onto the figure of a dinosaur or other uneven surface, thereby distorting the faces and bodies of the speakers portrayed (Figure 10.2) and introducing an additional element of grotesqueness to the proceedings. In *World Domination*, Beloufa as director presents us with a game-playing scenario. But this is not an abstract exercise in game theory.[17] Originally, game theory was largely concerned with zero-sum games, in which one person's gains result in losses for the other participants. And in *World Domination*, the participants seem to be involved in such a game, with the declaration of war the ultimate playing out of a "winner take all" ("America first") approach to international relations. However, the artist is employing a game scenario which has a minimal rule structure and no mathematical form. As he suggests, it is a dramatic device employed by a director to structure an improvisation conducted by ordinary people around a chosen theme. It's alarming how quickly seemingly rational political argument slips into a discourse of nationalism, racism and vicious international rivalry. Within this, warfare can appear as a feasible response to the multiple crises we face as capitalism and democracy drift apart and authoritarian solutions are sought.

Beloufa's interest in games takes a more concrete form in the other filmic element included in the *Sustainable Development* show. This piece, *Monopoly* first shown at the 7th Artesmundi Exhibition in Cardiff in 2016, was made in Ukraine. A group of Kiev teenagers film themselves playing a specially constructed game of Monopoly (Figure 10.3). In this, a range of Ukrainian real estate is available "on the market", including the Maidan Square where the so- called "Revolution of Dignity" took place in February 2014. Indeed, the entire Crimean Peninsula (now occupied by Russia) is available for purchase within the game.

Beloufa, conscious of the complex ideological history of Monopoly as a board game,[18] describes the project interview in a 2017 interview in Cardiff (Beloufa 2017) (video clips of the project can be seen at https://youtu.be/YUvrlhlpgA):

> I am interested in games, obviously, and so we had kids in Ukraine playing Monopoly and we project (images of) the cities and streets they are buying. But what interested me is that this game (is being played) in a former communist country which is capitalist now and wants to go into Europe. People play a capitalist game that was (originally) supposed to be a socialist criticism of capitalist greed, but (its values) became reversed because it worked commercially. And they (the young participants) made the jokes I wanted them to do without me asking them, making fun of speculation. So, they are all aware at fourteen years old of the system and they like to play it and they like to crush people. And someone is buying Karl Marx Street and now it's vastly more expensive than it used to be.

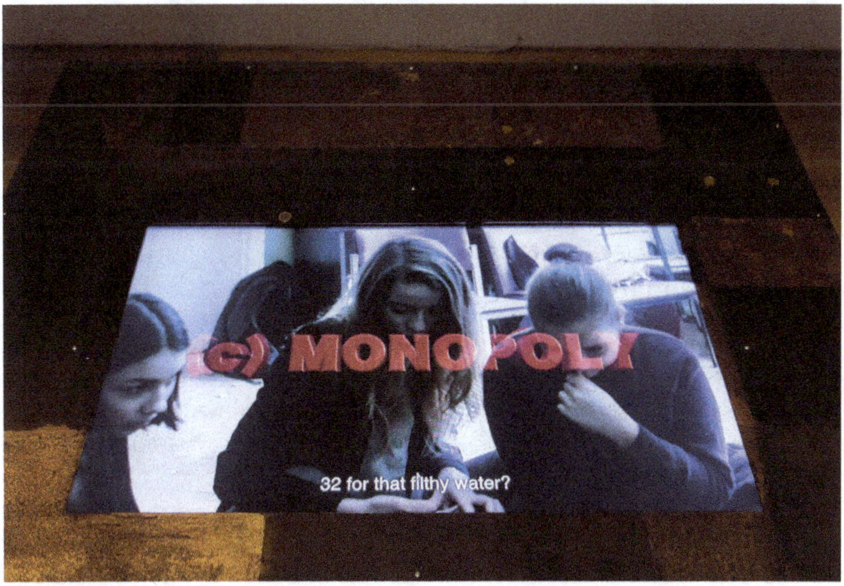

FIGURE 10.3 Neil Beloufa, *Ukrainian Monopoly* (2015), film still

Source: courtesy of the artist

Beloufa sets up a dramatic scenario through the game structure. Within this, the participants interact to explore what could be described as the performative character of capitalist exchange. This is not the manipulated gaming and performance of a television reality show such as *Big Brother* where the contestants, distinguished primarily by their narcissism and capacity for abject behaviour, are programmed by the producers to interact, often in conflictual ways, for the voyeuristic pleasures of the audience. Beloufa's scenarios are more like a Brechtian *Lehrstuck*, a form of drama that is specifically didactic in purpose and concerned with political and ethical lessons, though we'll find no *Mother Courage* to guide us through the moral maze with which Beloufa presents us.[19]

Beloufa's concern with how performative participation in the dominant discourse can structure our sensibilities and interactions with others is redolent of Marxist philosopher Louis Althusser's arguments (1971) about ideology and subjectivity,[20] although the artist does not reference this critical source. Beloufa's participants in his rhetorical games are invited to occupy a certain social position and set of discourses. The suggestion is that through the game or role playing, they eventually assume the identities offered to them in the games and "speak the speak" through a process that Althusser described as an "interpellation". As viewers, we witness this process and are invited to draw critical conclusions. However, it is unclear if the participants come to any reflective understanding about these interactions.[21]

The virtue of Beloufa's approach as an artist (rather than as a critical theorist) is its concreteness. The work is experimental and in its filmic form immediately revealing. The experimental dimension tracks how ordinary citizens navigate the rhetorical games constituting the forms of capitalist hegemony which seek to normalize the absurdities of an economic and political order that cannot serve the needs of the many, only the interests of the few. As an audience, we are confronted with the playing out of these absurdities as they move towards potentially disastrous consequences. Most of us sit up and take notice, and this, I think, is the artist's critical intent.

10.4.2 Niels Ackerman and Sebastian Gobert: Looking for Lenin

Other artists have also sought to engage with post-communist eastern Europe, employing different creative means and critical resources. In 2015 and 2016, photographer Niels Ackerman collaborated with journalist and writer Sebastian Gobert in a project, *Looking for Lenin.*[22] They travelled through war-torn Ukraine to locate and photographically document an extensive range of statues of Lenin, often toppled or vandalized in an outpouring of anti-Soviet hostility after 2013 (like the one seen in Figure 10.4 left, which lies by a Kremenchuk Housing Estate, or that in Teplivka, where they located a submerged statue buried to the waist and now painted blue, which arises in an overgrown plot like a ghostly spectre (Figure 10.4 right)). These remains are now in the possession of individuals, or hidden away

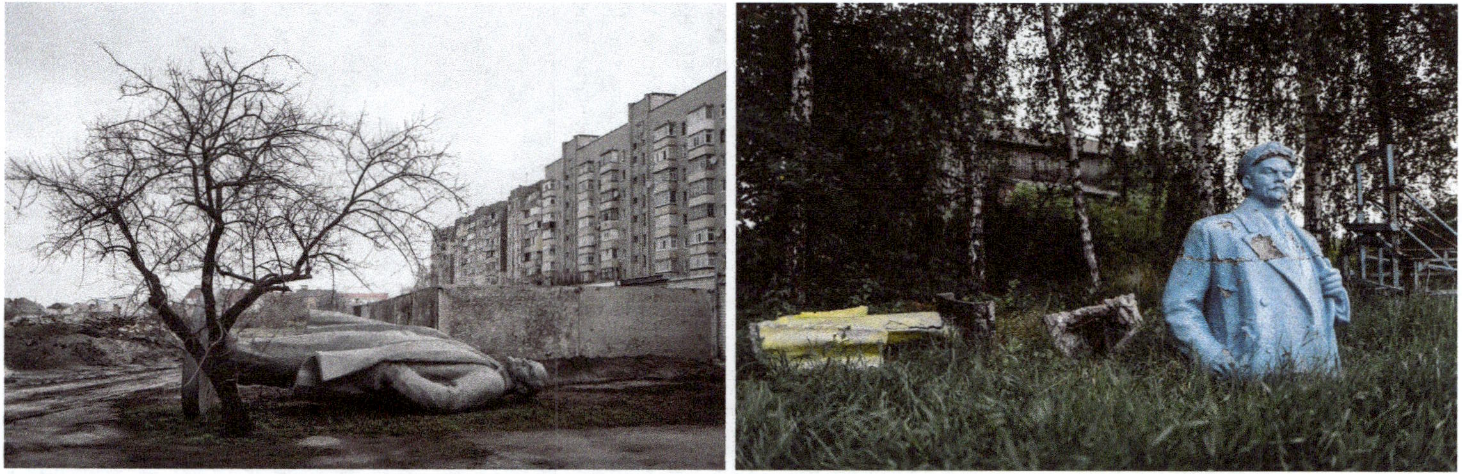

FIGURE 10.4 (left) Lenin in Kremenchuk; (right) Lenin in Teplivka

Source: courtesy of Niels Ackermann/Lundi13/Fuel Publishing 2017

in institutions, or lie abandoned in various overgrown locations. While Ackerman photographically documented the fallen statues, Gobert collected testimonies from those now watching over them.

In an email interview I conducted with Ackerman and Gobert in 2018, they describe the origins of their project.

A&G: The project's genesis was in the early days of the Maidan revolution. On the 8th of December 2013 nationalist protesters toppled the statue of Lenin on Bessarabska square in the centre of Kiev. Niels was there and saw the crowd trying to smash-up the monument, but also trying to take home a little souvenir. However, the statue's red quartzite (the same stone used in the Lenin mausoleum in Moscow and in Napoleon Bonaparte's tomb in Paris) was very resistant and only small fragments were broken off the statue.

However, the next morning the damaged statue was gone, completely vanished!

DB: How did your research proceed?

A&G: We started an investigation into what was happening to the Lenin monuments . . . scores of them across Ukraine. The hunt was a way to forget the violent upheavals and the civil war we had been covering as journalists and to deal with Ukraine related issues in a different, more nuanced, way. Through the fate of the Lenin monuments, we felt we could explain what had been going on in Ukraine since 2013.

In May 2015, the new Ukrainian government passed a Decommunization Law making the glorification of Soviet symbols illegal – like Nazi ones. So, all these monuments had to go. Ukraine had by far the highest density of Lenin statues of any country. In 1991, there were 5,500 statues in Ukraine, against 7,000 in the whole of Russia. Officially, not a single statue is left today. Where did they go? Could we locate them?

As they report, the search to locate the "decommissioned" statues of Lenin, was not without its difficulties.

A&G: This search was tough because Lenin moves rather quickly for a dead guy. Lenin monuments are moved, transformed, broken, stolen, sold. The Bessarabska Lenin, for instance, was removed by a private collector, quite illegally. Naturally, he is not anxious to talk about this. Sometimes we arrived at a reported site a few days too late. The monument had been removed and possibly sold on.

Officialdom could also be difficult. We knew that Melitopol, a city in south-east Ukraine, had three monuments stored somewhere. We asked to see them, called, wrote letters, tried high-ranked contacts . . . nothing worked. The answer back was, "we cannot allow you to see our Lenins. Your project does not depict our city and our country in a positive way". Taking pictures of abandoned Lenin statues in a warehouse was not acceptable for them.

However, we developed an extensive network of contacts via social media. People told us about Lenin locations and we began to collect testimonies from them, "I have a Lenin, come and have a tea with me and take a picture!" The quest to find each and every Lenin released different stories. Each, and everyone, has something to say about Lenin. Our project is never-ending, it seems.

Their research threw up a range of interesting questions. Gobert asks, "what is the meaning of this de-communization? How does it relate to the war in the east of the country?" Who, we might ask, are the winners and who the losers in this brave new post-communist world in eastern Europe? Why do the abandoned figures of Lenin still seem to haunt this fragmented land? What remains of any utopian Marxist vision after the collapse of "actually existing socialism"?

Ten years earlier, the editors of Jacques Derrida's *Spectres of Marx* (1994:ix) had posed this key question at a theoretical level.

> How will intellectuals in the Marxist tradition respond theoretically and politically to the global transformations now occurring? How has the crisis in Eastern Europe and the former Soviet Union affected the way intellectuals, scholars and government officials in those countries and around the world reconceive their intellectual and political projects?

Ackerman and Gobert pose these issues at the level of popular culture and memory and in the specific context of Ukraine's recent painful history. In Zaporizihia, the researchers come across the macabre sight of the headless torso of the Soviet leader lying on the ground by the original plinth which supported his statue and to which the legs still adhere. In Chernobyl, a 2-metre bust of Lenin formerly located on the site of nuclear facility is now preserved in a scruffy room now used by cleaning staff still working at the contaminated site (Figure 10.5 left).

In Pokrovska Bahachka, Ackerman photographs a tumbled statue lying prostrate in a clearing in a wood beside abandoned machinery that tokens a post-industrial Ukraine plunged into war and political uncertainty. In Kiev, a Lenin bust, now in the possession of a former nationalist volunteer who fought against the Russian-backed separatists in the Donbass region, erupts out of the soil of her back garden (she was give the piece as a birthday present by fellow soldiers who regarded it as a trophy of war). In a subterranean basement of the former Lenin Museum (now re-designated as Ukrainian House) Lenin's head and shoulders protrude from a packing case, a silent witness on a "revolution to free a ruling class" (Figure 10.5 right).

As researchers, Ackerman and Gobert were, however, anxious not to impose their own views but simply to listen and document.

A&G: We are not Ukrainians and judging the way the country and its people deal with their soviet past would be inappropriate. We simply aim to show to the rest of the world the complexity of this issue. We may think that it makes sense

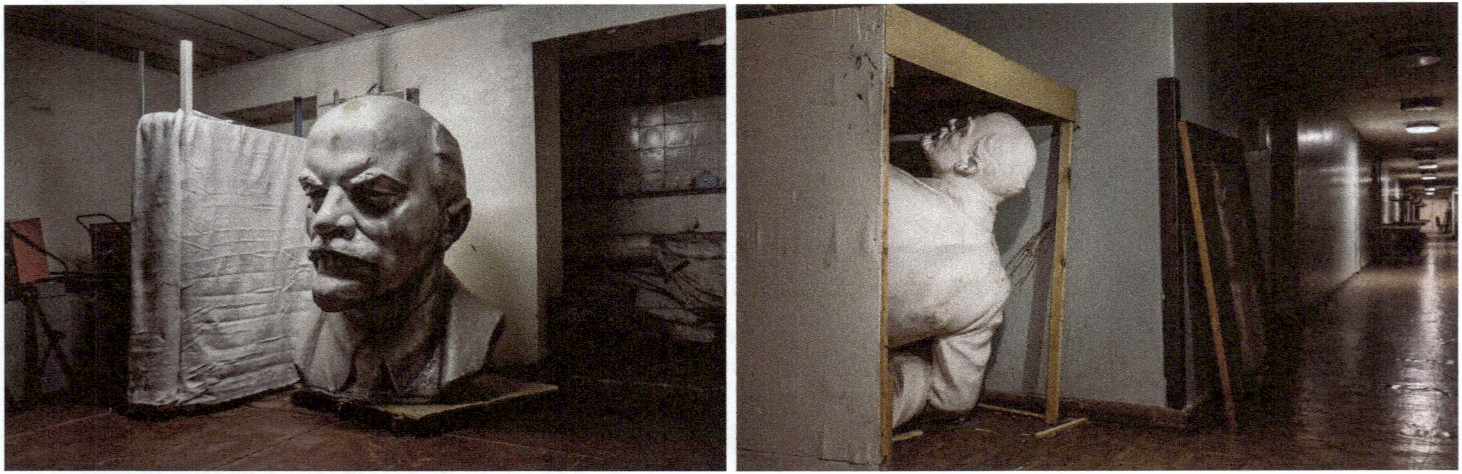

FIGURE 10.5 (left) Lenin in Chernobyl; (right) Lenin in former Lenin Museum, Kiev

Source: courtesy of Niels Ackermann/Lundi 13/Fuel Publishing 2017

> to take Lenin down. But it's not as consensual as this. The act of toppling in itself does not solve a number of issues. Where does Lenin go? What next? Which memory of the past will be preserved?

Gobert's wry observations and Ackerman's arresting images seem to testify to the continuing presence of this spectre.[23] As Gobert concludes (2017),

> the destruction of soviet imagery restores ideological meaning to objects that lost it a long time ago. We then witness a form of renaissance for these mosaics, paintings, sculptures and so on, as they regain their original ideological charge. In my opinion, nothing did more for the popularity of Soviet symbolism than the current process of de-communisation.

As Jacques Derrida mused in *Spectres of Marx* (1994:37),

> at a time when a new world disorder is attempting to install its neo-capitalism and neo-liberalism, no disavowal has managed to rid itself of all of Marx's ghosts. Hegemony still organizes the repression and thus the confirmation of the haunting. Haunting belongs to the structure of everyday hegemony.[24]

10.4.3 Mari Bastashevski, State Business

Somewhat more forensically, Mari Bastashevski, a Russian-born artist who now works in Sweden and the US, examines the brave new post-Cold War world characterized by fractured international relations, private security companies, illegal arms trading and the increased use of surveillance systems and data analytics by states against their citizens. Her ongoing project, *State Business* (2017),[25] is part investigative journalism, part photographic practice and part art project. Its uniqueness lies in this hybridity. Her images have to be read through the optic of her research.

Bastashevski draws on the work of the social theorist Keller Easterling and Easterling's concern with the politics of infrastructure in the contemporary capitalist world. Easterling rethinks our concept of political sovereignty around a notion of "extrastatecraft", seeking to identify the complex "operating system" constituted by the intermeshing of information networks and political and economic transactions that operate without any political accountability between states and international business. Bastashevski's focus as an artist is on the role that media representations play in obscuring these emergent, non-national forms of sovereignty and how this pernicious image ecology might be challenged. Her focus is on the shady world of arms sales and private security companies, glimpsed in Figure 10.6, and in an image of new recruits to leading Uganda private security firms SARACEN prior to their dispatch to Mogadishu in Somalia for security duties (Figure 10.7).[26]

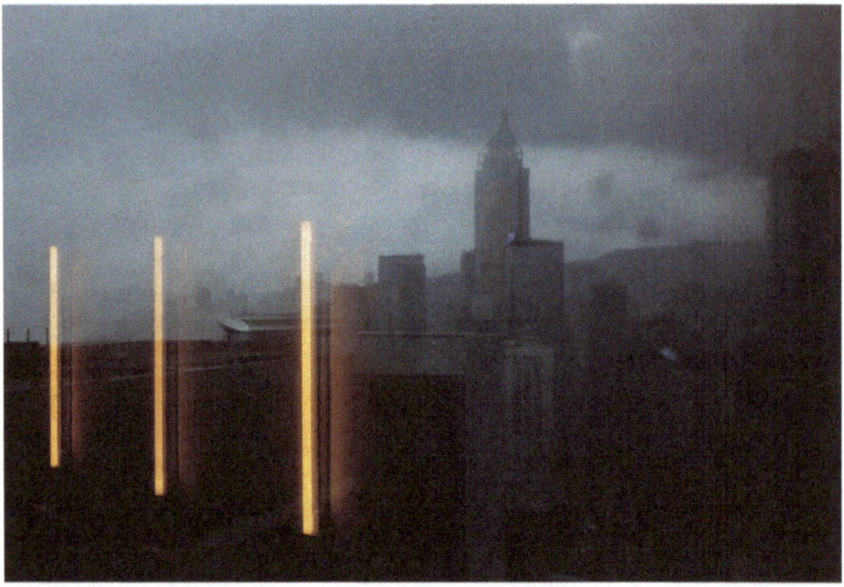

FIGURE 10.6 Mari Bastashevski, Erik Prince Frontier Service Group, view from the office in Hong Kong (2016), from *State Business*

Source: courtesy of the artist

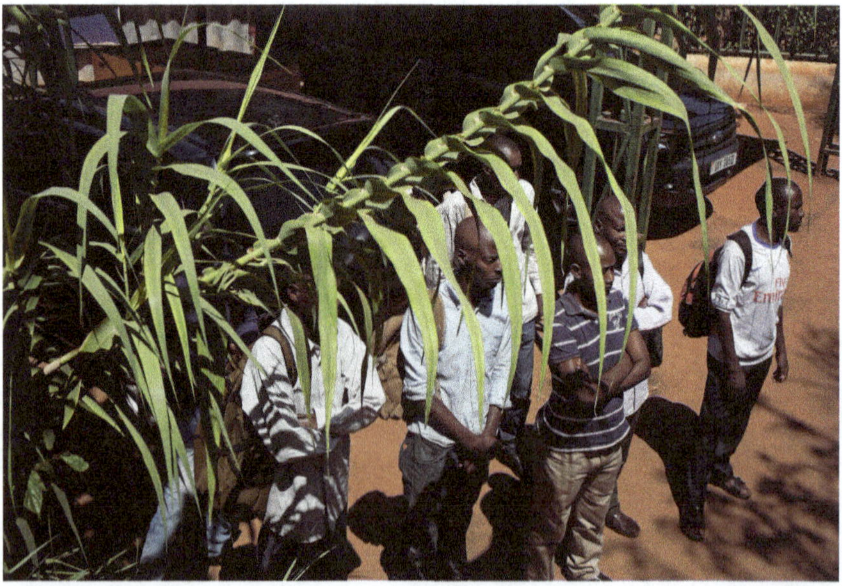

FIGURE 10.7 Mari Bastashevski, SARACEN Uganda, new recruits waiting for transport to Mogadishu, Kampala, Uganda (2017), *State Business*

Source: courtesy of the artist

Her focus is also on the merchandising of cutting-edge information technologies to various authoritarian governments which facilitate political actions, including seeking to influence elections – whether to shore up the status quo in crisis-prone polities or to destabilize other governments for economic and political gain. Her research is often conducted collaboratively with journalists and social researchers. It explores the cooperation between global corporations and authoritarian regimes in an economy of violence and political manipulation which she refers to as "necro power", adopting the term "necrocapitalism" used by the social theorist Subhabrata Bobby Banerjee (2008:1541). He defines necrocapitalism as the "contemporary forms of organizational accumulation that involve dispossession and the subjugation of life to the power of death".

As she says of her research practice (email interview with the author 20/08/2018)

> it's very transversal. I stumble on a specific transaction that teaches me something about the larger global information regimes we're inhabiting. I draw on heaps of media theory and on theories of power, then plunge into field research. My role as a photographer provides an opportunity for me be to be in these spaces of power as an observer. In my interactions with these political actors I perform a certain ritual and get to ask my questions. Then I turn the conversation around by adding another question when I'm summarily rejected in my initial enquiry.

I asked her about the difficulty of negotiating access to the various sites she wants photograph in and whether she seeks permissions to film.

MB: It's different every time. I probe, I dance around it. I file freedom of information requests. And I'm always surprised by what turns out to be accessible and what, in the end, inaccessible. For example, researching in Dubai I was able to gain access to the highly protected Jebel Ali power plant – just waving at the check point and driving right into the complex. On the other hand, I've never been able to go inside Siemens [headquarters in Munich]. Luckily, these regimes of corporate secrecy are difficult to maintain when those maintaining know them to be completely arbitrary.

My photographic method is to ask for a permission to take the image in a location knowing this will be denied. Then to ask from what distance photography would be permitted on a site. This not only allows the subject to draw out their perimeter of secrecy – be it legal or arbitrary – but creates a series of micro ruptures in the automated production of power. So suddenly there is this body in the middle of the room, entirely exposed and performing in a revealing way.

Bastashevski has developed a complex art practice in which her photographic work has to be seen in the context of her internet-driven social research. She acknowledges the influence of Allan Sekula's work[27] and particularly his recognition of the

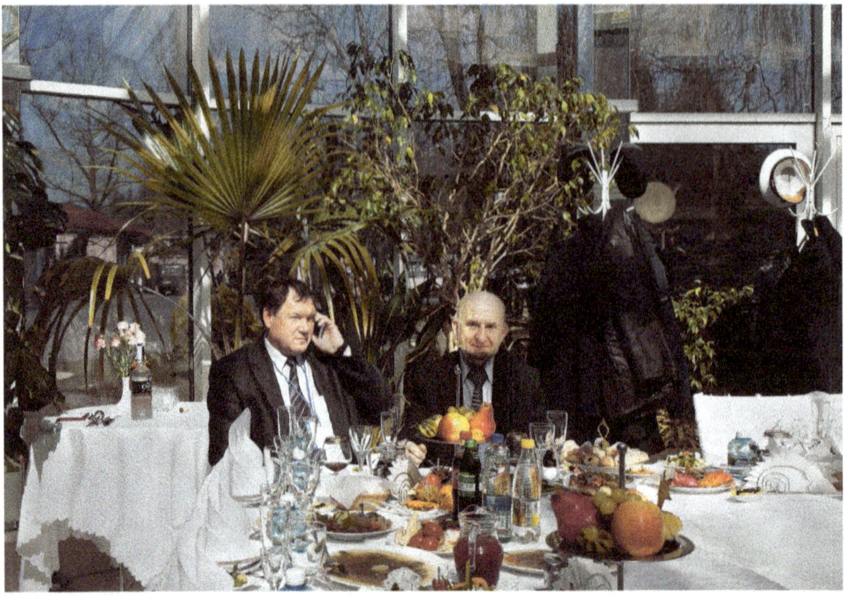

FIGURE 10.8 Mari Bastashevski, NATO Banquet, Pavlograd, Ukraine (2012), *State Business*

Source: courtesy of the artist

limitations of traditional documentary practice in "getting at" these covert filiations of power and corporate influence. An informational vacuum surrounds the mechanisms of power operative in disorganized capitalism, and photographers and journalists have limited access to these new sites of power. This means that her photographs seek to "photograph the unphotographable".[28]

MB: I find it ironic how Sekula's work as an artist is often oversimplified within photographic discourse where he is seen as a concerned, activist–documentarian. I want to develop Sekula's aspiration to go beyond a photographic aesthetic based on either the single image or photo series by transplanting some of the rules of performance art and the philosophy of language into the field of photography. In my case, this makes for a body of work that is initially hard to nail down but which accepts that it is meaningless without the provision of the research context from which it originated. So, you could say that I've spent the past decade making art that I've always known I'd not be able to sell. . . . I don't think many artists understand how empowering it is for an artist constantly broke to say no! to being rich.

At first glance, Bastashevski's images (such as the empty US State Department conference room or World Bank mail room seen in Figure 10.9) seem mundane.

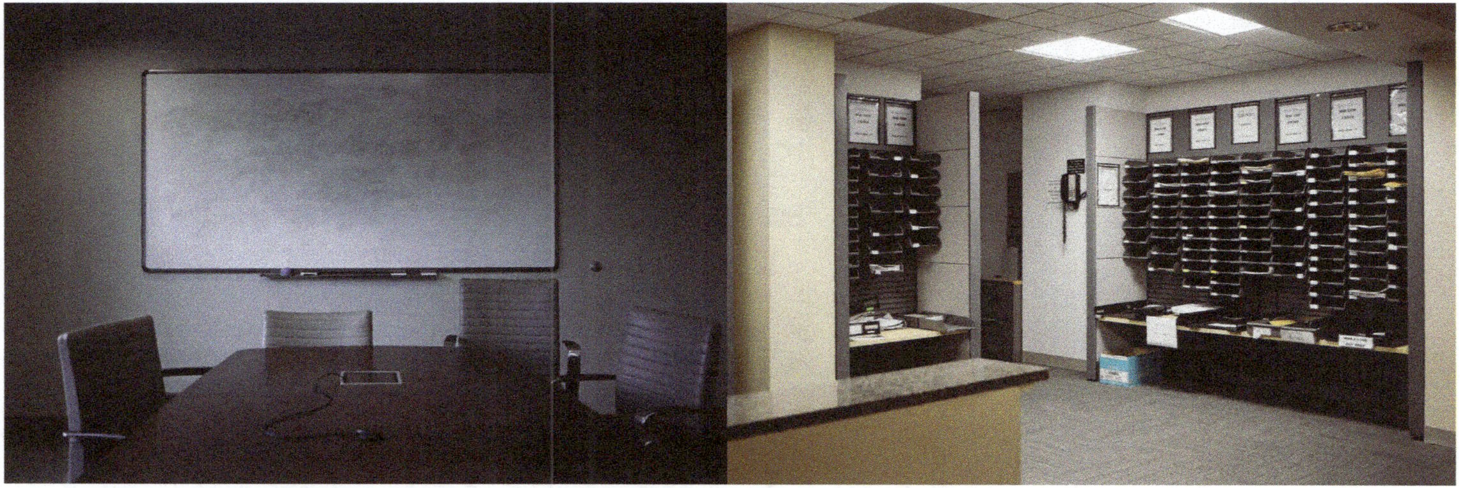

FIGURE 10.9 Mari Bastashevski, State Department conference room, Washington DC (2016); World Bank mailing room, Washington DC (2017), *State Business*

Source: courtesy of the artist

We are presented with pictures of empty meeting rooms in unidentifiable locations; anonymous corporate buildings; empty hotel corridors; distant figures huddled in conversation in the corners of café terraces (as seen in Figure 10.8). These images reveal little of the complex webs of clandestine power she is researching. In her project, there are no scenes of violence or military action, nor portrayals of the suffering of victims resulting from orchestrated violence and war. These are not images from the "front line", so beloved of the photojournalist, but from the "back rooms" and un-trackable spaces in which corporate business deals and undercover international economic and political transactions take place.

MB: My position as the author is somewhere at the edge of catastrophe and in situations where what is going on is not an event in the news journalism understanding of the word. My events lie on the surface, and at the same time they are not obvious. Photography allows for the creation of form for the invisible. This is something we pass by every day without taking notice. I am not looking for these situations on purpose, it is just that the decisive deals and meetings happen in these mundane spaces.

In the course of this project, Bastashevski travelled around dozens of countries in eastern Europe, Africa and Asia, researching hundreds of documents and information sources to produce a series of photographs in which, it seems, almost nothing is going on. Asked about the prosaic nature of her photographs in *State Business*, she recalls,

> Recently someone had a quick look at the publication and wrote on Facebook:
>
> > Just some military guys having lunch, some office with some examples of antique armour, a corridor, a business lunch on the roof, and so on and so forth. These are all rather simple situations in the places, which we have all seen. Without connection, these images do not convey the meaning.
>
> And this person accidentally nailed it.

I asked her how important the text accompanying her images was in amplifying the meaning of her work.

MB: The texts are not amplifying the meaning of the work. They are the work and the element making the dissonance between text and image more pronounced.

DB: How as artists can we visualize power relations and contest these?

MB: I don't think artists should visualize anything. Art and design can serve as modes of argumentation, or embodiments of policy propositions. They do so in their own right, rather than merely "illustrating" all the "real" thinking that happens elsewhere! For me art – and photography is no exception – is about conscious praxis and the organizing principles that inform the work.

So, contemporary artists in their research and practice can engage with a range of critical perspectives. For some of them, at least, their work has an emancipatory interest and intellectual roots in a body of critical social and political theory. In the work of these artists, Neil Beloufa, Neils Ackerman and Sebastian Gobert, and Mari Bastashevski, we witness a range of lens- and screen-based work produced within a critical paradigm of arts research.

Each of these artists employs a range of research activities – game playing and filmic performance, contemporary archaeology and photo-documentation, investigative journalism and reformulated documentary practice – to advance their art practice. The research approaches they adopt are closely related to the specifics of the project they are working on and directly inform their art process. The measure of the work's criticality lies not in its explicit reference to particular bodies of theory or to a set of intellectual precepts. However, in its commitment to develop artistic means to represent and understand the seismic economic, political and cultural shifts that have occurred in the global order we now live within, one senses that this work provides a searing critique of capitalist disorder.

10.5 Conclusion: artistic research and critical theory

To recap and conclude: Art and media students have been implored to *theorize* their work and are presented with exemplars of critically informed work as a guide to doing so. They are also presented with complex bodies of cultural theory primarily concerned with issues of cultural representation and the interlacing of signification, power and identity. This corpus of theory is only incidentally – if at all – concerned with issues of creativity and expression, or with artistic craft, art process and creative accomplishment. It often appears indifferent to questions of aesthetic evaluation. Certainly, a gap has arisen within the academy between this formulation of theory and current art practice.

Moreover, is it appropriate to align the dichotomy of theory and practice with that of text and art process? That's certainly not how the artists discussed here see the matter. Moreover, to do so seems to imply that the research value of a creative arts practice project is only realized when the "results" are delivered in written form and contextualized within a body of pre-formulated theory. In other words, it is this wordy intellection extrinsic to the creative process – rather than the forms of innovation, creative achievement, reflexivity and critical understanding present in the art processes and outcomes themselves – which generates an academically recognizable form of research.

Clearly, the practice-based model contests this position. In this text, I have argued that forms of critical insight and indeed engagement with a range of theoretical questions can arise directly through the art process and structured reflection on this informed by critical interlocution. Indeed, it must do so if the work is to count as artistic research.

This book seeks to plot the basis for a research culture for the creative and media arts by casting an eye over the genealogy of artistic research and seeking to elaborate the critical and methodological basis of contemporary research practice in the creative arts by the close study of a range of recent art projects.

The development of a research culture for the creative arts and media is likely to rest on our capacity to break from current formulations of theory at play within arts research. These are characterized by their postmodernist sophistry and seem to have little critical purchase on arts practice. However, we need to begin to develop a theory of practice attentive to the distinguishing features of the creative arts, namely their concern with invention and communication and with apprehending the world via artistic engagement.

So, to end where we began. The key focus of this book has been how practicing artists go about the business of research. Artists integrate various research strategies into their work, often collapsing the barriers between research, project development, art process and various forms of social engagement. Their work has a strong investigative basis.

As I have argued, the creative arts and media, with their focus on "ways of making and doing", have their own distinctive modes of apprehending and understanding the world. The knowledge they acquire is rooted in their creative encounters and critical reflections. Artistic research needs to start from this specificity. We have sought to identify a triangular dynamic within artistic research linking studio practice, critical thinking and professional and pedagogic practice. At the core of this dynamic is the autonomous creative arts and media activity conducted by individual artists, whether in the studio, field or classroom. In the first instance this is responsive both to its own "inner voice" and to forms of studio and field discipline focused on the invention of form, content and purpose. However, artists also need to engage with a range of critical interlocutors, interdisciplinary perspectives and methodologies while remaining focused on the core expressive tasks of art. Lastly, these contrasting elements are drawn together by means of an arts pedagogy (in the widest sense of that term). This, while facilitating the creative development of the student, acknowledges the central role of research within contemporary art practice, as it does a range of professional considerations and forms of social engagement.

Notes

1 Though it would be true to say that the new academic interest in the history of visuality represented, on the one hand, by the work of cultural historians like Jonathan Crary, William Mitchell and Martin Jay and, on the other, by a renewed interest in the phenomenology of embodied vision inspired by Maurice Merleau-Ponty's work, has had a significant impact on students and researchers in the visual arts field, including how they think about their creative practice.

2 The dyspraxic character of contemporary cultural theory vis-à-vis artistic research lies in its divorce from questions of both concrete arts practice and political challenges.
3 Noel Burch, the Paris-based film activist, used the term "theory of film practice" in his 1969 book of the same name to identify a form of film analysis that championed a type of descriptive analysis of the language and formal character of film drawn from both the intuition and speculations of the practicing film-maker as well as from the academic analysis of key film texts.
4 The most rigorous elaboration of the relation between art and language was provided by the American artist and theorist Joseph Kossuth, who was to argue in his essay "Art After Philosophy" (1993:7) that

> the propositions of art are not factual, but linguistic in *character* – that is, they do not describe the behaviour of physical, or even mental objects; they express definitions of art, or the formal consequences of definitions of art.

5 Fay's book sought to introduce the ideas of European critical theorists to an American readership, in particular the work of Jürgen Habermas.
6 An example of such research collaboration is the work of a joint team drawn from the School of Medicine in Trinity College Dublin; medical staff at St James Hospital, Dublin; and the National College of Art and Design. An art intervention, Open Window, was introduced into stem cell transplant unit in St James and its impact on the welfare of patients monitored using various quality of life indices. See Denis Roche, Philip Napier, Brian Maguire and Shaun McCann (2008).
7 The interpretative paradigm forges its own distinctive links between theory and practice by providing a grounded understanding of social relations and practices, a knowledge base that encourages various sorts of communicative action and social intervention. The most obvious application of such sociological knowledge is in the tutelary practices of the state – in social work and health and welfare policy, criminology and the management of deviance, in education and cultural provision. Ethnographies of the classroom and pupil subcultures inform educational policy. Studies of ethnic cultures inform social policy concerned with multicultural objectives. Police and criminological practice rests on theories and field analyses of deviancy. Political forecasting and marketing activity rests on focus group research. Design practice is increasingly user-oriented and rests on the application of ethnographic methods of research to understanding consumer preferences and needs, as does much professional media research. However, these forms of qualitative research are often employed in an uncritical manner driven by commercial and bureaucratic interests.
8 Those art theorists like Francis Halsall (2008) who follow the lead of Niklas Luhmann and seek to introduce a social systems approach into art historical study will, like a previous generation of social theorists such as Talcott Parsons, Edward Shils and Robert Merton, have to wrestle with the difficulties a systems approach has in addressing subjectivity and art as an expressive practice.
9 This is not to deny the essential ambiguity of the work of art itself, which in its polysemous form lends itself to multiple interpretations over which the artist has no monopoly.
10 It's not always clear that artists who employ a relational practice are alive to the methodological and ethical tensions that can arise when we elide the interpretative world of our subjects with our own as artists and naively assume a commonality of interests and values.
11 Such as Clare Bishop, whose seminal paper "Antagonism and Relational Aesthetics" (Bishop 2004) remains the most trenchant critique of Bourriaud's position.
12 Bourriaud concedes that the contemporary art world can be seen as a flow of commodities and is shaped by these market realities, but aspires to create a "haven in a heartless world" through the relational project. Indeed, contemporary art more generally often seems a heady mix of utopianism and market opportunism, lacking any critical backbone.

13 Which is to say he remained at that time under the sway of a Hegelian philosophy struggling to understand the emerging capitalist world by employing idealist constructions rather than by reference to a critique of political economy and by a materialist understanding of history.

14 For a recent account of the Frankfurt School which provides biographies of the leading figures, together with a social history of the movement and an imaginative exegesis of their main ideas, see Stuart Jeffries's *Grand Hotel Abyss* (2016).

15 Although many of his viewers seem to find his film pieces easier to understand than the other installation elements, not least because they are shot and edited in a professional manner with a degree of narrative coherence.

16 The film has been shown, for instance, at his MoMA and ICA shows and in his contribution to the Artesmundi exhibition at the National Museum of Wales in 2016.

17 Game theory is usually understood as "the study of mathematical models of conflict and cooperation between intelligent rational decision-makers" (Myerson 1991:3).

18 The game was initially designed and patented in 1904 by American radical Elizabeth Magie under the title "The Landlord's Game". Magie was significantly influenced by the economic thinking of Henry George and his critique of landownership and the rentier economy in the US. She intended her game to be a "practical demonstration of the present system of land grabbing with all its usual outcomes and consequences". The game was eventually acquired in dubious circumstances by a large toy manufacturer and remarketed as a celebration of US property capitalism.

19 Literally "learning-plays", through which Brecht sought to develop participative forms of drama which might provide an opportunity for both his actors and their audience to explore the possibilities of learning moral and political lessons through acting, role playing and adopting postures, attitudes and dispositions.

20 For Althusser, ideology in a capitalist society works by a process of "interpellation". Through this socializing process, the very nature of individual subjects' identities are constituted through the process of "hailing" them in social interactions and via their participation in various discourses which transform not only the individual's views but indeed transform them into subjects (Althusser 1971).

21 Indeed, in Althusser's account of ideology (and that of psychoanalytical theorists Jacques Lacan, from whom elements of his theory are derived), these processes of socialization and identity formation take place at a subconscious level.

22 Their work was exhibited at *Rencontres du Photographie* in Arles in the summer of 2017 and published in book form by Fuel (2017).

23 Derrida (1994:37) distinguishes the spectre from the "icon or the idol" and indeed from the simple phantasm. The spectre is "not an image or even an image of an image". It has a presence or, as Derrida puts it, "the tangible intangibility of a proper body without flesh, but still the body of some*one* as *someone other*". And, as he observes, this someone other is looking at us, indeed, "we feel ourselves being looked at by it".

24 Derrida assesses Marx's legacy not in terms of cold intellectual "capital" but rather through approaching the figure of Marx as a "revenant" still walking among us and posing difficult questions that unsettle a fragile new order.

25 Elements of Bastashevski's ongoing project were presented at the 2017 *Rencontres de la Photographie* in Arles, and a book reporting more fully on the project will be shortly published.

26 Sekula's *Fish Story* (1995) has been highly influential within recent photo-documentary practice in its constructivist commitment to get beyond the surface aspects of the photographic image in order to develop strategies of research and picturing which can reveal the character of global trade and capitalist transnational power relationships.

27 Brecht, reacting to the factographic approach to photography adopted by the proponents of the "new objectivity" in the 1920s, argued against any simple documentary realism within image making given the complexity and non-visual character of capitalist social relations.

28 As we have seen, in the case of the creative practice PhD, this written requirement becomes very exacting and the potential gap between theory and practice frighteningly wide. Indeed, the reference to bodies of theory in many practice-based arts PhD projects on occasion seems arbitrary. Too often the theoretical discourse employed appears decorative, designed to stake claims about the candidate's credentials and the intellectual claims of their work rather than explain anything much. Sometimes it seems to involve constructing a sort of protective intellectual carapace designed to shield the practice from critical interrogation rather than illuminate and evaluate it (Bell 2016).

BIBLIOGRAPHY

Abel, Theodore (1948) 'The Operation Called Verstehen' *American Journal of Sociology* vol.54, No.3 November, pp. 211–218, accessed on 01.01.2018 at www.rochelleterman.com/ir/sites/default/files/abel%201946.pdf.

Ackerman, Niels, and Gobert, Sebastian (2017) *Looking for Lenin*, London: Fuel Design and Publishing.

Aiken, Jane Andrews (1980) 'Leon Battista Alberti's System of Human Proportion' *Journal of the Warburg and Courtauld Institutes* Vol.43, pp. 69–96.

Alberti, Leon Battista (1991) *On Painting*, London: Penguin Classics.

Althusser, Louis (1971) 'Ideology and Ideological State Apparatus' in L Althusser (ed.) *Lenin and Philosophy and Other Essays*, New York: Monthly Review Press.

Amad, Paula (2010) *Counter Archive: Film, the Everyday, and Albert Kahn's Archives de la Planète*, New York: Columbia University Press.

Anderson, James (2012) *Media Research Methods*, London: Sage.

Anderson, Perry (1980) *Arguments within English Marxism*, London: Verso.

Andersen, Troels (ed.)/Malevich, Kasimir (1978) *The Artist, Infinity, Suprematism: Unpublished Writings* 1913–1933 Copenhagen: Borgen.

——— (ed.)/Malevich, Kasimir (2011) *K.S. Malevich: The Leporskaya Archive*, Aarhus: Aarhus University Press.

Arnheim, Rudolf (1966) *Towards a Psychology of Art*, Berkeley and Los Angeles, CA: University of California Press.

Arts and Humanities Research Council (AHRC) (2004) *Research Funding Guidelines*, Bristol: AHRC.

Aumont, Jacques (1987) *Montage Eisenstein*, Bloomington, ID: Indiana University Press.

Badiou, Alain (2012) *The Adventure of French Philosophy*, London: Verso.

——— (2017) *Conditions*, London: Bloomsbury.

Banerjee, Subhabrate Bobby (2008) 'Necrocapitalism' *Organization Studies* Vol.29, No.12, pp. 1541–1563.

Banks, Marcus (2001) *Visual Methods in Social Research*, London: Sage.

——— (1999) *Rethinking Visual Anthropology*, New Haven, CT: Yale University Press.

——— (2016) Preface to *Mind the Gap: Working Papers on Practice Doctoral Research in the Creative and Media Arts* in Desmond Bell (ed.), Dublin: Distillers Press.

Barbash, Illisha, and Taylor, Lucien (1997) *Cross Cultural Filmmaking: A Handbook for Making Documentary and Ethnographic Films and Videos*, Berkeley and Los Angeles, CA: University of California Press.

Barkan, Elazar and Bush, Ronald (1995) *Prehistories of the Future: The Primitivist Project and the Culture of Modernism*, Stanford, CA: Stanford University Press.

Barnes, Barry, Bloor, David, and Henry, John (1996) *Scientific Knowledge: A Sociological Analysis*, Chicago: University of Chicago Press.

Barr, Alfred (1936) *Cubism and Abstract Art*, New York: Museum of Modern Art.

Barret, Elizabeth, and Bolt, Barbara (ed.) (2007) *Practice as Research: Context Method, Knowledge*, London: IB Tauris.

Barthes, Roland (1972) *Mythologies* trans. Annette Lavers, London: Paladin.

Bazin, Andre (1967/2005) *What is Cinema?* Berkeley and Los Angeles, CA: University of California Press.

Beattie, Keith (2008) *Documentary Display: Reviewing Non-Fiction Film and Video*, London: Wallflower Press.

Bell, Desmond (Dir.) (1999) *Hard Road to Klondike/Rotha Mor an tSaoil*, RTE/TG4/BBC:NI, 55 mins, Faction Films, 1999 Venice Film Festival, Glass Machine Productions.

——— (2004) 'Shooting the Past: Found Footage Filmmaking and Popular Memory' *Kinema*, Spring, pp. 74–89.

——— (2006) 'Creative Film and Media Practice as Research: In Pursuit of that Obscure Object of Desire' *Journal of Media Practice* Vol.7, No.2, pp. 85–100.

——— (Dir.) (2012a) *The Enigma of Frank Ryan TG4*, 94 mins, Colour, 2013 Festival Films du Monde Glass Machine Productions, Montreal.

——— (2012b) 'Documentary Film and the Poetics of History' *Journal of Media Practice* Vol.12, No.1, pp. 3–26.

——— (2015) 'Sound, Music and Film – Towards a Total Sonic Fabric' *The New Sound Track* Vol.5, No.1.

——— (2016) *Mind the Gap! Working Papers on Practice Based Doctoral Research in the Creative Arts and Media*, Dublin: University College Distiller's Press.

Beloufa, Neil (2015) Skype interview Radicate.eu, accessed on 15.02.2018 at www.radicate.eu/world-domination-video-artist-neil-beloufa (transcribed DB).

——— (2017) Interview Artes Mundi 7/British Council, accessed on 24.08.2018 at https://uk.video.search.yahoo.com/yhs/search?fr=yhs-Lkry-SF01&hsimp=yhs-SF01&hspart=Lkry&p=neil+beloufa+interview#id=4&vid=0eceb9094fb225abd3128cbbd84439da&action=click (transcribed DB).

Berger, Peter L. and Luckman, Thomas (1967) *The Social Construction of Reality: A Treatise on the Sociology of Knowledge*, London: Allen Lane.

Bhaskar, Roy (2008) *A Realist Theory of Science*, London: Routledge.

Bishop, Clare (2004) 'Antagonism and Relational Aesthetics' *October* Vol.110 Autumn, pp. 51–79.

Boltanski, Luc, and Esquerre, Arnaud (2016) 'The Economic Life of Things' *New Left Review* Vol.98, March–April.

Borgdoff, Henk (2013) 'A brief Survey of Current Debates on the Concept and Practices of Research in the Arts' in Mick Wilson and Schelte van Ruiten *SHARE: Handbook for Artistic Research Education*, pp. 146–162, Amsterdam: ELIA.

Bourriaud, Nicolas (2002) *Relational Aesthetics*, Paris: Les Presses du Reel.

——— (2001) 'Public Relations: An Interview with Nicolas Bourriaud' by Bennet Simpson *Art Forum* April.

Bowlt, John E. (1976) *Russian Art of the Avant Garde: Theory and Criticism 1902–1934*, London: Thames and Hudson.

Bowlt, John, and Washton Long, Rose-Carol (1980) *The Life of Vasilii Kandinssky in Russian Art: A Study of 'On the Spiritual in Art*, Newtonville, MS: Oriental Research Partners.

Bruzzi, Stella (2000) *New Documentary: A Critical Introduction*, London and New York: Routledge.

Buchloh, Benjamin (1990) 'Conceptual Art 1962–1969: From the Aesthetic of Administration to the Critique of Institutions' *October* Vol.55 Winter, pp. 105–143.

Buckingham, Mathew (2011) *The Spirit and the Letter*. 15 mins Looped and Projected Video Installation New York: Brooklyn Museum.

Bürger, Peter (1976) *Theory of the Avant-Garde* trans. Micheal Shaw, Minneapolis, MN: University of Minnesota Press.

Burgin, Victor (1986) *The End of Art Theory: Criticism and Post-Modernity*, Atlantic Highlands, NJ: Humanities Press International.

Burgoyne, Robert (2008) 'The Balcony of History' *Rethinking History: The Journal of Theory and Practice* Vol.11, No.4, pp. 547–554.

Cabanne, Pierre (1987) *Dialogues with Marcel Duchamp*, Boston, MA: Da Capa Press.

Calle, Sophie (2009) Interview with Louise Neri, *Interview Magazine*, March 7th, accessed on 20.08.2018 at www.interviewmagazine.com/art/sophie-calle.

——— (2015) *Suite Vénitienne* Catskill, NY: Siglio.

Carter, Paul (2010) 'Interest: The Ethics of Invention' in Estelle Barrett and Barbara Bolt (ed.) *Practice as Research: Approaches to Creative Arts Enquiry*, pp. 15–26, London and New York: IB Tauris.

Casini, Sylvia (2010) 'The aesthetics of Magnetic Resonance Imaging (MRI): From the Scientific Laboratory to an Artwork' *Contemporary Aesthetics* Vol.8.

——— (2011) 'Magnetic Resonance Imaging (MRI) as Mirror and Portrait: MRI Configurations Between Science and Arts' *Configurations* Vol.19, No.1.

——— (2016) 'Displacing Concepts, Re-organising Knowledge: Beyond the Theory Versus Practice Dilemma in the Arts and Humanities PhD' in Desmond Bell (ed.) *Mind the Gap! Working Papers on Practice Based Doctoral Research in the Creative Arts and Media*, Dublin: University College Distillers Press.

Cetina, Karen Knorr, Savigny, Elke von, and Schatzki Theodore R. (2001) *The Practice Turn in Contemporary Theory* London and New York: Routledge.

Chasnik, Ilya (1922) 'Unovis 1920-1922' trans. from Russian by Sarah Bodine for the 1979 exhibition by Chasnik, 'Watercolours, Drawings, Reliefs,' New York: Leonard Hutton Galleries.

Chater, Cathy (1998) *Research for Media Production* London and New York Focal Press.

Chen, Nancy N. (1992) 'Speaking Nearby: A Conversation with Trinh T Minh-ha' *Visual Anthropology* Vol.8, No.1.

Chion, Micheal (1994) *Audio-Vision: Sound on Screen* (with foreword by Walter Murch) New York: Columbia University Press.

Clark, T.J. (2001) *Farewell to an Idea: Episodes from a History of Modernism*, New Haven, CT: Yale University Press.

Classen, Constance (1993) *Worlds of Sense: Exploring the Senses in History and Across Cultures*, London: Routledge.

Colleyn, Jean-Paul (2005) 'Jean Rouch: An Anthropologist Ahead of His Time' in Jeff Himpele and Faye Ginsburg (ed.) *Cine-Trance: A Tribute to Jean Rouch (1917–2004). American Anthropologist* Vol.107, No.1, pp. 108–129.

Combes, Annie (1991) 'Ethnography and the Formation of National and Cultural Identities' in Susan Hiller (ed.) *The Myth of Primitivism: Perspectives in Art*, London: Routledge.

Crary, Jonathan (1990) *Techniques of the Observer: On Vision and Modernity in the Nineteenth Century*, New York: October Books.

Crittenden, Roger (2010) *Fine Cuts: The Art of European Film Editing*, New York and London: Focal Press.

Damisch, Hubert (1995) *The Origins of Perspective*, Cambridge, MA: MIT Press.

Dancyger, Ken (2006) *The Director's Idea: The Path to Great Directing*, Burlington, MA: Focal Press.

——— (2010) *The Technique of Film and Video Editing*, Burlington, MA: Focal Press.

Davies, David (1994) *Art as Performance*, Oxford: Blackwell.

Deleuze, Gilles (2003) *Francis Bacon: The Logic of Sensation*, Minneapolis, MN: University of Minnesota Press.

Deleuze, Gilles, and Guattari, Felix (1987) *A Thousand Plateaus*, Minneaolis, MN: University of Minnesota Press.

——— (2013) *A Thousand Plateaus*, London: Bloomsbury.

Derrida, Jacques (1994) *Specters of Marx*, London: Routledge.

Dewey, John (1998) *The Essential Dewey: Pragmatism, Education, Democracy* (ed.) Larry A Hickman and Thomas M Alexander, Bloomington, ID: Indiana University Press.

Douglas, Charlotte (1980) *Malevich*, New York: Harry Abrams.

Dray, William (1957) *Laws and Explanation in History*, Oxford: Oxford University Press.

Drutt, Mathew. (ed.) (2003) *Kasimir Malevich: Suprematism*, Berlin: Deutsche Guggenheim.

Durkheim, Emile (1897/2002) *Suicide: A Study in Sociology*, Abingdon Oxon: Routledge.

——— (1912/1964) *The Elementary Forms of the Religious Life* (trans. Joseph Swain), London: Allen and Unwin.

Easterling, Keller (2014) *Extrastatecraft: The Power of Infrastructure Space*, London: Verso.

Edgerton, Sam (1975) *The Renaissance Rediscovery of Linear Perspective*, New York: Basic Books Inc.

——— (2009) *The Mirror, the Window, and the Telescope: How Renaissance Linear Perspective Changed Our Vision of the Universe*, Ithaca, NY: Cornell University Press.

Edwards, Elizabeth (1992) *Anthropology and Photography, 1860–1920*, New Haven, CT: Yale University Press.

Eisenstein, Sergei (1971) 'A Strange Provincial City' *Selected Works in 5 vols*, Moscow: Iskusstvo.

——— (1976) 'Notes for a Film of *Capital*' Translated by Maciej Sliwowski, Jay Leyda and Annette Michelson *October* No.2, pp. 115–138, accessed on 04.01.2017 at https://thecharnelhouse.org/wp-content/uploads/2015/03/sergei-eisenstein-notes-for-a-film-of-capital-1927.pdf.

——— (1983) *Immoral Memories*, Boston, MA: Houghton Mifflin.

Ellingson, Laura L. and Ellis, Carolyn (2008) 'Autoethnography as Constructionist Project' in James A. Holstein and Jaber F. Gubrium (ed.) *Handbook of Constructionist Research*, New York: Guilford Press.

Elliot, David (1975) *Susan Hiller* (ed.) Ann Gallagher, London: Tate Publishing.

Ellis, Carolyn, and Flaherty, Micheal (ed.) (1992) *Investigating Subjectivity: Research on Lived Experience*, Newbury, CA: Sage.

Fay, Brian (1975/2015) *Social Theory and Political Practice*, London: Allen and Unwin.

Feld, Stephen (2005) 'Introduction' in Jean Rouch, *Ciné-Ethnography* (ed. and trans. Steven Feld) Minneapolis, MN: University of Minnesota Press.

Feyerabend, Paul (1995a) 'Science as a Product of Nature as a Work of Art' in Gavroglu, K. et al. (eds.) *Science, Mind and Art*, Dordrecht-Boston-London: Kulwer Academic Publishers.

——— (1995b) *Killing Time: The Autobiography of Paul Feyerabend*, Chicago: University of Chicago Press.

——— (2010) *Against Method: Outline of an Anarchistic Theory of Knowledge*, London: Verso.

Finch, James (2017) 'A Wistful Dream of Far-Off Californian Glamour: David Sylvester and the British View of American Art' *Tate Papers* No.27 Spring, accessed on 04.01.2018 at www.tate.org.uk/research/publications/tate-papers/27/david-sylvester.

Foster, Hal (1996a) 'The Artist as Ethnographer' in *The Return of the Real*, Cambridge, MA: MIT Press.

——— (1996b) 'The Archive Without Museums' *October* Vol.77, pp. 97–119.

Foucault, Michel (1977) 'What Is an Author?' in *Language, Counter-Memory, Practice: Selected Essays and Interviews by Michel Foucault*, edited with an introduction by Donald F. Buchard, Ithaca, NY: Cornell University Press.

Frayling, Christopher (1994) 'Research in Art and Design' *Royal College of Art Research Papers Series* Vol.1, No.1, London: RCA.

Fruitmarket Gallery (Edinburgh) (October 2016) Publicity leaflet produced for Damian Ortega exhibition, *States of Time.*

Gage, John (1999) *Color and Meaning: Art Science and Symbolism*, Berkeley and Los Angeles, CA: University of California Press.

Gale, Mathew, and Sidlina, Natalia (eds.) (2017) *Red Star Over Russia a Revolution in Visual Culture 1905–55 a Celebration of the David King Collection*, London: Tate Publishing.

Gell, Alfred (1998) *Art and Agency: An Anthropological Theory*, Oxford: Oxford University Press.

Gibbons, Luke (2005) '"We Knew Their Plight Well" Race and Immigration in Some Recent Irish Films' *Third Text* Vol.19, No.5.

Ginsburg, Faye (2005) 'Dans le Bain Avec Rouch: A Reminiscence' in Jeff Himpele and Faye Ginsburg (eds.) *Cinetrance: A Tribute to Jean Rouch (1917–2004). American Anthropologist* Vol.107, No.1, pp. 108–129.

Gombrich, Ernst (1960/2002) *Art and Illusion, a Study in the Psychology of Pictorial Representation*, London: Phaidon.

Grey, Camilla (1962) *The Great Experiment: Russian Art 1863–1922*, London: Thames and Hudson.

Grey, Carole, and Malins, Julian (2004) *Visualising Research: A Guide to the Research Process in Art and Design*, London and New York: Routledge.

Gunning, Tom (1990a) *Cinema of Attractions: Early Film, Its Spectator and the Avant-Garde*, London: BFI Publishing.

Gunning, Tom (1990b) 'The Cinema of Attraction: Early Cinema, its Spectator and the Avant-Garde' in Thomas Elsaesser et al. (ed.) *Early Cinema: Space, Frame, Narrative*, pp. 56–62, London: BFI Publishing.

Habermas, Jürgen (1972) *Knowledge and Human Interest*, London: Heineman.

Hahl-Koch, Jelena (1993) *Kandinsky*, London: Thames and Hudson.

Hall, Stuart, and Jefferson, Tony (ed.) (1975/2006) *Resistance Through Rituals: Youth Subcultures in Post-War Britain*, London: Routledge.

Halsall, Francis (2008) *Systems of Art: Art, History and Systems Theory*, Oxford: Peter Lang.

Hammond, John H. (1981) *The Camera Obscura, a Chronicle*, London and Cleveland, OH: CRC Press.

Harrison, Charles, and Wood, Paul (eds.) (2003) 'The Realistic Manifesto' in *Art in Theory, 1900–2000*, Malden: Blackwell.

Healy, John (2016) 'The Components of the "Crit" in Art and Design Education' *Irish Journal of Academic Practice* Vol.5, Article 7, accessed at http://arrow.dit.ie/ijap/vol5/iss1/7.

Hempel, Carl (1942) 'The Function of General Laws in History' *The Journal of Philosophy* Vol.39, No.2, pp. 35–48.

Hiller, Susan (ed.) (1991) *The Myth of Primitivism: Perspectives on Art*, London: Routledge.

——— (1996) *Thinking About Art: Conversations with Susan Hiller* (ed.) Barbara Einzig, Manchester: Manchester University Press.

——— (2011) *Susan Hiller* (ed.) Ann Gallagher, London: Tate Hockings, Paul (1975).

Hockney, David (2001) *Secret Knowledge* 1hr 15 mins, BBC4 Omnibus, Television Documentary Film.
——— (2006) *Secret Knowledge: Rediscovering the Lost Techniques of the Old Masters*, London: Thames and Hudson.
Holm, Gunilla (2014) *Photography as a Research Method*, Oxford: Oxford University Press.
Horniman Museum (2019) Introductory statement on the museum's web page on Museums of London, accessed on 25.02.2016 at https://www.museumslondon.org/museum/130/free-museums-in-london.
Howes, David (ed.) (1991) *The Varieties of Sensory Experience: A Sourcebook in the Anthropology of the Senses*, Toronto: University of Toronto Press.
Itten, Johannes (1974) *The Art of Color: The Subjective Experience and Objective Rationale of Color*, New York: John Wiley.
Jameson, Frederick (1975) *The Prison-House of Language: A Critical Account of Structuralism and Russian Formalism*, Princeton, NJ: Princeton University Press.
Janacek, Gerard (1996) *Zaum: The Transrational Poetry of Russian Futurism*, San Diego, CA: University of San Diego Press.
Jay, Martin (1999) *Downcast Eyes: The Denigration of Vision in Twentieth-Century French Thought*, Berkeley and Los Angeles, CA: University of California Press.
Jeffrey Charles Henry Peacock (JCHP) (January 2011) *Sic, Sic, Sic: A Mutual Disillusion*, Supplementary Material produced to accompany exhibition at All Hallows On the Wall, London.
——— (2014) Critical Décor, Exhibition Lanchester Gallery Coventry, together with web publication of a range of art critical documents and statements (no pagination).
——— (2017) 'Bad Goods' Supplement in *Art Monthly* No.412, December-January 2017–2018, 10/10, accessed on 23.02.2019 at https://www.badgoods.org/project-10.
Jeffries, Stuart (2016) *Grand Hotel Abyss: The Lives of the Frankfurt School*, London: Verso.
Jesse, Heike (ed.) (2016) *Fashion Studies: Research Methods, Site and Practices*, London: Bloomsbury.
Kandinsky, Wassily (1947) *Point and Line to Plane* trans. Howard Dearstyne and Hilda Rebay New York: Solomon Guggenheim Foundation.
——— (1977) *On the Spiritual in Art* trans. Micheal Sandler, Toronto: Dover Publications, accessed on 13.01.2018 at www.gutenberg.org/cache/epub/5321/pg5321-images.html.
——— (1994) *Complete Writing on Art* (ed.) Lindsay, K and Vergo, P., Boston, MA: Da Capo Press.
Kant, Immanuel (2007) *Critique of Pure Reason*, London: Penguin Classics.
——— (2009) *Critique of Judgement*, Oxford: Oxford University Press.
——— (2015) *Critique of Practical Reason*, Cambridge: Cambridge University Press.
Kemp, Martin (2006a) *Leonardo da Vinci: The Marvellous Works of Nature and Man* Oxford: Oxford University Press.
——— (2006b) *Leonardo da Vinci: Experience, Experiment and Design*, London: Victoria and Albert Museum Publications.
——— (2006c) *Seen/Unseen: Art, Science and Intuition from Leonardo to the Hubble Telescope*, Oxford: Oxford University Press.
Kemp, Martin, and Walker, Margaret (2001) *Leonardo on Painting: An Anthology of Writings by Leonardo da Vinci: With a Selection of Documents Relating to his Career as an Artist*, New Haven, CT: Yale University Press.
Kentridge, William (2012) 'A Natural History of the Studio' Lecture, Tel Aviv University 10 June 2012, accessed on 20.08.2018 at https://uk.video.search.yahoo.com/yhs/search;_ylt=AwrIRln5V31bGCcAkil3Bwx.;_ylu= (transcribed DB).

——— (2013) 'Listening to the Image' Neubauer Collegium Inaugural Lecture at the University of Chicago, accessed on 03.04.2018 at www.youtube.com/watch?v=WO5_FZiB950.

——— (2014) *Six Drawing Lessons* (the Charles Eliott Norton Lectures), Cambridge, MA: Harvard University Press.

——— (2015) *More Sweetly Play the Dance* (ed.) Jaap Guldermond, Amsterdam: Nai010 Publishers.

Kentridge, William, and Vivienne Koorland (2017) *Conversation in Letters and Lines* (with foreword by Tamar Garb, Briony Fer, Joseph Leo Koerner, Ed Krcma, Griselda Pollock), Chicago: University of Chicago Press.

Knight Powell, Amy (2016) 'Squaring the Circle: The Telescopic View in Early Modern Landscapes' *Art History* Vol.39, No.2, pp. 282–301.

Kogan, Nina (1920) 'The Principles of Painterly Abstraction,' *Unovis* No.1. trans. Nicoletta Misler, published in Experiment 1, 1995 (University of California) pp. 85–88.

Kokkori, Maria, and Bouras, Alexander (2014) 'Charting Modernism: Kazimir Malevich's Research Tables' in Malevich (ed.) *Achim Borchardt Hume*, pp. 164–195, London: Tate Publishing.

Kossuth, Joseph (1993) *Art after Philosophy and after, Collected Writings 1966–1990*, Cambridge, MA: MIT Press.

Kuhn, Thomas (1962) *The Structure of Scientific Revolutions*, Chicago: University of Chicago Press.

Kwint, Marius (1999) *Material Memories: Design and Evocation*, London and Oxford: Berg.

Latour, Bruno (1979) *Laboratory Life: The Construction of Scientific Facts*, Princeton, NJ: Princeton University Press.

——— (1988) *Science in Action: How to Follow Scientists and Engineers Through Society*, Cambridge, MA: Harvard University Press.

Le Breton, David (2017) *Sensing the World: An Anthropology of the Senses*, London: Bloomsbury.

Lenin, Vladimir Illich (1923/1964) 'Better Fewer, But Better' in *Collected Works* Vol.33, pp. 487–502, Moscow: Progress Publishers (original source Pravda, 4 March 1923).

Lennon, Peter (1968) *The Rocky Road to Dublin* 99 mins, black and white, New York: Icarus Films.

LeWitt, Sol (1967) 'Paragraphs on Conceptual Art' *Artforum* June, accessed on 03.01.2018 at www.corner-college.com/udb/cproVozeFxParagraphs_on_Conceptual_Art._Sol_leWitt.pdf.

Leyda, Jay (1964) *Films Beget Films: A Study of the Compilation Film*, New York: Hill and Wang.

Lippard, Lucy (1997) *Six Years: The Dematerialization of the Art Object from 1966 to 1972*, Berkeley and Los Angeles, CA: University of California Press.

Lodder, Christine (1983) *Russian Constructivism*, New Haven, CT: Yale University Press.

Luccia, Cynthia (1994) 'Revisiting *High School* – An Interview with Frederick Wiseman' *Cinéaste* Vol.20, No.4, pp. 5–11.

Luhmann, Niklas (2000) *Art as Social System*, Los Angeles, CA: Stanford University Press.

Lund, Christian (2016) 'Bill Morrison: The Film Archaeologist' Interview with Bill Morrison Louisiana Museum of Modern Art, Denmark, accessed on 01.01.2018 at http://channel.louisiana.dk/video/bill-morrison-film-archaeologist (transcribed DB).

Lütticken, Sven (2014) 'Hito Steyerl: Post-Cinematic Essays after the Future' in Nick Aikens (ed.) *Too Much World: The Films of Hito Steyerl*, Eindhoven and Berlin: Van Abbemuseum and Sternberg Press.

MacGowan, Micheal (1962/2014) *Hard Road to Klondike*, Cork: The Collins Press.

Macleod, Lucy, and Holdridge, Lin (2007) *Thinking Through Art: Reflections on Art as Research*, London and New York: Routledge.

Madoff, Steven Henry (2009) *Art School: Proposition for the 21st Century*, Cambridge, MA: MIT Press.

Malevich, Kasimir (December 1915) 'From Cubism to Suprematism in Art, to the New Realism of Painting' trans. Charlotte Douglas, included in her *Swans of Other Worlds: Kasimir Malevich and the Origins of Abstraction in Russia* Michigan: Ann Arbor 1976.

Malevich, Kasimir (1959) *The Non-Objective World*, Chicago: Paul Theobold.

——— (1978) *The World as Non-Objectivity: Unpublished Writing* Vols I–III (ed.) Troels Anderson, Copenhagen: Borgen.

Mannoni, Laurent (2000) *The Great Art of Light and Shadow: Archaeology of the Cinema*, Exeter: University of Exeter Press.

Marcus, George, and Myers, Fred (ed.) (1995) *The Traffic in Culture: Refiguring Art and Anthropology*, Berkeley and Los Angeles, CA: University of California Press.

Markov, Vladimir (1968) *Russian Futurism: A History*, Berkeley and Los Angeles, CA: University of California Press.

Marks, Laura (2000) *The Skin of the Film: Intercultural Cinema, Embodiment, and the Senses*, Durham, NC: Duke University Press.

Martindale, Andrew (1972) *The Rise of the Artist in the Middle Ages and Early Renaissance*, London: Thames and Hudson.

Marx, Karl (1968) 'Theses on Fuerbach' in *Marx and Engels Selected Works in One Volume*, London: Lawrence and Wishart.

Mauss, Marcel (1990) *The Gift: Forms and Functions of Exchange in Archaic Societies*, London: Routledge.

McGarry, Fearghal (2015) *Frank Ryan*, Dublin: University College Dublin Press.

Mead, Margaret (1975) 'Visual Anthropology in a Discipline of Words' in Paul Hockings (ed.) *Principles of Visual Anthropology*, Paris: Mouton.

Merton, Robert (1968) *Social Theory and Social Structure*, New York: Free Press.

Milner, John (1983) *Vladimir Tatlin and The Russian Avant-Garde*, New Haven, CT: Yale University Press.

Milton, Alex, and Rodgers, Paul (2013) *Research Methods for Product Design*, London: Lawrence King.

Minh-ha, Trinh T. (1989) *Woman, Native, Other: Postcoloniality and Feminism*, Bloomington, ID: Indiana University Press.

——— (1992) *Framer Framed*, London: Routledge.

Mink, Louis (1966) 'The Autonomy of Historical Understanding' *History and Theory* Vol.5, No.1, pp. 24–47.

Mitchell, W.J.T. (1996) 'Response to October "Visual Culture Questionnaire"' *October* Vol.77 Summer.

Moholy-Nagy, Lazlo (1925/1969) *Painting Photography and Film*, Cambridge, MA: MIT Press.

MRAC (2017) (director Sandra Patron) Exhibition leaflet produced to accompany the exhibition by Neil Beloufa 'Sustainable Development', Serignan: MRAC.

Myerson, Roger (1991) *Game Theory: Analysis of Conflict*, Cambridge, MA: Harvard University Press.

Nekes, Werner (1986) *Film before film* (Was geschah wirklich zwischen den Bildern), 1hr 23 mins, Colour, Documentary Film London: BFI Publishing.

Nelson, Robin (2006) *Practice as Research in the Arts*, London: Palgrave Macmillan.

Newbury, Darren (1996) *Designing and Managing a Research Project: A Research Guide*, Birmingham: Research Training Initiative, University of Central England.

Nicholls, Bill (1992) *Representing Reality: Issues and Concepts in Documentary*, Bloomington, ID: Indiana University Press.

——— (1994) *Blurred Boundaries: Questions of Meaning in Contemporary Culture*, Bloomington, ID: Indiana University Press.

——— (2010) *Introduction to Documentary*, 2nd edition. Bloomington, ID: Indiana University Press.

Nicholls, Bill, and Renov, Micheal (ed.) (2011) *The Cinema's Alchemist: The Films of Péter Forgács*, Minneapolis, MN: University of Minnesota Press.

Obrist, Hans Ulrich (2001a) 'Public Relations: Bennet Simpson Talks with Nicolas Bourriaud' *Art Forum* April, accessed on 06.01.2017 at http://web.mit.edu/allanmc/www/simpson1.pdf.

——— (2001b) 'Battery, Kraftwerk and Laboratory' in Carin Couni (ed.) *Words of Wisdom: A Curators' Vade Mecum on Contemporary Art*, New York: Independent Curators International.

——— (2008) *A Brief History of Curating*, Zurich: JRP and Ringier.

O'Neill, Paul, and Mick Wilson (eds.) (2015) *Curating Research*, London: Open Editions.

Panofosky, Erwin (1991) *Perspective as Symbolic Form*, New York: Zone Books.

Parsons, Talcott (1951/1991) *The Social System*, London: Routledge.

Peucker, Brigette (2007) *The Material Image: Art and the Real in Film*, Stanford, CA: Stanford University Press.

Phelan, Peggy (1993) *Unmarked: The Politics of Performance*, London and New York: Routledge.

Pilkington, A. (2015) 'Unfamiliar Terrain- An Investigation into how the Uncanny can Occur in Painting' PhD thesis National College of Art and Design, Dublin.

Pink, Sarah (2013) *Doing Visual Ethnography*, London: Sage.

Pollock, Griselda (2003) 'Visual Culture and its Discontents: Joining in the Debate' *Journal of Visual Culture* Vol.2, No.2.

Popper, Karl (2002) *Conjectures and Refutations: The Growth of Scientific Knowledge*, London and New York: Routledge.

Powell, Amy Knight (2016) 'Squaring the Circle: The Telescopic View in Early Modern Landscapes' *Art History* Vol.39, No.2, pp. 282–301.

Priest, Susanne (2009) *Doing Media Research: An Introduction*, London: Sage.

Rabinger, Micheal (2006) *Developing Story Ideas*, New York: Focal Press.

——— (2014) *Directing the Documentary*, New York: Focal Press.

Railing, Patricia (1989) *From Science to Systems of Art: On Russian Abstract Art and Language 1910/1920 and Other Essays*, Sussex: Artists Bookworks.

Rainsfort, Sue (2014) 'Maleventos: Alison Pilkington at Pallas Projects/Studios February 28th -March 8th', accessed on 01.08.2018 at www.suerainsford.com/2014/03/malevolentos-ailson-pilkington-at-pallas-projectsstudios-february-28th-march-8th-2014.

Rains, Stephanie (2011) 'The Ideal Home (Rule) Exhibition: Ballymaclinton and the 1908 Franco British Exhibition' *Field Day Review* Vol.7, pp. 5–21.

Rancière, Jacques (1994) *Names of History: On the Poetics of Knowledge*, Minneapolis, MN: University of Minnesota University Press.

——— (2004) *The Politics of Aesthetics: The Distribution of the Sensible* trans. Gabriel Rockhill, London: Continuum.

——— (2011) *Althusser's Lesson* trans. Emiliano Battista, London: Continuum.

Read, Herbert (1968) *A Concise History of Modern Painting*, London: Thames and Hudson.

Rescher, Nicholas (1973) *The Primacy of Practice: Essays Towards a Pragmatically Kantian Theory of Empirical Knowledge*, Oxford: Blackwell.

Rilke, Rainer Maria (2002) *Letters on Cezanne* trans. Joel Agee, New York: North Point Press.

Robinson, Denise (2004) 'Scarce Stains the Dust. . .: Freud's Museum – The Work of Susan Hiller' in James Lingwood (ed.) *Susan Hiller: Recall, Selected Works 1969–2004, Exhibition Catalogue*, Gateshead: Baltic Centre for Contemporary Art, Gateshead.

Roca, J. (2007) 'Colombia, Botany and Classification: Fake Flowers and Post-Colonialism' *Frieze* June, accessed on 16.07.2018 at https://frieze.com/article/alberto-barraya.

Roche, Denis Napier, Philip Maguire, Brian, and McCann, Shaun (2008) 'A Clinically Useful Artwork? Between a Dialogical and a Relational Approach to art in a Clinical Environment' *The International Journal of the Arts in Society* Vol.3.

Rodchenko, Aleksandre (2008) *Revolution in Photography*, Moscow: Moscow House of Photography.

Rodchenko, Varvara, and Lavrentiev, Aleksandr (1989) *Rodchenko-Stepanova Family Workshop*, London: Serpentine Gallery.

Ronai, Carol (1992) 'The Reflexive Self Through Narrative: A Night in the Life of an Exotic. Dancer/Researcher' in Ellis and Flaherty (ed.) *Investigating Subjectivity: Research on Lived Experience*, London: Sage.

Rony, Fatimah T. (1996) *The Third Eye: Race, Cinema, and Ethnographic Spectacle*, Durham, NC: Duke University Press.

Rosenstone, Robert (2012) *History on Film: Film on History*, London: Routledge.

Roses, Gillian (2001) *Visual Methodologies: An Introduction to the Interpretation of Visual Materials*, London: Sage.

Rouch, Jean, and Feld, Steven (ed.) (2003) *Cine-Ethnography*, Minneapolis, MN: University of Minnesota Press.

Royal Institute of British Architects (RIBA) (2017) *How Architects Use Research*, London: RIBA.

Ruby, Jay (1998) *The Death of Ethnographic Film* (paper presented at the American Anthropological Association meetings, December 2, in Philadelphia), accessed on 16.02.2018 at www.researchgate.net/publication/265025213_The_Death_of_Ethnographic_Film.

——— (2000) *Picturing Culture: Explorations of Film and Anthropology*, Chicago: University of Chicago.

Russell, Catherine (1999) *Experimental Ethnography: The Work of Film in the Age of Video*, Durham, NC: Duke University Press.

Rye, Caroline (2002) 'Living Cameras a Study of Live Bodies and Mediatized Images in Multi-Media Performance and Installation Art Practice' PhD thesis Napier University, Edinburgh.

Sansi, Roger (2013) *Art, Anthropology and the Gift*, London: Bloomsbury.

Schneider, Arnd, and Wright, Christopher (ed.) (2005) *Contemporary Art and Anthropology*, London: Berg.

——— (ed.) (2010) *Between Art and Anthropology: Contemporary Ethnographic Practice*, London: Berg.

——— (ed.) (2013) *Anthropology and Art Practice*, London: Bloomsbury.

Schon, Donald (1983) *The Reflective Practitioner: How Professionals Think in Action*, New York: Basic Books Inc.

Science Gallery (Dublin) (2017) Publicity posters for the exhibition *In Case of Emergency*, 2017.

Schutz, Alfred (1967) *The Phenomenology of the Social World*, Chicago: Northwestern University Press.

Scriviner, Stephen 'The Art Object Does not Embody a form of Knowledge' *Working Papers in Art and Design Coventry University* No.2, pp. 25–32, accessed on 09.01.2017 at www.herts.ac.uk/__data/assets/pdf_file/0008/12311/WPIAAD_vol2_scrivener.pdf.

Sekula, Alan (1995) *Fish Story*, Düsseldorf: Richter Verlag.

Sekula, Alan, and Birch, Noel (2010) *The Forgotten Space* 112 mins, Colour, Essay Film English, Amsterdam: Doc-Eye Film.

Sell Tower, Beeke (1981) *Klee and Kandinsky in Munich and at the Bauhaus*, Michigan: University of Michigan Press.

Shanken, Edward A. (2003) *Telematic Embrace: Visionary Theories of Art, Technology and Consciousness*, Berkeley and Los Angeles, CA: University of California Press.

Sheik, Simon (2009) 'Objects of Study or Commodification of Knowledge? Remarks on Artistic Research' *Art & Research: A Journal of Ideas, Contexts and Methods* Vol.2, No.2 Spring, accessed on 08.01.2017 at www.artandresearch.org.uk/v2n2/pdfs/sheikh.pdf.

Shklovsky, Viktor (2016) *Viktor Shklovsky: A Reader* (ed.) Alexendra Berlina, London: Bloomsbury.

Sobchack, Vivian (2004) *Carnal Thoughts: Embodiment and Moving Image Culture*, Berkeley: University of California Press.

Sloan, Luke, and Quaane-Hase, Anabel (2017) *The Sage Handbook of Social Research Methods*, London: Sage.

Smith, Hazel, and Deane, Roger (eds.) (2009) *Practice-led Research, Research-led Practice in the Creative Arts*, Edinburgh: Edinburgh University Press.

Smith, Mark A. (2014) *From Sight to Light: The Passage from Ancient to Modern Optics*, Chicago: University of Chicago Press.

Smolik, Noemi (2006) 'Kandinsky – Resurrection and Cultural Renewal' in Hartwig Fischer and Sean Rainbird (eds.) *Kandinsky: The Path to Abstraction*, pp. 139–158, London: Publishing.

Sohn-Rethel, Alfred (1978) *Intellectual and Manual Labour: A Critique of Epistemology*, London: Macmillan.

Spangler, Tina (1993) 'Interviewer Interviewed: A Discussion with Trinh T. Minh-ha' *Latent Image* Winter.

Stafford, Barbara Maria (1991) *Body Criticism: Imaging the Unseen in Enlightenment Art and Medicine*, Cambridge, MA: MIT Press.

——— (1996) *Good Looking: Essays on the Virtues of Images*, Cambridge, MA: MIT Press.

——— (2006) *Devices of Wonder: From the World in a Box to Images on a Screen*, Los Angeles, CA: Getty Publications.

Steadman, Ralph (2001) *Vermeer's Camera: Uncovering the Truth Behind the Masterpieces*, Oxford: Oxford University Press.

Steyerl, Hito (2010a) 'Aesthetics of Resistance? Artistic Research as Discipline and Conflict' *eipcp* January, accessed on 18.02.21018 at http://eipcp.net/transversal/0311/steyerl/en.

——— (2010b) *In Free Fall: A Thought Experiment* Recorded Public Lecture, November 06, accessed on 18.02.2018 at https://vimeo.com/81109235.

——— (2010c) *In Free Fall* 32 mins Colour, Essay Film, German/English Vienna: Sixpackfilm.

——— (2010d) Hito Steyerl *Picture This Interview on the Making of 'In Free Fall'* accessed on 18.02.2018 at www.picture-this.org.uk/worksprojects/works/by-date/2010/in-free-fall (transcribed DB).

——— (2011) 'In Free Fall' *e-flux* No.24 April.

Strathearn, Margaret (1988) *The Gender of the Gift: Problems with Women and Problems with Society in Melanesia*, Berkeley and Los Angeles, CA: University of California Press.

Sullivan, Graham (2005) *Art Practice as Research: Inquiry in the Visual Arts*, London: Sage.

Sweeney, Moira (2016) 'Dublin Docks: Visualising Changing Identities, Communities and labour Practices' in, Desmond Bell (ed.) *Mind the Gap! Working Papers on Practice-Based Doctoral Research in the Creative and Media Arts*, Dublin: University College Distillers Press.

Sylvester, David (1987) *The Brutality of Fact: Interviews with Francis Bacon*, London: Thames and Hudson.

Tall, Emily (1987) 'Eisenstein on Joyce: Sergei Eisenstein's Lecture on James Joyce at the State Institute of Cinematography, November 1, 1934' *James Joyce Quarterly* Vol.24, No.2 Winter, pp. 133–142.

Taylor, Richard, and Ian Christie (eds.) (1988) *The Film Factory: Russian and Soviet Cinema in Documents, 1896–1939*, Cambridge, MA: Harvard University Press.

Thompson, Edward P. (1978) *The Poverty of Theory and Other Essays*, London: Merlin Press.

Tretyakov, Sergei (1929/2006) 'The Biography of the Object' *October* Vol.118 Fall, pp. 57–62.

Trotsky, Leon (2008) *History of the Russian Revolution*, Vol.1 Chicago: Haymarket Press.

Tupitsyn, Margarita (2002) *Malevich and Film*, Newhaven, CT: Yale University Press.

Udaltsova, Nadezhda and Larissa Zhadova (eds.) (1980) *Woman Artists of the Russian Avant-Garde 1910–1930*, Zurich: Galerie Gmurzynska.

Vaughan, Dai (1999) *For Documentary: Twelve Essays*, Berkeley and Los Angeles, CA: University of California Press.

Vaughan, Laurene (2017) *Practice-Based Design Research*, London: Bloomsbury.

Volans, Kevin (1986) *White Man Sleeps: Composer's Statement*, accessed on 04.01.2017 at http://kevinvolans.com/essays/white-man-sleeps-composers-statement/.

Vygotsky, Lev (1925/1971) *The Psychology of Art*, Cambridge, MA: MIT Press.

Warwick, Genevieve (2016) 'Looking in the Mirror of Renaissance Art' *Art History* Vol.39, pp. 254–281.

Weber, Max (1905/2011) *The Protestant Ethic and the Spirit of Capitalism*, Oxford: Oxford University Press.

——— (1949) *The Methodology of the Social Sciences* (trans and ed. Edward Shils and Henry Finch), New York: The Free Press.

Weiss, Peter (2005) *The Aesthetics of Resistance* Vol.1. trans. Joachim Neugroschel, Durham, NC: Duke University Press.

Werner, Gosta (1990) 'James Joyce and Sergei Eisenstein' *James Joyce Quarterly* Vol.27, No.3, pp. 491–507.

White, Hayden (1987) *The Content of the Form: Narrative Discourse and Historical Representation*, Baltimore: The Johns Hopkins University Press.

Wicks, Rainer, and Gabriele Graw (2001) *Teaching at the Bauhaus*, Berlin: Hatje Cantz Verlag.

Williams, Raymond (1976) *Keywords: A Vocabulary of Culture and Society*, London: Fontana Paperbacks.

Wilson, Mick, and van Ruiten, Schelte (eds.) (2013) *SHARE: Handbook for Artistic Research Education*, Amsterdam: ELIA.

Wilson, Sarah, and Ssorin-Chaikov, Nikolai (2012) 'Ethnographic Conceptualism: Performing Methodological Experiments' Abstract of a workshop at the Courtauld Institute, London, January 31, referenced in Schneider and Wright (2013).

Witkovsky, Mathew S. (ed.) (2017) *Revolintsiia! Demonstratsiia!* Chicago: Art Institute of Chicago.

Wittgenstein, Ludwig (1961) *Tractatus Logico-Philosophicus*, London: Routledge and Kegan Paul.

——— (1998) *Culture and Value* (ed.) Georg Henrik von Wright, London: Wiley-Blackwell.

Wood, Christopher (1995) 'Review of *The Origin of Perspective* by Hubert Damisch' *Art Bulletin* Vol. LXXVII, No.4, pp. 677–682.

Zhadova, Larissa A. (1982) *Malevich: Suprematism and Revolution in Russia Art 1910–1930*, London: Thames and Hudson.

INDEX

Note: Page numbers in *italic* indicate figures and page numbers in **bold** indicate tables on the corresponding pages.

abacchi 130
abduction 297
Abwehr operatives 243, *245*
action 3, 84–85
Adorno, Theodor 33n14, 91n17, 306
aesthetics 83–84; administrative 4, 29, 249; relational 73–80, *74*, *77–78*, 90n13, 90n15, 304; of resistance 250, 252
Africa 43, 46, 283, 296n14; African art 277–280; South Africa 15, 45, 67; West Africa 264, 271, *287*
aesthetic research 38, 250
AKhRR 219
Alberti, Leon Battista 129, 133–134, *134*, 136–137
Althusser, Louis 255n5, 312, 326n20, 326n21
anthropology 2, 76, 232, 255n6, 257–258, 303–304; and auto-ethnography 290–294, *292*, 296n16; and Baraya 267–271, *269–270*; as conceptual art 295n3; and an ethnographic turn in the visual arts 258–260; fusion between art and 294–295n2; and Haddon and Browne 260–261, *261–262*; and Hillen 263, *266*; and Hiller 277–283, *281*; and Hinde *265*; and Kindness *267*; and lens-based ethnography 272–276, *273*, *275*; and method 295n10; and Minh-ha 283–290, *287*, 296n13, 296n14; and Rouch 264–267, *268*, 271–272, 295n7; and surrealism 295n4
Appelt, Dieter 80
applied art, research 15, 29, 58, 147, 183, 206, 221
architecture 37–38, 129–131, 136–138
archive 52–54, 231–235, *233*
art education 10, 12, 26, 27, 63, 100, 176, 184, 199, 293, 297, 300
artefact 10, 20, 25, 28, 29, 69, 231–232, 234, 256n11, 257, 259, 277, 279, 280, 282, 283
art history 18–20, 167–168, 181n28, 223n14, 257–259; and the avant-garde 198–199, 295n4
artistic reflection 43, 46
artistic research 1–5, 9, 35–38, 88–89, 105–106, 133–137; for art and design 28–29; auto-ethnographic impulse in 290–294; into the creative arts 18–20; through creative arts practice 20–28, *25*; and critical theory 323–324, 325n2; distinguishing characteristics of 15–18; founding texts and axioms of 173–178; and history 227–231, 248–250; key terms in 9–15, *11*, *14*; methodological drive in 92–94; modelling of 29–32, *30–31*; and the modernist avant-garde 145–148, 171–173; PhD research route *21*; Renaissance contribution to 142–144; and the revolution 183–184,

186–187, 221–222; the specificity of 97–101; *see also specific artists*
artist interview 46–49
art making 93, 98, 105, 148, 259, 294
art process 23–24, 28–29, 36–37; autonomy of 180n15; and critical theory 323–324; and a research turn in visual arts 58–59, 75; and the revolution 206–207, 222n1
art system 27, 232, 303
atonality 164
Austria 254; Linz 246–249, *247*
auto-criticism 45–46, 48
autonomy 4, 34n14, 56n2, 62, 180n15, 208, 254
avant-garde 1–2, 26–28, 48–49, 100–101, 145–147, 181n19; and abstraction 223n16; and art as expression 168–173, **172**; eclipse of 218–221; and modernist artistic research 173–178; and *Narkompros* 222n3; and progressivism 181n25; and the revolution 198–200, 221–222, 222n2; and a science of art 147–149, 179n9, 182n35; women in 180n12; and the Zaum poetry movement 182n30; *see also specific artists*
axioms 85, 173–178

background research 28, 35–38, 42, 49, 54, 98
Bacon, Francis 22, 46–47, 69, *70*
Badiou, Alain 85, 121n7, 146
Balthus Inventory 71, *72*, 114
Banks, Marcus 110, 296n20
Baptistry of San Giovani *132*
Baraya, Alberto 267–271, *269–270*, 282
Baroque, the 45, 126, 164, 198
Barr, Alfred 150, 180n13, 193–196, *194*
Barthes, Roland 12, 60, 62, 71
Bastashevski, Mari 317–323, *318*, *320–321*, 326n25
Bauhaus, the 27, 150, 152, 179n8, 186; and art as expression 169, 171; and Constructivism 206, 215; and the eclipse of the avant-garde 218, 221; and Kandinsky 164; and modernist artistic research 173, 175–176; and non-objective art 196; and the UNOVIS experiment 200, 203
beaux-arts 187
Bell, Desmond *25*, *30–31*; *The Enigma of Frank Ryan* 242–246, *245*; *Hard Road to Klondike* 46, 54–55, *55*, 57n9, 57n14
Beloufa, Neil 2–3, 307–312, *308–309*, *311*, 323
Bishop, Clare 76, 90n13
Blau Reiter group 151, 169, 224n25
Bourriaud, Nicolas 75–76, 90n13, 90n15, 304, 325n12
Bridgehead building 246–248, *247*
Brunelleschi, Filippo 104, 125, 129–134, *132*, 136–138, 143, 144n6
Buchloh, Benjamin 33–34n14, 59, 89n1, 299
Buckingham, Mathew 228, 233–234, *234*
Burch, Noel 325n3; *Forgotten Space, The* 272–273, *273*

Cabanne, Pierre 47, 59–60
Calle, Sophie 80, *82*, 83, 296n16
Caouette, Jonathan 19
capitalism 12, 33n14, 295n6, 326n13, 326n18, 326n20; critical theory and 300–302, 304–308, 310–312, 317, 319–320; history and 229, 249–252, 254; the modernist avant-garde and 146; necrocapitalism 319; neo-capitalism 217; and a research turn in visual arts 79, 86; science and 101
Carter, Paul 15
Casini, Sylvia 115, 117–118, 121
Cezanne, Paul 33n10, 47, 128, 156, 158, 193
Chater, Kathy 37
Chion, Michael 42
choreography 274
cinema, of attractions, archeology of 126, 144n3
cinematography 242, 284, 288
cinéma verité 295n7
collaboration 40, *81*, 271–272, 278–279; art–science 101, 106, 110–120; research 302, 325n6
Colleyn, Jean-Paul 272–273
Collins, Robbie 109, *109*
communication 12, 19, 169–171, **172**, 189, 299, 304
Comolli, Jean Luc 48–49
Conceptualism 34n14, 58–65, 299–300
Constructivism 26–27, 179n5, 179n6, 182n34, 206–214, 218–219
Constructivist theory of learning 26
contemporary art 9–10, 28, 58–59, 75–76, 257–258, 325n12; *see also specific artists*
contextual/contextualization, critical contextualizaion 2, 19–20, 30, 56, 86, 88, 167, 172, 223n10, 256n12, 259, 282, 299, 303, 323
counter-relief 181n22, 224n23
counter rules 86, 276

Coutard, Raoul 39, 56n3
covering law 96, 301
Cozarinsky, Edgardo 52, 57n10
craft 98–99, 137, 228–231, 282
cranial measurement 261, *261*
Crary, Jonathan 101, 125, 324n1
creative and performing arts 1–5, 9–32, 36–38, 42, 49, 56, 58–59, 62–63, 83–84, 86, 88, 89, 92, 93, 98, 103, 105, 112, 125, 130, 136, 145, 147, 148, 151, 156, 164, 173, 178, 184–187, 217, 219, 222, 227, 228, 231, 241, 249, 254, 255, 257–259, 272, 277, 279, 283, 290, 291, 293, 294, 297–300, 302, 323, 324
creative arts 1–5, 36–38, 58–59, 62–63, 83–84, 88–89; and critical theory 297–300, 323–324; and history 227–228, 254–255; and the human sciences 257–259, 290–291, 293–294; key terms and 9–18; and modelling artistic research 29–32, *30–31*; and the modernist avant-garde 147–148; and research for art and design 28–29; research into 18–20; research through 20–29; and the revolution 184–187; and science 92–94
creative arts and media 3–4, 30, 58, 254, 297, 324; theory in 298–300; *see also* creative arts; media
critical dialogues 28, 36, 42, 43, 46, 48, 258
critical theory 297–298, 323–324; and Ackerman and Gobert 312–317; and Bastashevski 317–323, *318*, *320–321*; and Beloufa 307–312, *308–309*, *311*; in the creative arts and media 298–300; and the critical paradigm 305–307; and the interpretative paradigm 303–305; media theory 16, 19, 298, 306, 319; and positivism 301–302; theory–practice relation 300; *see also* theorization
criticism 12–13, 47–48, 167–169
critique (crit) 13, 27, 33n14, 59, 77, 84, 88, 92, 103, 105, 110, 219, 249–251, 259, 264, 278, 279, 283, 289, 294, 300, 306, 323
Cubism 145, 179n7, 180n13, 181n20; and the revolution 188, 193–196, *194–195*, 199–201, 203–204; and twentieth-century artistic research 156, 158, 160
cultural studies 298–300
curation 59, 65–73, 75–76, 106, 238, 240–241

Damisch, Hubert 101, 133, 144n6
dance 13–15, 176–177, *177*
Davies, David 83–84
da Vinci, Leonardo 104, 125, 129–130, 137–143, *139*, 207–208, *211*
Davis, Whitney 50–51
de-communization 315
Deleuze, Gilles 12, 16, 47, 105; and Felix Guattari 16, 85–86, 98, 129, 138, 298
Delpeut, Peter 52, 57n9
dematerialization 59, 62
demonstration 17, 104, 131, 133, 134, 136, 137, 138, 143, 223n9
Derrida, Jacques 12, 315, 317, 326n23, 326n24
design 1–3, 18–19, 24–26, 33n9, 98–100; and the modernist avant-garde 145–148, 175–176; and the Renaissance 129–131, 140–143; research for 28–29; and the revolution 183–189, 201–204, 206–209, 211–214, *212*
Dewey, John 26, 28
dialogue 42–48, 88–89, 245, 304–305
Didou, Marc 101, 117–121, *119–120*
digital revolution 62, 126
discovery 97–100, 230–231
discovery led 16, 97, 231, 298
docklands 272–276
docu-drama 243
Doherty, Willie 63–64, *64*, 90n10
drama 80–83, 242–244; *Lehrstuck* 312
Dublin 24, 48, 56n3, 69, 108–109, 243, 263; documenting the docklands of 272–276, *273*, *275*; Irish Museum of Modern Art (IMMA) 75, 112, 114; National College of Art and Design 20, 231; Trinity College 122n8, 260, 325n6
Duchamp, Marcel 47, 56n5, 59–60, 89n4, 89–90n6; Duchamp effect 89n3
Durkheim, Emile 178, 223n13, 255n8
dyspraxis 297, 325n2

economics 270–272, 301–302, 305–307, 310, 322–323
Edgerton, Sam 101, 130, 134, 143
Eisenstein, Sergei 27, 47–49, 56n6, 182n34; and artistic research 150, 173; and Kandinsky 164; and the revolution 214, 219
ekstasis 43
England 108, 232–233, 295n6
episteme 142
ethno-fiction 264
ethnography 36–38, 267–268, *268*, 270–272, 295n10, 304, 325n7; art ethnography 295n2; auto-ethnography

23, 290–291, 293, 296n16, 296n17; as conceptual art 295n3; ethnographic turn 258–260; and Hiller 277–279, 282–283; lens-based 272–276, *273*, *275*; and Minh-ha 283–290, *287*; and the tarnished legacy of anthropology 260–261
ethnomusicology 45, 283, 296n13
Europe 3, 75, 90n15, 141, 156; and critical theory 306, 311; eastern Europe 312, 315, 322; European world-view 296n15; and history 230, 244, 250; and the human sciences 271, 279–280; and the revolution 193, 199
exhibition 75–79, 158–160, 240–241, 307–308; exhibition practice 2, 75, 187–188
experimental, film practice 16, 33n7, 40, 45, 49, 54, 88, 151, 164, 252, 260, 271, 283, 284, 286, 288
explanandum 96
explanans 96
expression 49–50, 167–168, 177–178, 227–228, 236–238; art as 168–173, **172**
Expressionism 145, 150–152, 196; abstract 33n11, 150; German 180n15, 199, 223n16
extrastatecraft 317

fascism 243–244, 250, 306
Fay, Brian 300–301, 303, 305–306
Feyerabend, Paul 103–105, 121n4, 121n5, 121n7, 276
feminism 86, 277–278, 299–300
film 1–4, 15–17, 19–20, 27–28; and auto-ethnography 291–294; avant-garde 57n14; bio-pic 50; *cinéma vérite* 295n7; *ciné-provocation* 274; *ciné-trance* 273; and critical theory 297–299, 306–311; docu-drama 243; film theory 57n12, 181n28, 182n34, 298, 325n3; flick book 126, *128*; found footage 42, 46, 49–56, *51–52*, *55*, 57n8–9, 254; and history 228–229, 240–246, 250–255, 256n14; and the human sciences 260–261, 271–274; and the interview 46–49; and the modernist avant-garde 172–173; and painting 181n23; point-of-view style of 296n18; reflections and dialogues as research sources for 42–46, *44*; and the Renaissance 125–126; and research 35–42, *39*, *41*, 63–64, 67–69, 81–83; and the revolution 214–215; and science 112–115; tableau in 32n6; *see also specific film-makers*
Fitzgibbon, Margaret 228, 231, *233*, 234, 256n13
Forgács, Péter 42–43, 49–52, *51*, 57n8
Foster, Hal 199, 223n17, 258–259, 283, 295n4, 296n17
Foucault, Michel 12, 60
Frayling, Chris 18, 20, 28–29, 33n8
Freud, Sigmund 22
Fruitmarket Gallery 65–68, *66*, *68*, 75
funding 33n13, 110, 231; public 3, 29, 106
Futurism 71, 145, 158–162, *161*, 179n7; and the revolution 185, 188, 193, *195*, 196, 201
fuzzy logic 42

Gage, John 181n25, 222n4
Gedankenexperimenten 103
Geisteswissenschaften 148
Generalbass 164
geocentrism 103, 143
Germany 12, 27, 101, 183, 221; and Constructivism 206, 214; and critical theory 306; and history 229, 243–245, 248, 250; and the modernist avant-garde 147–148, 150, 152, 165, 175; and non-objective art 196; and positivism 255n1; women in 180n12
Gestalt psychology 169
GINKhUK 186–187, 189, *190*, 199, 219, 221
Godard, Jean Luc 38–40, 56n3
Gordon, Douglas 40, *41*, 56n4, 76, 90n15
Guattari, Felix *see under* Deleuze, Gilles
Gunning, Tom 126, 144n3
Guston, Philip 23

Habermas, Jürgen 101–102, 105, 121n3, 306, 325n5
Hempel, Carl 230
Hillen, Sean 97, 263, *266*
Hiller, Susan 258, 277–283, *281*, 289, 296n15
Hinde, John 263, *265*
Hirst, Damien 106–108, *107*
historiography 49–50, 88, 181n25, 227, 229, 256n10; and art–science collaboration 114; causal attribution in 255n2; and the family archive 233; Marxist 198, 255n5; and the Spanish Civil War 234–246
history 4–5, 51–52, 145–148, 227–228, 254–255, 260–261; and contemporary

art 246–254, *247*, *253*; and figurative devices 255n3; film history 242–246, *245*; history writing 255n4; and material memory 231–234, *233*; of mentalities 255n6; method, craft and science in 228–231; of science 103–104, 108, 126; *see also* art history; historiography
Holocaust 51, *247*
human sciences 4, 10, 18, 37, 84–86, 88; and critical theory 299–300, 302, 305–306; Dilthey and 179n2; and history 227, 229; Kandinsky and 179n10; and the modernist avant-garde 148, 171; *see also* anthropology
hypothesis led 16, 94–95, 97–99, 198, 200, 231, 298

Ideal type 229, 230, 255n7
imaging 3, 20, 69, 115–117; Constructivist 214–218; and history 242–246
immanent critique 59, 84, 89
impressionism 179n7, 188
induction 95, 98, 297
internal necessity 180n15
Institute for the Arts (INKhUK) 148, 151, 167, 171, 185–187, 204; and Constructivism 213; and the eclipse of the avant-garde 219; and psychology 222n4
Institute for Social Research 12, 91n17, 306
interactive media 1
interpretative, sociology, paradigm 178, 227, 229, 255n8, 260, 290, 303–305, 325n7
interview 36–38, 42–43, 53–54, 59–60, 234–236; with Ackerman and Geobert 314–315; with Bastashevski 322; with Hillen 263; with Jeb 71–73; with McGarry 244–246; as a research tool 46–49; with Zenk 236–238; with Ziff 240–241
intuition 16, 18, 98, 173, 178, 242, 303, 325n3
invention 13–16, 129–131, 145–147, 187–188
Ireland 3, 52, 54, 57n14, 90n12, 122n10; and history 232, 245; and the human sciences 261, *265*, 291, 295n6; Northern 49, 63, 263; *see also* Dublin
Italy 80; Florence 104, *128*, 130, *132*, 133–137, *135*

Jebb, Katerina 69, 71–73, *72*, 114
Jeffrey Charles Henry Peacock (JCHP) *11*, 77–79, *78*, 108
Joyce, James 47–48, 56n6
Jugenstil 152

Kandinsky, Wassily 49, 180n15; and abstraction 180n18; and affect 182n31; and Constructivism 206–207, 213–214; and critical theory 299; and the eclipse of the avant-garde 218, 221; and the modernist avant-garde 146, 149–156, *153–155*, 160–162, 168–177, *174*; and non-objective art 192, 196; and progressivism 181n25; and psychology 222n4; and the revolution 183–186, 188–189; and a science of painting 223n14; and the social sciences 179n10, 180n16; teaching notes of 182n33; theory and practice in the work of 162–168, *165*; and Theosophy 179n9; and the UNOVIS experiment 200, 203
Kant, Immanuel 12, 32n2; Kantian thought 85, 255n1
Kemp, Martin 87, 101, 129, 131, 144n4, 207; and da Vinci 138, 140–141
Kentridge, William 14–17, *14*, 33n7, 43–44, *44*, 67–69, *68*
Khondji, Darius 40
Kindness, John 263, *267*
Klee, Paul 26, 182n33; and artistic research 150, 152, 173, 175, 178; and Kandinsky 164
Klutsis, Gustav 189, *192*
knowledge 4, 35–36, 82–87, 90–91n16; and the characteristics of artistic research 16–18; critical theory and 300–301, 325n7; history and 249–250; the human sciences and 269–271; and key terms in artistic research 9–10; and the main types of artistic research 23–27; the Renaissance and 133–134, 136–138, 140–143; science and 93–94, 100–102, 104–106
Kogan, Nina 180n12, 203–204, 224n22
Koorland, Vivienne 67
Kossuth, Joseph 60–62, *61*, 90n7, 325n4
Kubrick, Stanley 38–40, *39*, 56n2
Kuhn, Thomas 102, 149, 298
Kunstwissenschaft 223n14

language 47–48, 50–51, 86–87, 99–100; Minh-ha and 286; the modernist avant-garde and 146–148, 168–170, **172**, 182n30; non-objective art and 198–199; of representation 285; visual 145,

166–168, 173–175, 202; the UNOVIS experiment and 202–204
Lenin, Vladimir 198, 221, 312–317, *313*, *316*
LeWitt, Sol 60, 88, 90n9
linguistics 57n12, 60, 62, 87, 89, 156, 167, 168, 171, 182n30, 227, 229, 299, 325
Lippard, Lucy 62, 282–283
Lizzitsky, El 151, 175, 180n11, 183–184, 189; and Constructivism 207, *209*, 214, 216, *216*; and the eclipse of the avant-garde 219, 221; and the UNOVIS experiment 204–205, *205*
Lodder, Christine 179n8, 207, 222n3
logical empiricism 230

Madoff, Steven Henry
magic lantern 126
Malevich, Kasimir 26, 49, 146–147, 182n35, 223n9–10; and art as expression 171–173; *Black Square* 115, 158–159, *161*, 181n21, 181n23, 224n23; and Constructivism 206–207, 213–215, 217–218; and critical theory 299; and the eclipse of the avant-garde 218–221, *220*; Jakobson and 223–224n19; Kogan and 224n22; language of 223n13; Leporskaya and 223n18; and modernist artistic research 173–175, *174*; and non-objective art 192–200, *194–195*, *197*; and progressivism 181n25; and the revolution 183–184, 186–187, 189–192, *190*; as a revolutionary artist 224n21; Richter and 224n25; and a science of painting 223n14; and Steyerl 251; and style 223n15; and Theosophy 179n9; and twentieth-century artistic research 150–151, 156–161, *157*, *159–160*; and the UNOVIS experiment *191*, 200–204, *202*; writings of 179n8, 223n12; and the Zaum poetry movement 182n30
maquettes 35, 40, 98, 207, 231
Marks, Laura 88, 246
Martin, Rosy *81*
Marx, Karl 12, 48, 50, 305, 311, 326n24
Marxism 85, 148, 180n19, 198, 229, 255n5; and contemporary critical arts practice 315; Hiller and 278; and the theory–practice relation 300, 306
Masaccio, Tommaso 125, 129, 133–136, *135*, 143, 144n6
material memory 227, 231–234
Mauss, Marcel 76, 79
Mayakosky, Vladimir *192*
Maybury, John 47
McGarry, Fearghal 243–244
Mead, Margaret 260, 277
media 1–4, 9–10, 12–13, 15–19, 29–30, 37–38; and critical theory 297–299, 306–307, 323–324; new media 125, 256n15; and the Renaissance 125–126; and a research turn 62–63; and the revolution 214–215
media arts 1, 3, 10, 12, 19, 297, 324
memento mori 71, 118
method 1–2, 228–231, 258–260; anthropological 276–283, 295n10, 296n13; art historical 295n4; scientific 10, 94, 101–102, 104–105, 301–302
methodological unity science 99, 230
methodology 9–10, 18–20, 23–26, 28–32, 85–86, 88–89; critical theory and 298–300; history and 227–228, 254–255; the human sciences and 258–259, 289–290, 293–294; interview and 47–48; methodological drive in artistic research 92–94; the modernist avant-garde and 145–146; non-objective art and 199–200; science and 97–100, 103–104
middle range theory 89, 91n18
Milner, John 156, 181n20
Milton, Alex: and Paul Rodgers 25, 33n9, 37
Minh-ha, Trinh T. 88, 246, 258, 283–290; *Naked Spaces 287*; *Reassemblage* 283–289, *287*, 296n13; and Rouch 296n14
models 10–12, 23–24, 27, 35–36, 184, 186–187; Constructivist 26, 150, 179n5–6, 183, 207–209, 213–214; critical theory and 299–302, 304–305; deductive–nomological (DN) 96–97, *96*; history and 229–231; the human sciences and 293–294; hypothetical–deductive *95*; method and 92–94, 101–102; modelling artistic research 29–32, *30–31*; the modernist avant-garde and 145, 149, 169–172; of space 180n11; and the specificity of artistic research 97–101; the Renaissance and 135–138, 140–144
modernism 4, 10, 15, 28, 43, 128; and Constructivism 206, 213; Cozarinsky and 57n10; and critical theory 299; and history 229, 249–250; and the human sciences 257, 259, 279, 283; and the interview 46, 48; and non-objective art

193, *194*, 196, 198–199; and reflexivity 295n4; and the revolution 187–189, 221; and science 100–101, 103, 106; and the UNOVIS experiment 200–201; *see also* avant-garde; *specific artists*
Moholy-Nagy, Lazlo 26–27, 150–151, 164, *197*; and Constructivism 214–218, *218*; and modernist artistic research 173, 175
Morin, Edgar 38–39, *39*, 295n7
Morrison, Bill 52–53, *52*, 56n7, 57n9
MRAC 235, 307, *308*
MRI 69, 101, 115, 117–118, 121
Murch, Walter 42
music 42–43, 45–46, 54, 163–164, 168–170; Minh-ha and 283, 288–289

Narkompros 184, 188, 222n3
nationalism 243–244, 310, 314–315
Nazism 50, 152, 238, 243–248, 250, 254, 314
neo-liberalism 300, 308, 317
Neo-Platonist 121n5
neo-positivism 103
newsreel 57n8, 57n10
Nicholls, Bill 19; and Michael Renov 42
nomadic science 16, 87, 88, 102, 129, 138, 143
non-objective art 62, 115, 183–184, 188–189, 192–200, 224n19; and artistic research 151, 156, 159; and Constructivism 206–207, 214, 217–218; and history 251; Kandinsky and 164, 167–168; and the modernist avant-garde 147, 178; and the UNOVIS experiment 204
nous 18, 244

Obrist, Hans 47, 56n5, 75, 90n15, 106, 187
observation 10, 24, 94, 103, 104, 121, 126, 129, 137, 138, 140–142, 176, 259, 291, 317
O'Neill, Paul: and Mick Wilson 65, 90n11
optical, optical science 171
originality 10, 29, 35, 242
Ortega, Damian 65–66, *66*

painting 22–24, 59–60, 133–137, 142–143; and history 234–238; and the modernist avant-garde 149–152, 158–162, 168–169; and progressivism 181n25; and the revolution 185–189, 192–193, 200–204, 206–207; and science 179n3, 182n35, 223n14; Suprematist 214–218, 222n7; *see also specific painters*
Pan-Africanism 264
Participation 75, 80, 110, 259, 276, 277, 302, 304, 312, 326n20
Parreno, Philippe 40, *41*, 75–77, 77, 90n15
pedagogy 12, 27, 30, 146–147, 150, 151, 167, 175, 182n33, 183, 186, 188, 196, 201, 249, 324
Pelucca *177*
performance 3–4, 15–17; *see also* performativity
performativity 80–84, 271–272
perspective, linear 104, 130, 131, 133, 134, 252
phantasmagoria 126
Phd, doctorate, double PhD 3, 20, *21*, 24, 110, 115, 144n2, 196, 227–228, 231, 232, 272, 290, 291, 294, 296n14, 296n20, 302, 327n28
photography 3–4, 15–16, 19–20, 35–38, 57n12, 125–126; calotype 112, 114, 122n10; camera obscura 87, 126–128, *127*, 144n2; and critical theory 314–315, 319–320, 322–323; and history 235–241; and the human sciences 260–262, 272–276, 291–294; photo-documentation *261–262*, 323; pin-hole cameras 126; and a research turn in visual arts 79–83; and the revolution 214–217; and science 114–115, *116*, 122n10, 122n13; *see also specific artists*
Picasso, Pablo 106, 156, 158, *158*, 196, 203, 279–280
Pilkington, Alison 20, *22*, 23
political economy 12, 151, 179n10, 180n19, 273, 278, 300, 306, 326n13
Popper, Karl 94, 98, 103, 121n2
positivism 10, 84, 92, 101–104, 121n3, 143, 229, 255n1, 301–302
postcard 263, *265*, 283
post-colonialism 282, 290
Post-impressionism 152, 156, 179n3, 196
postmodernism 83, 92, 106, 227, 246, 324
poststructuralism 12, 19, 76, 87, 151, 299
practice-based research 33, 35, 46, 56, 92, 121, 143, 146, 151, 171, 178, 200, 219, 258, 272, 302
practice-led research 11
primitive art 151, 199, 279, 282
probabilistic 95, 97, 98, 230
problem solving 100, 140
process *see* art process; research process
Productivism 152, 168, 207, 213, 218–219

professional practice 16, 32, 33n13, 37, 38, 115, 137, 138, 298
project development 2, 37, 40, 49
Prouns 204, 207
provocation 16, 17, 23, 40
pseudo-intellectualism 108

radiogram 232
Railing, Patricia 198, 223n15
Rancière, Jacques 85, 115, 255n4
Read, Herbert 4, 150–151, 180n14
readymade 59–60, 89–90n6
re-appropriation 73, 231–232
re-enactment 242, 244
reflection 42–47, 97–98, 112–114, 323–324; reflective practice 24–28, *25*, 33n12, 42–43, 134
reflexivity 27–28, 258–259, 285n4
Renaissance 1, 15, 87, 122n13, 125–128, 144n6; Hiller and 277; and the modernist avant-garde 145, 149, 162, 166, 179n1; and non-objective art 198; Renaissance vision 128–137; and science 101, 104; *see also* da Vinci, Leonardo
Renov, Michael *see under* Nicholls, Bill
relational aesthetics 73–80, 90n13, 304
research *see* artistic research; background research; research culture; research process
research culture 1, 3, 30, 42, 88–89; and critical theory 324; and history 228–229; and the modernist avant-garde 146, 148, 178; and the revolution 184
research process 16, 24–27, 179n1, 183, 302; and science 93–94, 97, 122n14
revolution: and artistic research 184–192, 221–222; Bolshevik revolution 151, 184; Conceptualist 90n9; digital 62, 126; and Constructivism 206–214, 224n23; and the eclipse of the avant-garde 218–221; February revolution 224n21; Maidan revolution 311, 314; in music 163; and non-objective art 192–200; October Revolution 146–147, 179n7, 189; political 145; of Quattrocento Florence 137; in Russia 156, 159–160, 175, 183–184, 250; and science 102, 104, 108, 128, 140, 143, 149; and Suprematist painting 214–218; Tatlin and 179n5; and the UNOVIS experiment 200–206
Rilke, Rainer Maria 33n10, 47
Rodchenko, Aleksandr 26, 146, 179n6, 183–184, 186, 188–189; and artistic research 150–151, 175; and Constructivism 207–213, *210*; and Kandinsky 164; and non-objective art 192, 200; and Suprematist painting 214–217, *217*
Rodgers, Paul *see under* Milton, Alex
Ronai, Carol Rambo 290–291
Rosanova, Olga *161*
Rosenstone, Robert 242, 246, 256n14
Rouch, Jean 38–40, *39*, 45, 258, 295n7, 296n14; and anthropology 264, 266–268, *268*, 271–272; and auto-ethnography 293; and ethnography 272–274
Ruby, Jay 271, 296n13
Rudoph, Charlotte *177*
Russia 1, 26–28, 49, 182n35, 250; and art as expression 168, 171; artistic research and 183–192, 221–222; Association of Artists of Revolutionary Russia (AKhRR) 219; and Constructivism 206–214; and critical theory 311, 314–315; and diagrammatic representations 223n9; and the eclipse of the avant-garde 218–221; and Kandinsky 162–163, 166; and Marxism 181n19; and modernist artistic research 175; and the modernist avant-garde 146–148; Moscow *158*, 165–166, 179n7, 185–186, 213; and non-objective art 192–200; and progressivism 181n25; and reflex psychology 222n5; renaissance in 179n1, 179n4; and science 101, 103; and Suprematism 214–218; and twentieth-century artistic research 150–152, 156, 158–159, 160; and the UNOVIS experiment 200–206; women and 180n12; *see also specific artists*

scannography 69–71
Schneider, Arnd: and Christopher Wright 256, 276, 294n2, 295n3, 296n20
Schön, Donald 24, 26
science 1–2, 85–88, 92, 101–105, 121; of art 145–150, 178; and art as expression 168–173; artists and 106–110; art–science collaboration 110–119; colonial 269–271; and critical theory 301–302; da Vinci and 137–142; Dilthey and 179n2; Feyerabend and 121n4, 121n7; Kemp and 144n4; the logic of scientific discovery 94–97, *95–96*; and the methodological drive in artistic research 92–94; nomadic 16, 87–88, 102, 129, 138; of painting

182n35, 223n14; photography and 122n13; and positivism 255n1; and practice-based research 121n1; and the Renaissance 125–126, 128–130, 133–134, 143–144; and the revolution 183–187, 219–222; science–art nexus 101, 106, 207; scientific method 10, 94, 101–102, 104–105, 301–302; and the specificity of artistic research 97–101; Steiner and 179n9; *see also* human sciences
Science Gallery 108–109, *111*, 122n8
sculpture 117–118, 231–232
sean-nós 46, 54
Sekula, Allan 295n8, 319–320, 326n26; *Forgotten Space, The* 272–273, *273*
semiotics 167–168, 170–171, **172**, 182n29, 223n13, 257, 300
senses, anthropology/phenomenology of 171
Sergei Schukin Mansion *158*
Sheikh, Simon 62–63
Shklovsky, Viktor 200, 224n20
social engagement 2, 30, 73, 79, 103, 304, 324
social sciences *see* human sciences
Spain: Madrid 40; *see also* Spanish Civil War
Spanish Civil War 57n8, 228, 234–243
spectatorship 79, 118, 131, 167, 181
Spence, Jo 80, *81*
Stafford, Barbara Maria 87, 126, 167, 182n29
Stepanova, Varvova 146, 179n6, 184, 188, 208–212, *212*, 222n3
Steyerl, Hito 228, 246–254, *247*, *253*, 256n15
Stockhausen, Karl Heinz 43
storytelling 230, 232, 244, 273, 291
studio 4, 10, 12, 13, 16, 17, 19, 27–30, 32, 35, 38, 42, 43, 56, 62–63, 69, 71, 73, 93, 97–99, 108, 114, 117, 129, 136, 145–148, 164, 174, 175, 186, 187, 200, 203, 204, 208, 213, 228, 231, 271, 289, 294, 324
style: and non-objective art 192–200
Suprematism 179n7, 181n23, 182n36, 223n19, 224n23, 224n25; and art as expression 168; and Constructivism 206–207, 214–218; and the eclipse of the avant-garde 219; and modernist artistic research 173; and the modernist avant-garde 145; and non-objective art 193, *195*, 196, 198–200; and the revolution 186, 188–189; and twentieth-century artistic research 158–160; and the UNOVIS experiment 200–201, 203–204
Sweeney, Moira 272–276, *275*
Sylvester, David 46–47
symbolic interactionist 42, 279
Symbolism 158, 167, 182n30, 196, 201

Tatlin, Vladimir 179n5, 181n20, 181n22, 182n30; and the modernist avant-garde 146, 150–151, 156, 168, 175; and the revolution 183–184, 192–193, 207–208, *208*, *211*
techne 142
theoreticism 84, 300
theory of practice 147, **172**, 173, 297, 299, 324
theoretico-critical 18
theorization 19–20, 48–49, 140–141, 147–148, 185–187
theory *see* critical theory; theorization
Theosophy 163, 179n9
Tiravanija, Rirkrit 2, 73–76, *74*, 304
Trotsky, Leon 180–181n19, 221

United Kingdom (UK) 3, 29, 49, 108, 112; London 47, 69, 90n8, 232
United States (US) 3, 144n3, 278, 301, 326n18; anthropologists in 295–296n11; and found footage film-making 52, 54; and history 230, 233; and non-objective art 196
universities 3, 92–93, 110–112, 133
UNOVIS *see under* Malevich, Kasimir
Utopian 207, 221, 250, 308, 315
utopianism 207, 221, 250, 308, 315, 325n12

value free 85, 295n11, 301
viajero 269
Vietnam War 278, 295–296n11
visual methodologies 296n20
VKhUTEMAS 27, 175, 206, 209–213, 224n21
Volans, Kevin 43, 45–46, 54

Wallis, Ruby 291–293, *292*, 296n18
Walls, Howard 53
Weber, Max 255n7, 255n9
Weir, Grace *113*, *116*
Weiss, Peter 250
weltanschauung 149, 193
Williams, Raymond 13, 32n3, 32n5

Wilson, Mick *see under* O'Neill, Paul
Wiseman, Frederick 42
Wittgenstein, Ludwig 49–52, 178, 182n36
Wright, Christopher *see under* Schneider, Arnd

Young, Catherine Sarah 110–111, *111*

Zaum, poetry 182n30, 224
Zenk, Pauline 228, 235, *235*, 237–238, *237*
Zidane, Zinédine 40, *41*
Ziff, Trisha 228, 238–241, *239*
zoetrope 126

For Product Safety Concerns and Information please contact our EU
representative GPSR@taylorandfrancis.com
Taylor & Francis Verlag GmbH, Kaufingerstraße 24, 80331 München, Germany

www.ingramcontent.com/pod-product-compliance
Lightning Source LLC
LaVergne TN
LVHW021117300926
849160LV00010B/762

* 9 7 8 1 1 3 8 5 8 9 0 6 3 *